Feeding and Processing Affect Meat Quality and Sensory Evaluation

Feeding and Processing Affect Meat Quality and Sensory Evaluation

Guest Editor

Sandra Rodrigues

MDPI

Basel • Beijing • Wuhan • Barcelona • Belgrade • Novi Sad • Cluj • Manchester

Guest Editor
Sandra Rodrigues
CIMO
LA SusTEC
Instituto Politécnico
de Bragança
Braganca
Portugal

Editorial Office
MDPI AG
Grossspeteranlage 5
4052 Basel, Switzerland

This is a reprint of the Special Issue, published open access by the journal *Foods* (ISSN 2304-8158), freely accessible at: https://www.mdpi.com/journal/foods/special_issues/PK6R2833EV.

For citation purposes, cite each article independently as indicated on the article page online and as indicated below:

Lastname, A.A.; Lastname, B.B. Article Title. *Journal Name* **Year**, *Volume Number*, Page Range.

ISBN 978-3-7258-6932-9 (Hbk)
ISBN 978-3-7258-6933-6 (PDF)
https://doi.org/10.3390/books978-3-7258-6933-6

Contents

About the Editor

Sandra Rodrigues

Sandra Rodrigues is an Adjunct Professor at the School of Agriculture of the Polytechnic Institute of Bragança (ESA-IPB). Holding a PhD in Animal Science, her research within the Mountain Research Centre (CIMO) centers on meat quality, sensory analysis, and innovative processing of traditional small ruminants and pig breeds. Her scientific output features over 50 indexed journal articles, 11 book chapters, and more than 100 papers in conference proceedings. Dr. Rodrigues has contributed to 16 research projects, notably serving as a key investigator for the BISOLIVE project, which explores olive pomace as a dietary strategy to improve meat quality. She also leads research into healthier meat products by investigating alternatives to salt and animal fat. Beyond her research, she is a member of the Editorial Board for *Foods* (MDPI) and has served as a Guest Editor for three Special Issues.

foods

Editorial

Feeding and Processing Affect Meat Quality and Sensory Evaluation

Sandra S. Q. Rodrigues

CIMO, LA SusTEC, Instituto Politécnico de Bragança, Campus de Santa Apolónia, 5300-253 Bragança, Portugal; srodrigues@ipb.pt

Paul D. Warriss [1] aptly described the evolution of agricultural systems as progressing through three distinct phases: first, meeting basic needs; second, improving production efficiency; and finally, enhancing product quality. Meat production follows this trajectory, and we are now in the phase where delivering high-quality products is paramount.

Meat quality [2] encompasses a set of desirable attributes that align with consumer expectations—appropriate conformation, nutritional value, sensory appeal, and safety. However, perceptions of meat quality vary widely depending on cultural background, personal experience, and the stakeholder's role in the production chain—be it producer, retailer, or consumer. To meet these diverse expectations, meat production must strive for maximum output at minimal cost (efficiency), while simultaneously ensuring the highest standards of quality. However, this pursuit must never compromise animal welfare, health, or environmental sustainability.

Growth performance, carcass yield, and meat quality are influenced by a range of animal-related factors, breed and genetics [3–5], sex [6,7], age [8,9], and slaughter weight [10,11]. These elements must be carefully considered in management practices to optimize outcomes. External factors, including diet or feeding strategies [3,4,12–14], as well as processing techniques [15]—individually or in combination—can significantly affect meat quality and consumers' sensory perception [16].

Feeding practices play a pivotal role in shaping various quality traits, from carcass composition and commercial value to nutritional content and organoleptic properties. Across species—from fish and monogastrics (including poultry) to ruminants—diverse strategies can be employed to produce meat that satisfies consumer expectations in terms of both nutrition and sensory appeal, while adhering to sustainable and animal-friendly principles. Approaches such as grass-fed systems, organic feed, and targeted supplementation are commonly used to modulate meat quality.

Since ancient times, humans have developed preservation techniques to extend the shelf life of food. In the absence of such methods, degradation due to microbial activity, chemical and enzymatic reactions, and physical changes become inevitable. Traditional preservation methods such as dehydration, smoking, brining, canning, fermentation, and refrigeration have increasingly been complemented or replaced by innovative techniques, including chemical and biological preservation and non-thermal technologies. Beyond extending shelf life, processing can also affect the technological and organoleptic characteristics of meat, influencing its flavor, texture, and overall consumer acceptance.

Meat quality and sensory appeal are central to consumer satisfaction, nutritional value, and innovation across the food industry. This Special Issue of *Foods*, entitled "Feeding and Processing Affect Meat Quality and Sensory Evaluation", presents a diverse and insightful collection of studies that explore how feeding strategies and processing technologies influence the final attributes of meat products.

The published papers span a wide range of approaches—from dietary interventions using saline-grown oats and fruit powders to alternative protein sources like fly maggots and algae extracts. These studies underscore the intricate interplay between animal nutrition, biochemical composition, and post-harvest processing in determining meat texture, flavor, shelf life, and consumer acceptance. The findings presented offer practical insights for producers, researchers, and policymakers aiming to elevate meat quality through sustainable and science-driven methods. A summary of the published papers is presented below.

1. Natural Antioxidants as Nitrite Alternatives in Processed Meats

A major research front in this field is the search for natural alternatives to synthetic additives in processed meats. Two studies featured in this Special Issue advance the clean-label agenda by exploring plant-based antioxidants to replace synthetic nitrites in meat products.

Ferreira et al. investigated Bougainvillea spectabilis bracts, processed via three drying methods, as natural additives in cooked pork ham. Bougainvillea powder enhanced the antioxidant stability of the ham across all drying methods, as reflected by higher total polyphenol levels and improved results in DPPH, ABTS, and FRAP assays. At 0.1% inclusion, the powder significantly reduced lipid oxidation over eight weeks of storage, with no compromise in flavor, texture, or appearance. These findings position Bougainvillea spectabilis as a viable, consumer-friendly alternative to nitrites.

Manea et al. evaluated the use of blackcurrant, lingonberry, and sour cherry powders as natural antioxidants in nitrite-free salami formulations. Among them, blackcurrant powder showed the strongest oxidative protection, positioning it as a promising clean-label alternative to synthetic nitrites. The antioxidant efficacy varied by fruit type and was closely linked to phenolic content, with higher concentrations yielding better results. Additionally, smoked and cooked salami exhibited greater oxidative stability than scalded variants, underscoring the need to align antioxidant strategies with both ingredient functionality and processing conditions.

Together, these studies exemplify how targeted processing and phytochemical strategies can meet consumer demand for safer, additive-free meat products without sacrificing technological or sensory performance.

2. Sustainable Feeding and Meat Quality Enhancement

Sustainable feeding systems are increasingly recognized as key to reconciling animal performance with environmental stewardship. Two studies explore how alternative feed sources can improve meat quality while promoting environmental sustainability.

Xin et al. demonstrated that saline-grown oats enhance the nutritional profile of forage and improve meat quality in Qinghai Tibetan sheep. Using metabolomics, this study links elevated levels of protein, amino acids, and sugars in saline oats to improved meat texture and sensory quality, maintaining water-holding capacity. These findings suggest that saline-adapted crops can modulate muscle metabolism and offer a viable strategy for livestock production on marginal lands.

Leite et al. evaluated the inclusion of olive cake by-products in Bísaro pig diets. Incorporating centrifuged and pressed olive cake into dry-cured meat products did not alter their overall physicochemical properties, but notable differences emerged between product types in fat, protein, moisture, water activity, and haem pigments. Interactions between diet and product also affected parameters like aw, ash, and NaCl, indicating that products from the same muscle (LTL) can develop distinct profiles. Olive cake inclusion—especially at 25%—significantly increased n-3 fatty acid content and lowered the PUFA n-6/n-3 ratio, enhancing nutritional value without affecting other fatty acid fractions or quality indices. These distinctions may boost consumer appeal and support circular feed strategies.

3. Functional Feeds and Packaging Technologies

Two complementary studies highlight how feed innovation and smart packaging can synergistically improve meat quality and safety.

Fehri et al. assessed diets enriched with extruded linseed and *Padina pavonica* algae in meat rabbits. The functional feed improved the fatty acid profile of rabbit meat by reducing saturated fatty acids and the n-6/n-3 ratio, while increasing n-3 PUFA levels across two production cycles. Nutritional gains included higher levels of vitamin E, coenzyme Q10, and essential minerals, contributing to health-oriented meat products aligned with the One Health framework.

Castrica et al. investigated modified atmosphere packaging (MAP) with an active absorbent pad (aPAD) for omega-3-enriched rabbit meat. The aPAD reduced microbial growth and oxidative degradation over 21 days. Oxidative stability improved, especially in algae-fed groups, suggesting a synergistic effect of polyphenols and bioactive compounds in *Padina pavonica*, combined with the physical and chemical properties of aPAD. Sensory quality and consumer acceptability decreased over time, demonstrating the value of integrating functional feeds with smart packaging to extend shelf life, ensuring safety, and consumer acceptability.

4. Aquatic Protein Innovation

Liang et al. present a novel approach to crustacean nutrition by partially replacing conventional feed with housefly maggot larvae (HML) in adult Chinese mitten crabs. Over a 40-day trial, HML supplementation improved edible yield, antioxidant capacity, and sensory quality—particularly in female crabs. Nutritional enhancements included elevated essential amino acids, astaxanthin, and inosine monophosphate (IMP), contributing to better texture and flavor. These findings support insect-based feed strategies as sustainable, high-performance alternatives in aquaculture.

5. Phytogenic Additives in Poultry

Shu et al. investigated the effects of dietary curcumin (CUR) supplementation on chicken meat quality. At 150 mg/kg, CUR improved amino acid deposition, increased 5′-nucleotide levels, and enhanced the PUFA/SFA ratio by reducing saturated fats. Volatile compound analysis revealed elevated aldehyde concentrations, contributing to improved flavor. This study confirms curcumin's potential as a natural additive to enhance both nutritional and sensory attributes in poultry meat.

Closing Remarks

As Guest Editor, I extend my sincere gratitude to all authors and reviewers and to the editorial team for their contributions to this Special Issue. The research presented here reflects a shared commitment to advancing meat science through sustainable feeding strategies and innovative processing techniques. It is my hope that this Issue will serve as both a reference and a catalyst for future innovations in food quality and sensory evaluation.

Funding: This research received no external funding.

Data Availability Statement: Not applicable.

Conflicts of Interest: The author declares no conflicts of interest.

References

1. Warris, P.D. *Meat Science: An Introductory Text*, 2nd ed.; Modular Texts; CABI: Wallingford, UK, 2010; pp. 77–96.
2. Purslow, P.P. *New Aspects of Meat Quality*, 2nd ed.; From Genes to Ethics; Woodhead Publishing: Cambridge, UK, 2022; pp. 1–22.

3. Mwangi, F.W.; Charmley, E.; Gardiner, C.P.; Malau-Aduli, B.S.; Kinobe, R.T.; Malau-Aduli, A.E.O. Diet and Genetics Influence Beef Cattle Performance and Meat Quality Characteristics. *Foods* **2019**, *8*, 648. [CrossRef] [PubMed]
4. Sakowski, T.; Grodkowski, G.; Gołebiewski, M.; Slósarz, J.; Kostusiak, P.; Solarczyk, P.; Puppel, K. Genetic and Environmental Determinants of Beef Quality—A Review. *Front. Vet. Sci.* **2022**, *9*, 819605. [CrossRef] [PubMed]
5. Romero, J.V.; Olleta, J.L.; Resconi, V.C.; Santolaria, P.; del Campo, M.M. Genetic Markers Associated with Beef Quality: A Review. *Livest. Sci.* **2024**, *289*, 105583. [CrossRef]
6. AbdElfattah, A.; Ahmed, H.; Maky, M. Effect of sex on meat quality characteristics. *SVU-Int. J. Vet. Sci.* **2022**, *5*, 165–173. [CrossRef]
7. Dai, Z.; Feng, M.; Feng, C.; Zhu, H.; Chen, Z.; Guo, B.; Yan, L. Effects of sex on meat quality traits, amino acid and fatty acid compositions, and plasma metabolome profiles in White King squabs. *Poult. Sci.* **2024**, *103*, 103524. [CrossRef] [PubMed] [PubMed Central]
8. Chen, X.; Cao, J.; Chang, C.; Geng, A.; Wang, H.; Chu, Q.; Yan, Z.; Zhang, X.; Zhang, Y.; Liu, H.; et al. Effects of Age on Compounds, Metabolites and Meat Quality in Beijing-You Chicken Breast Meat. *Animals* **2023**, *13*, 3419. [CrossRef] [PubMed]
9. Rivera-Bautista, C.; Grajales-Lagunes, A.; Relling, A.; Chay-Canul, A.; Vazquez-Valladolid, A.; Vargas-Bello-Pérez, E.; Lee-Rangel, H.A. Effect of age and diet on carcass and meat quality in ewes. *Heliyon* **2023**, *9*, e22350. [CrossRef] [PubMed]
10. Li, J.; Yang, Y.; Zhan, T.; Zhao, Q.; Zhang, J.; Ao, X.; He, J.; Zhou, J.; Tang, C. Effect of slaughter weight on carcass characteristics, meat quality, and lipidomics profiling in longissimus thoracis of finishing pigs. *LWT-Food Sci. Technol.* **2021**, *140*, 110705. [CrossRef]
11. Nikolic, A.; Babic, M.; Jovanovic, J.; Cobanovic, N.; Brankovic-Lazic, I.; Milojevic, L.; Parunovic, N. Effect of broiler slaughter weight on meat yield and quality. *Sci. J. Meat Technol.* **2019**, *60*, 17–23. [CrossRef]
12. Muzolf-Panek, M.; Zaworska-Zakrzewska, A.; Czech, A.; Lisiak, D.; Kasprowicz-Potocka, M. Antioxidative Status and Meat Quality Traits as Affected by Dietary Supplementation of Finishing Pigs with Natural Phenolics. *Antioxidants* **2024**, *13*, 1362. [CrossRef] [PubMed] [PubMed Central]
13. Liao, J.; Zhang, P.; Yin, J.; Zhang, X. New insights into the effects of dietary amino acid composition on meat quality in pigs: A review. *Meat Sci.* **2025**, *221*, 109721. [CrossRef] [PubMed]
14. Teodora, P.; Evgeni, P.; Desislava, V.-V.; Nikolay, K.; Desislav, B.; Stefan, D.; Krasimir, D. Meat quality and fatty acid profile in broilers as affected by low-fat Tenebrio molitor meal in the diet. *Front. Anim. Sci.* **2025**, *6*, 1629411. [CrossRef]
15. Gómez, I.; Janardhanan, R.; Ibañez, F.C.; Beriain, M.J. The Effects of Processing and Preservation Technologies on Meat Quality: Sensory and Nutritional Aspects. *Foods* **2020**, *9*, 1416. [CrossRef] [PubMed]
16. Rodrigues, S.S.Q.; Leite, A.; Vasconcelos, L.; Teixeira, A. Exploring the Nexus of Feeding and Processing: Implications for Meat Quality and Sensory Perception. *Foods* **2024**, *13*, 3642. [CrossRef] [PubMed]

foods

MDPI

Article

Preliminary Study on the Mechanism of the Influence of Saline Oat Pasture on Muscle Metabolism and Meat Quality of Tibetan Sheep

Xiaoming Xin, Lijuan Han *, Shengzhen Hou, Linsheng Gui, Zhenzhen Yuan, Shengnan Sun, Zhiyou Wang, Baochun Yang and Chao Yang

College of Agriculture and Animal Husbandry, Qinghai University, Xining 810016, China;
hlj880105@163.com (X.X.); 1987990009@qhu.edu.cn (S.H.); 2017990039@qhu.edu.cn (L.G.);
2017990038@qhu.edu.cn (Z.Y.); 2016980007@qhu.edu.cn (S.S.); 1992990011@qhu.edu.cn (Z.W.);
1989990021@qhu.edu.cn (B.Y.); yangchao@qhu.edu.cn (C.Y.)
* Correspondence: 2016990034@qhu.edu.cn; Tel.: +86-155-9746-0033

Abstract: This study aimed to examine the effects of oats cultivated in saline and non-saline environments on the meat quality and muscle metabolism of Qinghai Tibetan sheep. First, targeted and untargeted metabolomics were used to examine oat quality and metabolites. Second, sheep muscle quality and metabolites were analyzed. Finally, a combined examination of the quality of the oats and their metabolites, as well as that of the muscles, was compared with saline oats. This study hypothesizes that, compared with non-saline environments, soil salinization can improve the nutritional quality of oats, thereby enhancing the meat quality and muscle metabolism of Qinghai Tibetan sheep. Saline-grown oats were shown to have higher levels of crude protein, crude fat, free amino acids, and simple sugar. The meat quality of the sheep fed on saline oats was higher due to free amino acid and carbohydrate metabolism, resulting in improved texture, color, water-holding capacity, and cooked meat percentage, with lower steaming loss. The findings of this study confirm the hypothesis that salinization improves Tibetan sheep meat quality by optimizing oat composition, providing a reference for agricultural and animal husbandry production in saline areas.

Keywords: saline land; growing oats; Tibetan sheep; meat quality; metabolomics

1. Introduction

The formation of muscle quality is regulated by feed nutrition. As the core feed for ruminants, forage has metabolomic characteristics (such as amino acids, fatty acids, and secondary metabolites) that enter the animal body through digestion and absorption, directly participating in material synthesis and energy metabolism in muscle cells and ultimately affecting the flavor, tenderness, and nutritional composition of meat [1].

Saline soils represent a specialized habitat in terms of biological resources and biodiversity [2]. Saline–alkaline soils are typically rich in soluble salts, such as Na^+, Cl^-, SO_4^{2-}, and exchangeable Na^+, with a total salt content significantly higher than that of non-saline–alkaline soils. High concentrations of Na^+ can cause an increase in pH, triggering an imbalance of mineral elements. This further leads to elevated contents of mineral elements such as Na, Cl, Ca, and Mg in saline–alkaline soils while reducing the availability of trace elements such as Fe, Mn, and Zn [3]. Changes in the availability of mineral elements in the soil can affect the optimal growth and functional status of plants. The saline–alkaline land in Qinghai Province constitutes 3.2% of the national saline–alkaline land area. The grass in these regions is suitable for the soil conditions, exhibiting enhanced tolerance to cold, drought, and salinity, and serves as feed for animals such as horses, cattle, and sheep [4]. As the primary forage in the saline–alkaline land of Qinghai Province, oats (*Avena sativa* L.) possess a high relative forage value (RFV). Oats grown in saline–alkaline land are rich in essential nutrients, including fats, proteins, and vitamins. They are renowned for their tolerance to cold, drought, and saline–alkaline conditions and are widely regarded as a viable option for the improvement of saline–alkaline land [5]. Liu et al. [6] studied different oat varieties grown in saline and alkaline soils of the Songnen Plain, observing that these oats had high stem-to-leaf ratios, high RFV values, and good relative feeding values. For this reason, they can serve as pioneer plants for the improvement of saline–alkaline soils. Bai et al. [7] investigated the effects of alkaline stress on oats. They found that under alkaline stimulation, the activities of superoxide dismutase (SOD) and peroxidase (POD) in oats increased, along with an increase in the content of soluble sugars. These responses can enhance the alkaline tolerance of oats. Moreover, Yang et al. [8] found that drought stress (DS) can induce peroxidation and osmotic stress in plants, which in turn respond to drought stress by synthesizing osmoprotectants to regulate osmotic pressure. There is insufficient understanding of the disparities in the quality of oats grown on saline compared with non-saline soils in Qinghai.

Tibetan sheep (*Ovis aries*), one of China's three major basic sheep breeds, is an endemic species to the Qinghai–Tibet Plateau [9]. These sheep are indigenous to Qinghai Province, where they are reared by farmers and herders. They constitute a fundamental basis of the animal husbandry industry in the province [10]. Liang et al. [11] showed that feeding saline forage to sheep can improve the growth performance and feed-to-weight ratio, serum protein metabolism, immunity, antioxidant capacity, and mineral- and flavor-associated compounds in the meat. Qiu et al. [12] investigated the effect of saline alfalfa (*Medicago sativa* L.) on the meat quality, organ development, and serum biochemical indices of meat goats, finding that saline alfalfa was more palatable than conventional alfalfa, with better color, freshness, and tenderness of the meat, and reduced values of cooking and dripping loss. A study by Moreno et al. [13] showed that the meat of lambs fed on saline oats had higher contents of ash, total saturated fatty acids, and polyunsaturated fatty acids, together with lower n-6:n-3 ratios. Pearce et al. [14] discovered that the carcasses of goats fed with halophytes, such as saltbush (*Atriplex* spp.), can yield a higher proportion of lean meat and a lower amount of fat. Furthermore, this method of feeding can increase vitamin E levels, which helps to maintain the meat's color. Therefore, it was thought that the high metabolite content of oat grasses growing in salt water could improve the quality of Tibetan lamb by encouraging the sheep to store more nutrients and metabolites in their muscles.

There is currently no research on the impact of oat grasses cultivated in saline–alkaline parts of Qinghai on the quality of Tibetan sheep meat. Furthermore, there is no association between saline oats and meat quality. To clarify the mechanism by which oat cultivation in saline soil affects the quality of Tibetan sheep meat, and to identify the key components in oats that influence this change, we conducted the following experiments. Samples were obtained from both saline and non-saline areas in Qinghai Province to investigate nutrient levels and metabolites in oat plants. After a 120-day feeding period, the longissimus dorsi muscles of Tibetan sheep were collected to analyze their eating quality, nutritional quality, and targeted and untargeted metabolites. The relationships between the differential metabolites of the saline oats and the quality of Tibetan sheep, as well as among the differential metabolites, were also examined. The principal metabolites and metabolic pathways by which saline oats affect the quality of Tibetan sheep were identified. The findings serve as a standard for advancing the animal husbandry ecosystem in the Qinghai region, offering extensive data and technical assistance for integrating specific nutrients into standardized feeding protocols.

2. Materials and Methods

The Committee of Experimental Animal Care approved all experimental procedures involving animals, while the Qinghai University of Animal Care approved the handling techniques (QUA-2022-0515).

2.1. Samples Collection

2.1.1. Oat Sample Collection

Following the standards set by the National Forage Testing Association (NFTA), oats were collected from the saline–alkaline land of Gonghe County, Hainan Prefecture, Qinghai Province, China (GX; latitude: 36°28′2″ N; longitude: 99°16′26″ E; and altitude: 3168.1 m), and non-saline land of Haiyan County, Haibei Prefecture, Qinghai Province, China (YX; latitude: 36°59′36″ N; longitude: 100°55′5″ E; and altitude: 3111 m), and were harvested and preserved at −80 °C for subsequent analysis of oat quality across the distinct regions. The soil sampling areas were consistent with the oat-growing areas, and soil samples were collected in the oat-growing fields of the two locations.

2.1.2. Experimental Design of Oat Feeding in Different Areas of Cultivation

Sixty healthy male sheep, 2 months old and with similar body conditions, were selected and randomly allocated to the Gonghe-housed group (YB, n = 30) and the Hai-yan-housed group (B0, n = 30). Group YB sheep were raised at Xiangka Meiduo farm in Gonghe County, Hainan Prefecture, Qinghai Province, China, and were fed local saline-cultivated oats. Group B0 sheep were raised at Jinzang Farm in Haiyan County, Haibei Prefecture, Qinghai Province, China, and were fed local non-saline-cultivated oats. Referring to the literature with minor modifications [15,16], both groups were provided with identical feed concentrates and intake, as detailed in Table 1. The two groups of animals (30 in each group) were housed in enclosures with wind-sheltered exercise areas that were also sunny, dry, and well ventilated. The animals were fed twice daily at 08:30 am and 4:30 pm with unrestricted access to feed and water; any feed remaining from the previous feeding time was collected and weighed before the next feeding. Furthermore, the housing was swept, gutters were cleaned daily, and the housing and exercise yards underwent weekly disinfection and sterilization. The fences were maintained in a clean and hygienic condition. All Tibetan sheep underwent immunization, and the transmission of internal and external parasites was systematically prevented and managed. The official 120-day experiment was conducted after a 7-day adaptation period, and slaughtering was initiated at 6 months of age.

Table 1. Diet composition and nutritional level (dry matter basis, %).

Dietary Composition	B0 (%)	YB (%)
Corn	27.6	27.6
Soybean meal	3.6	3.6
Canola meal	6.6	6.6
Cottonseed meal	9.6	9.6
Wheat	7.8	7.8
Sodium chloride	0.6	0.6
Limestone	0.6	0.6
Sodium bicarbonate	0.6	0.6
Premix	3	3
Total	60	60
Roughage	40 (Haiyan oats)	40 (Gonghe oats)

Note: YB represents feeding with oats grown in saline–alkali land, and B0 represents feeding with oats grown in non-saline–alkali land. The same below.

2.1.3. Tibetan Sheep Meat Sample Collection

At the end of the feeding trial, six experimental animals were randomly selected from each group and transported to a nearby commercial abattoir. The animals were fasted for 12 h (no food or liquid) and were humanely slaughtered according to animal welfare procedures; i.e., the lambs were stunned and bled. After slaughtering, the Longissimus dorsi lumborum was also removed from one side of each carcass. Slaughtering and sampling were performed together by professionals following uniform standards. All samples were placed in dry ice and transferred to the laboratory for storage at −80 °C for subsequent analysis. Six replicates per group were used for all meat and metabolomics analyses.

2.2. Determination of Soil Mineral Elements

Following the method described by Song et al. [17], the concentrations of mineral elements in soil samples were determined using inductively coupled plasma optical emission spectrometry (ICP-OES, Optima 8300, Perkin Elmer, Waltham, MA, USA).

2.3. Oat Quality Analysis

2.3.1. Nutritional Values Analysis of Oats

The oats were dried in a blast-drying oven, pulverized, and passed through a 40-mesh sieve to determine their value. Then, according to the method described by Han et al. [18], the moisture content was determined using the oven method (DHG-9070A, Shanghai Bluepard Instruments Co., Shanghai, China), the crude protein content was determined using the Kjeldahl method (K9840, Hanon Advanced Technology Group Co., Jinan, Shandong, China), and the crude fat content was determined using the Soxhlet extraction method (SOx406, Shandong Haineng Scientific Instrument Co., Jinan, Shandong, China). Near-infrared (NIR) spectroscopy (INFRAMATIC 8620) was used to determine acid detergent fiber (ADF) and neutral detergent fiber (NDF) [19]. The acid–base fractionated hydrolysis method was used to determine the crude fiber [20].

2.3.2. Determination of Oats Quality Indices

The relative feeding value (RFV) was calculated as described by Gao [21] using the following formula:

$$\text{RFV} = \text{DMI (\%DW)} \times \text{DDM (\%DW)}/1.29 \quad (1)$$

where DMI represents the dry matter intake, and DDM indicates the digestible dry matter;

$$\text{DMI} = 120/\text{NDF} \quad (2)$$

where NDF indicates the neutral detergent fiber;

$$\text{DDM} = 88.9 - 0.779 \times \text{ADF} \quad (3)$$

where ADF represents the acidic detergent fiber.

2.3.3. Free Amino Acid-Targeted Metabolomics Determination of Oats

The samples were extracted from storage at −80 °C and accurately weighed to 60 mg utilizing an electronic balance (AL104, Mettler Toledo, Greifensee, Zurich, Switzerland). Then, 50 μL of water homogenate was incorporated into each sample, followed by vortexing for 60 s with a vortex mixer (QT-1, Shanghai Kit Analytical Instrument Co., Shanghai, China). A solution of methanol (≥99.0%, Fisher Chemical, Pittsburgh, PA, USA) and acetonitrile (≥99.0%, Fisher Chemical, Pittsburgh, PA, USA) (1:1, *v*/*v*) was then introduced in a volume of 400 μL, along with 50 μL of a 50 μM internal standard mixture containing 16 isotopes. The samples were incubated at −20 °C for one hour to precipitate proteins after the mixture was vortexed for 60 s and then subjected to low-temperature sonication using an ultrasonic instrument (JP-100, Shenzhen Jiemeng Cleaning Equipment Co., Shenzhen, Guangdong, China) for two 30 min intervals. Centrifugation was performed at 14,000 rcf and 4 °C for 20 min using a centrifuge (5430R, Eppendorf, Hamburg, Germany). The resulting supernatant was freeze-dried using a vacuum freeze dryer (FD-IC-50, Shanghai Bilang Instrument Co., Shanghai, China) and stored at −80 °C.

Chromatographic separation was conducted using a UHPLC system (1290 Infinity, Agilent, Santa Clara, CA, USA). Standards (≥99.0%, Sigma-Aldrich, St. Louis, MO, USA) were maintained in an autosampler at 4 °C, with the column temperature set to 35 °C. Mass spectrometry analyses were conducted using a mass spectrometer (6500/5500 QTRAP, SCIEX, Framingham, MA, USA) operating in positive ion mode. Quality control (QC) samples were produced by combining aliquots from all samples to evaluate data stability and reproducibility. The relative standard deviation (RSD) for the analyte in the QC samples was under 10%, signifying that the results were stable and reliable.

The distribution diagram of the RSD of free amino acids in the QC samples is presented in the Supplementary Materials, Figure S1A. Table S1 and Formula (S1) of the Supplementary Materials present the relevant standard curves and formulas.

2.3.4. Fatty Acid-Targeted Metabolomics Determination of Oats

The quantitative method was strictly validated following the relevant standards of the International Organization for Standardization (ISO), including validation items such as relative standard deviation (RSD), limit of detection, limit of quantification, and linear range. Detailed data on the method performance characteristics are provided in the Supplementary Materials.

Following the gradual thawing of the sample at 4 °C, 60 mg of the sample was accurately weighed using an electronic analytical balance (AL104, Mettler Toledo, Greifensee, Zurich, Switzerland) and combined with 5 mL of dichloromethane (≥99.0%, Sigma, St. Louis, MO, USA)–methanol (≥99.0%, Fisher Chemical, Pittsburgh, PA, USA) solution (2:1 *v*/*v*). The mixture was thoroughly vortexed, and 2 mL of ultrapure water was added to wash it. The lower phase of the solution was then isolated and evaporated to dryness using a nitrogen stream. Following this, 2 mL of n-hexane was introduced, along with the internal standard, and the mixture underwent methyl esterification for 30 min. Following methylation, 2 mL of ultrapure water was introduced, and 2000 μL of the supernatant was aspirated and evaporated under nitrogen.

The residue was re-dissolved in n-hexane, and the supernatant was transferred into an injection vial for gas chromatography–mass spectrometry (GC-MS) analysis (Agilent, Santa Clara, CA, USA). The samples were separated on a capillary column (19091S-433UI: HP-5ms, 30 m × 250 μm × 0.25 μm, Agilent, America) using a gas chromatography system, with helium as the carrier gas at a flow rate of 1.0 mL/min. Mass spectrometric analysis was conducted using a triple quadrupole mass spectrometer (5977B MSD, Agilent, Santa Clara, CA, USA), and the detection mode was selected ion monitoring (SIM). The QC samples were produced by combining aliquots from all samples to evaluate data stability and reproducibility. The relative standard deviation (RSD) of the analyte in the QC samples was below 10%, signifying reliable and stable results. The internal standard method was used for quantitative analysis, with methyl nonadecanoate (≥99%, NU-CHEK Prep, Inc., Elysian, MN, USA) as the reference material. The calibration process adopted matrix-matched calibration; that is, a series of concentration standard solutions was prepared using blank sample matrices. The linear correlation coefficients (R^2) of the plotted calibration curves were all greater than 0.99, meeting the requirements of quantitative analysis.

The distribution diagram of the RSD of fatty acids in the QC samples is presented in the Supplementary Materials, Figure S1B. Table S1 and Formula (S2) of the Supplementary Materials present the relevant standard curves and formulas.

2.3.5. Monosaccharide-Targeted Metabolomics Determination of Oats

In this experiment, detection was performed using GC-MS (8890-5977B, Agilent, Santa Clara, CA, USA) with a triple quadrupole mass spectrometer, and the detection mode was selected ion monitoring (SIM). The quantitative method was validated following the relevant standards of the International Organization for Standardization (ISO), including validation items such as relative standard deviation (RSD), limit of detection, limit of quantification, and linear range. Monosaccharide standards were used as calibration standards, and a series of concentration standard solutions was prepared using matrix-matched calibration. The correlation coefficient (R^2) of the calibration curve was greater than 0.99.

The samples underwent vacuum freeze-drying using a vacuum freeze dryer (CentriVap LABCONCO, Kansas City, MO, USA). They were ground into a powder using a ball mill (MM400, Retsch, Haan, North Rhine-Westphalia, Germany) operating at 30 Hz for 1.5 min. In total, 20 mg of the resultant powder was then measured into appropriately labeled centrifuge tubes. A solvent mixture consisting of methanol (chromatographically pure, Merck, Darmstadt, Hesse, Germany), isopropanol (Merck, Kenilworth, NJ, USA), and water in a volumetric ratio of 3:3:2 ($v/v/v$) was prepared, and 500 μL of this extract was added to each sample. The samples were vortexed for 3 min and sonicated at 4 °C for 30 min using a multi-tube vortex mixer (MIX-200, Shanghai Jingmei, Shanghai, China). Following this, centrifugation was performed at 4 °C and 12,000 rpm for 3 min using a centrifuge (5424R, Eppendorf, Hamburg, Germany). In total, 50 μL of the supernatant was aspirated, to which 20 μL of an internal standard solution at a concentration of 1000 μg/mL was added. The mixture was subjected to nitrogen evaporation and lyophilization. Subsequently, 100 μL of pyridine methoxide ammonium salt (99%, Sigma-Aldrich, St. Louis, MO, USA) (15 mg/mL) was added, and the samples were incubated at 37 °C for 2 h. Following this, 100 μL of BSTFA (99%, Shanghai Aladdin Biochemical Technology Co., Shanghai, China) was added, and the incubation continued at 37 °C for another 30 min (Thermo Scientific Forma 311, Thermo Fisher Scientific, Waltham, MA, USA).

Over 80% of the compounds in the QC samples had coefficient of variation (CV) values below 0.3, signifying the stability of the experimental data. Moreover, the proportion of compounds exhibiting CV values below 0.2 in the QC samples surpassed 80%, indicating substantial data stability.

Table S2 and Formula (S3) of the Supplementary Materials present the relevant standard curves and formulas. The parameters of GC-MS are shown in the Supplementary Materials, Table S3.

2.3.6. Untargeted Metabolomics Determination of Oats

After the samples were slowly thawed at 4 °C, 60 mg of each sample was weighed using an electronic balance and added to pre-chilled methanol (≥99.0%, Fisher Chemical, Pittsburgh, PA, USA)–acetonitrile (≥99.0%, Fisher Chemical, Pittsburgh, PA, USA)–water solution (2:2:1, v/v). The mixture was homogenized using vortexing with a vortex mixer (QT-1, Shanghai Kit Analytical Instrument Co., Shanghai, China) and then subjected to low-temperature sonication for 30 min using an ultrasonic cleaner (KQ5200E, Kunshan

Shumei, Kunshan, Jiangsu, China). Subsequently, the samples were incubated at −20 °C for 10 min. After that, the samples were centrifuged at 14,000 rpm for 20 min at 4 °C using a centrifuge (5430R, Eppendorf, Hamburg, Germany). The supernatant was collected and dried under a vacuum (FD-IC-50, Shanghai Bilang Instrument Co., Shanghai, China). For mass spectrometry analysis, the dried residue was reconstituted in 100 μL of an acetonitrile–water solution (1:1, *v*/*v*), vortexed again, and then centrifuged at 14,000× *g* for 15 min at 4 °C. Finally, the supernatant was injected for analysis.

The samples were separated using a UHPLC system (1290 Infinity LC, Agilent, Santa Clara, CA, USA) with a HILIC column (ACQUITY UPLC BEH Amide 1.7 μm, 2.1 mm× 100 mm column, Waters, Milford, MA, USA). The column temperature was maintained at 25 °C, the flow rate was set at 0.5 mL/min, and the injection volume was 2 μL. The mobile phase consisted of two components: mobile phase A consisted of a mixture of water, 25 mM ammonium acetate (≥99.0%, Sigma-Aldrich, St. Louis, MO, USA), and 25 mM Ammonia solution, while mobile phase B consisted of acetonitrile (≥99.0%, Fisher Chemical, Pittsburgh, PA, USA). The gradient elution schedule was 0.5–7 min, and the fraction of mobile phase B linearly reduced from 95% to 65%. From 7 to 8 min, it declined linearly from 65% to 40%. From 8 to 9 min, the proportion of mobile phase B remained constant at 40%. From 9 to 9.1 min, the proportion of mobile phase B produced linearly from 40% to 95%. From 9.1 to 12 min, the proportion of mobile phase B was sustained at 95%. The samples were maintained in an autosampler at 4 °C during the complete analysis process.

Mass spectrometry analysis was conducted using Q Exactive-series mass spectrometers (Thermo Fisher Scientific, Waltham, MA, USA), with detection performed in both the positive and negative electrospray ionization (ESI) modes. The parameters for the ESI source and mass spectrometry settings were as follows: nebulizing gas and auxiliary heating gas 1 (Gas1): 60; auxiliary heating gas 2 (Gas2): 60; curtain gas (CUR): 30 psi; ion source temperature: 600 °C; and spray voltage (ISVF): ±5500 V (for both positive and negative modes). The mass spectrometry acquisition mode was full scan. The primary mass-to-charge ratio detection range was 80–1200 Da with a resolution of 60,000 and a scan accumulation time of 100 ms. The secondary level adopted a segmented acquisition method, with a scanning range of 70–1200 Da; a secondary resolution of 30,000; and a scan accumulation time of 50 ms.

Raw data were converted to the mzXML format using ProteoWizard, followed by peak alignment, retention time correction, and peak area extraction via the XCMS software (version 3.14.0, Scripps Research, La Jolla, CA, USA). The extracted data underwent initial metabolite annotation through the combined use of these two tools, with subsequent structural confirmation referencing the Human Metabolome Database (HMDB) and Kyoto Encyclopedia of Genes and Genomes (KEGG). Fragmentation spectra obtained using liquid chromatography–high resolution tandem mass spectrometry (LC-HRMS/MS) were utilized, where MS/MS data assisted in accurate metabolite annotation to ensure the reliability of identification results. The identified metabolites were further subjected to data preprocessing, and their functions and involved metabolic pathways were determined using HMDB and KEGG.

2.4. Determination of Meat Quality

2.4.1. Determination of Carcass Traits

Carcass segmentation involved measuring the thickness of rib meat, abdominal wall, backfat, and the eye muscle area, as outlined by Ma et al. [22].

The area of the eye muscle was determined at the cross-section between the 12th and 13th ribs of the Tibetan sheep during carcass segmentation. This cross-section was outlined using sulfuric acid paper, and then, the area was calculated with a 1 cm × 1 cm grid. The rib thickness was measured as the tissue thickness 110 mm from the 12th and 13th ribs to the midline of the spine in the Tibetan sheep. The thickness of the abdominal wall was assessed at a point 127 mm from the 12th and 13th ribs. Furthermore, the backfat thickness was measured as the fat layer directly above the center of the eye muscle between the 12th and 13th ribs of the Tibetan sheep.

2.4.2. Determination of Meat-Eating Quality

Qualities associated with eating, including pH, color, thawing loss, cooking loss, cooked meat percentage, and texture, were determined using the method of Zhang et al. [23] with slight modifications.

Briefly, the pH_{45min} and pH_{24h} were determined by inserting a portable pH meter (PHS-3C, Shanghai Leici Instrument Factory, Shanghai, China) into the meat samples at a depth of 2–3 cm. We calibrated the pH meter using pH 4.0 and 6.86. An automatic colorimeter (ADCI-60-C, Beijing Chentaike Instrument Technology Co., Beijing, China) was used to measure the values of L^* (lightness), a^* (redness), and b^* (yellowness) on the meat surface. The colorimeter was equipped with a standard xenon lamp within the close aperture of 8 mm set to Illuminant D65 with an observer angle of 2°. In addition, the meat samples were heated in a water bath at a constant temperature of 80 °C until the internal temperature reached 70 °C. Next, using an iron ruler and a scalpel, the samples were cut into 3 cm × 1 cm × 1 cm meat columns in the direction of the muscle fibers. A muscle tenderness meter (RH-N50, Guangzhou Runhu Instrument Co., Guangzhou, Guangdong, China) was then used in the direction of the vertical muscle fiber to evaluate the shear force.

The water-holding capacity was calculated as the ratio of the amount of water lost by the meat samples to the initial weight of the samples, which was determined after the meat samples (1 cm × 1 cm × 1 cm) were subjected to a pressure of 350 N by a water-holding capacity tester (RH-1000, Guangzhou Runhu Instrument Co., Guangzhou, Guangdong, China). The cooked meat percentage was calculated as the ratio of the weight of the meat samples after being boiled in a water bath at 80 °C for 40 min to the initial weight of the samples (average weight of 60 g). For cooking loss analysis, the meat samples (2 cm × 3 cm × 2 cm) were cooked in a water bath at 80 °C for 30 min, and cooking loss was calculated as a percentage of the weight change in the samples from the initial weight of the samples. In the same way, the thawing loss was calculated as the ratio of the weight of the meat samples after being unfrozen in a refrigerator at 4 °C for 12 h to the initial weight of the samples (average weight of 30 g). Next, a texture profile analysis (TPA) analyzer (TA.XTC-18, Shanghai Baosheng Industrial Development Co., Shanghai, China) was used to measure the hardness, elasticity, adhesion, cohesion, and chewiness of the samples (1 cm × 1 cm × 1 cm). The probe model utilized was TA3/100, while the fixture model was TA-RT-KIT.

2.4.3. Determination of Meat Sensory Evaluation

This study references the standards ISO 13299:2016 (en) and ISO 5492 for the sensory evaluation methodology, specifically focusing on establishing a sensory profile for two groups of Tibetan sheep samples.

Approximately 100 g of Tibetan sheep meat samples from each group were weighed, and the two groups were individually labeled. These samples were then cooked in a thermostatically regulated water bath (HH-6, Changzhou Langyue Instrument Manufacturing

Co., Nanjing, Jiangsu, China) at a constant temperature of 100 °C. Once the sheep's internal temperature fell between 60 and 70 °C, the samples were taken from the water bath. The cooked sheep was cut further into equally sized and shaped cubes, each weighing roughly 3 g, guaranteeing the objectivity of the evaluation by the panelists. A sensory evaluation panel of 15 members assessed the color, aroma, juiciness, taste, texture, and general acceptability of the Tibetan sheep samples in each group. The more detailed sensory evaluation is shown in the Supplementary Materials, Table S4.

2.4.4. Determination of Meat Nutritional Quality

The method of the AOAC (2005) was used to determine the moisture, crude fat, and crude protein contents of the meat samples [24]. In brief, the moisture content of the muscle was determined using direct drying in an oven (DHG-9070A, Shanghai Bluepard Instruments Co., Shanghai, China) at 105 °C until a constant weight was obtained. The crude fat content was assessed using the Soxhlet extraction method (SOx406, Shandong Haineng Scientific Instrument Co., Jinan, Shandong, China). Moreover, the content of crude protein was measured using the Kjeldahl nitrogen method (K9840, Hanon Advanced Technology Group Co., Jinan, Shandong, China).

2.4.5. Free Amino Acid and Fatty Acid Contents in Tibetan Sheep Meat

The free amino acid and fatty acid contents in the Tibetan sheep meat were determined using the methods described in Sections 2.3.3 and 2.3.4, respectively.

2.4.6. Untargeted Metabolomics Determination of Tibetan Sheep Meat

The untargeted metabolomics analysis of Tibetan sheep meat was performed using the methods described in Section 2.3.6.

2.5. Data Processing and Analysis

The data were expressed as mean ± standard deviation (SD) and analyzed using independent samples t-tests in SPSS version 26.0 (IBM Corp., Armonk, NY, USA), with $p < 0.05$ considered statistically significant. Following sum-normalization, the data were processed and then subjected to multivariate data analysis using the R package (ropls) (version 1.26.0, maintained by Etienne Thevenot, Lyon, Auvergne-Rhône-Alpes, France), which included partial least squares discriminant analysis (PLS-DA), orthogonal partial least squares discriminant analysis (OPLS-DA), and Pareto-scaled principal component analysis (PCA). To determine each variable's contribution to the classification, the variable importance in the projection (VIP) value was computed for each variable in the OPLS-DA model. The associations between meat quality and meat and oat metabolites were assessed using Pearson's correlation coefficients, with $p < 0.05$ considered statistically significant.

3. Results

3.1. Analysis of Soil Mineral Elements

As shown in Table 2, both types of soils had relatively high contents of elements such as Al, Ca, Fe, K, Mg, and Na, while the contents of trace elements, including Cu, V, Co, Rb, Se, and Ni, were all less than 100 mg·kg^{-1}. Specifically, the contents of Al, Ca, K, Mg, Na, and S in saline–alkaline soils were significantly higher than those in non-saline–alkaline soils, whereas the contents of Fe, P, and Mn were significantly lower ($p < 0.001$). In addition, all trace elements in non-saline–alkaline soils were significantly higher than those in saline–alkaline soils.

Table 2. Differences in mineral element contents between saline–alkali soils and non–saline–alkali soils (mg·kg^{-1}).

Mineral Elements	Groups		*p*
	Saline-Alkali Soils	Non-Saline-Alkali Soils	
Al	68,035.35 ± 4736.96	34,008.04 ± 1424.82	<0.001
Ca	57,564.72 ± 209.07	13,656.04 ± 222.85	<0.001
Fe	26,630.92 ± 71.74	32,134.5 ± 44.50	<0.001
K	20,612.75 ± 182.08	16,377.34 ± 81.11	<0.001
Mg	13,457.35 ± 299.94	6690.74 ± 99.85	<0.001
Mn	691.79 ± 7.89	884.94 ± 6.80	<0.001
Na	15,369.20 ± 87.31	11,658.48 ± 5.98	<0.001
P	718.92 ± 1.80	1370.62 ± 26.10	<0.001
S	1071.58 ± 64.08	661.8 ± 28.56	<0.001
V	6.49 ± 0.30	99.45 ± 0.46	<0.001
Cr	65.84 ± 1.68	104.77 ± 0.85	<0.001
Co	0.86 ± 0.01	15.25 ± 0.10	<0.001
Ni	2.25 ± 0.19	38.21 ± 0.20	<0.001
Cu	15.56 ± 0.44	18.22 ± 0.28	<0.001
Zn	73.48 ± 0.13	109.20 ± 0.45	<0.001
Se	0.04 ± 0.01	0.50 ± 0.01	<0.001
Rb	5.49±0.35	18.97±0.82	<0.001

3.2. *Regional Variations in Oat Nutritional Quality and Metabolite Composition*

3.2.1. Analysis of Oat Nutritional Values

There were significant differences in the nutritional quality of oats grown in different regions. Table 3 demonstrates that the crude protein levels in the GX group (9.03%) were significantly elevated compared with the YX group (7.16%) ($p < 0.01$). The crude fat content in the YX group was 7.45 g/kg, which is significantly lower than that of the GX group (14.92 g/kg). The neutral detergent fiber content showed a significant difference between the two groups ($p < 0.01$), which was significantly lower in the GX group than in the YX group.

Table 3. Differences in oat quality between regions.

Items	Groups		*p*
	GX	YX	
	Analysis of oat nutritional values		
Moisture (%)	73.94 ± 0.01	72.97 ± 0.01	0.520
Crude protein (%)	9.03 ± 0.02	7.16 ± 0.04	<0.001
Crude fat (g/kg)	14.92 ± 0.87	7.45 ± 0.88	0.005
Crude fiber (%)	6.37 ± 0.06	6.57 ± 0.15	0.139
NDF (%)	53.37 ± 0.38	57.30 ± 0.30	<0.001
ADF (%)	31.37 ± 0.24	32.33 ± 0.05	0.073
DMI (%)	2.25 ± 0.01	2.09 ± 0.01	0.010
DDM (%)	63.84 ± 0.19	63.07 ± 0.40	0.147
RFV (%)	111.28 ± 0.51	102.38 ± 0.49	0.001
	Amino acid composition (mg/100 g)		
Serine	1130.88 ± 16.86	512.06 ± 1.32	<0.001
Valine	961.42 ± 10.77	549.16 ± 21.54	<0.001
Glycine	1178.60 ± 8.73	398.49 ± 29.40	0.002
Lysine	46.34 ± 0.53	34.43 ± 0.96	0.002
Arginine	1458.35 ± 95.18	613.77 ± 16.49	0.003
Alanine	208.71 ± 0.95	159.47 ± 3.46	0.004

Table 3. *Cont.*

Items	Groups		*p*
	GX	YX	
Amino acid composition (mg/100 g)			
Phenylalanine	1474.83 ± 87.32	749.18 ± 31.37	0.004
Tyrosine	368.95 ± 16.62	195.97 ± 3.90	0.004
Asparagine	20.99 ± 2.66	7.09 ± 0.85	0.006
Glutamate	238.90 ± 13.04	85.02 ± 20.90	0.009
Methionine	107.60 ± 7.53	53.66 ± 2.05	0.010
Spermidine	6.00 ± 0.73	2.82 ± 1.28	0.020
Choline	310.25 ± 25.01	231.43 ± 8.31	0.024
Citrulline	1.50 ± 0.32	0.51 ± 0.24	0.030
Threonine	1549.18 ± 50.01	1149.47 ± 79.41	0.031
Glutamine	0.27 ± 0.02	0.22 ± 0.01	0.040
Hydroxyproline	0.15 ± 0.01	0.13 ± 0.00	0.040
Aspartate	105.13 ± 2.35	68.86 ± 12.54	0.043
Aminoadipic Acid	0.30 ± 0.02	0.36 ± 0.01	0.044
Creatinine	0.01 ± 0.00	0.01 ± 0.00	0.070
Histidine	224.10 ± 10.95	198.97 ± 24.16	0.072
Proline	387.29 ± 10.60	353.17 ± 10.22	0.091
Isoleucine	445.08 ± 27.86	402.20 ± 3.80	0.120
Creatine	0.08 ± 0.00	0.29 ± 0.18	0.150
Leucine	412.86 ± 33.75	384.23 ± 3.80	0.220
Taurine	0.06 ± 0.01	0.18 ± 0.09	0.220
Tryptophan	4.44 ± 1.13	2.88 ± 0.56	0.340
Ornithine	0.49 ± 0.08	0.44 ± 0.14	0.410
Putrescine	1.49 ± 0.16	1.85 ± 1.11	0.690
Cystine	4.74 ± 2.89	5.21 ± 2.85	0.910
Cysteine	0.03 ± 0.00	0.03 ± 0.01	0.950
EAA	5118.25 ± 173.30	3470.51 ± 52.00	0.012
NEAA	5530.76 ± 131.29	2691.04 ± 58.33	0.002
TAA	10,649.00 ± 214.13	6161.56 ± 106.44	0.003
Fatty acid composition (mg/100 g)			
Decanoic acid (c10:0)	0.01 ± 0.00	0.01 ± 0.00	0.001
Heneicosanoic acid (c21:0)	0.09 ± 0.01	0.17 ± 0.01	0.002
Nervonic acid (c24:1n9)	0.17 ± 0.01	0.25 ± 0.00	0.006
Eicosenoic acid (c20:1n9)	0.56 ± 0.10	1.01 ± 0.04	0.010
Linoleic acid (c18:2n6)	41.49 ± 4.66	23.61 ± 0.33	0.018
Docosapentaenoic acid (c22:5n3)	0.05 ± 0.00	0.06 ± 0.01	0.019
Palmitoleic acid (c16:1n7)	4.72 ± 0.41	2.71 ± 0.35	0.021
Heptadecanoic acid (c17:0)	3.53 ± 0.55	1.89 ± 0.38	0.022
Eicosatrienoic acid (c20:3n3)	0.25 ± 0.00	0.32 ± 0.02	0.027
Arachidic acid (c20:0)	0.33 ± 0.04	0.23 ± 0.01	0.028
α-Linolenic acid (c18:3n3)	1.09 ± 0.17	0.77 ± 0.05	0.035
Palmitic acid (c16:0)	259.90 ± 19.93	156.37 ± 20.82	0.042
Eicosadienoic acid (c20:2n6)	0.16 ± 0.05	40.63 ± 1.27	0.060
Eicosapentaenoic acid (c20:5n3)	0.70 ± 0.07	0.07 ± 0.01	0.134
Lignoceric acid (c24:0)	0.90 ± 0.08	0.69 ± 0.04	0.178
Pentadecenoic acid (c15:1n5)	0.61 ± 0.02	0.79 ± 0.07	0.180
Pentadecanoic acid (c15:0)	0.14 ± 0.02	0.80 ± 0.15	0.195
Heptadecenoic acid (c17:1n7)	0.06 ± 0.02	0.14 ± 0.00	0.232
Myristoleic acid (c14:1n5)	0.09 ± 0.03	0.05 ± 0.01	0.260
Docosadienoic acid (c22:2n6)	0.05 ± 0.01	0.12 ± 0.01	0.318
Tricosanoic acid (c23:0)	0.13 ± 0.02	0.05 ± 0.00	0.380

Table 3. *Cont.*

Items	Groups		p
	GX	YX	
Fatty acid composition (mg/100 g)			
Myristic acid (c14:0)	0.92 ± 0.22	0.16 ± 0.02	0.440
Lauric acid (c12:0)	0.51 ± 0.14	0.72 ± 0.13	0.451
Octanoic acid (c8:0)	0.01 ± 0.00	0.45 ± 0.05	0.480
Tridecanoic acid (c13:0)	0.01 ± 0.00	0.01 ± 0.00	0.570
Stearic acid (c18:0)	3.75 ± 0.39	0.01 ± 0.01	0.600
Docosahexaenoic acid (c22:6n3)	0.13 ± 0.02	4.11 ± 0.77	0.930
ΣPUFA	302.81 ± 13.24	181.21 ± 17.73	0.020
ΣSFA	55.37 ± 1.22	48.29 ± 1.57	0.040
ΣPUFA: ΣSFA	5.48 ± 0.35	3.77 ± 0.47	0.062
ΣMUFA	7.73 ± 0.05	6.83 ± 0.55	0.100
Σn-3	261.11 ± 17.30	157.49 ± 18.02	0.035
Σn-6	41.71 ± 4.06	23.72 ± 0.29	0.018
EPA	0.70 ± 0.06	0.69 ± 0.03	0.452
DHA	0.13 ± 0.02	0.14 ± 0.02	0.466
TFA	365.91 ± 12.09	236.33 ± 16.91	0.016
Carbohydrate content (mg/100g)			
Sucrose	4643.76 ± 78.41	1787.41 ± 158.78	<0.001
Levoglucosan	4.55 ± 0.03	3.45 ± 0.02	0.001
Mannose	4.81 ± 0.09	3.35 ± 0.13	0.002
Xylulose	1.18 ± 0.01	0.93 ± 0.01	0.003
Fructose	2934.81 ± 51.76	2743.11 ± 38.97	0.003
Glucuronic Acid	1.22 ± 0.03	1.45 ± 0.02	0.006
Galactose	43.86 ± 0.74	40.51 ± 0.44	0.012
Arabinitol	3.31 ± 0.37	1.61 ± 0.01	0.014
Maltose	61.09 ± 5.51	32.99 ± 0.83	0.015
Galacturonic Acid	0.85 ± 0.08	1.12 ± 0.09	0.019
D-Arabinose	4.33 ± 0.21	3.56 ± 0.06	0.026
Xylose	4.21 ± 0.16	3.71 ± 0.09	0.029
Trehalose	10.91 ± 0.78	8.47 ± 0.12	0.036
Ribose	8.56 ± 0.40	7.57 ± 0.20	0.036
2-Ace-2-Deo-D-Glucosamine	1.46 ± 0.05	1.63 ± 0.08	0.093
Cellobiose	44.77 ± 6.96	36.76 ± 3.10	0.121
Rhamnose	1.49 ± 0.26	1.22 ± 0.00	0.164
Inositol	32.38 ± 3.19	29.04 ± 3.79	0.183
Ribose-5-pho-Ba	1.34 ± 0.13	1.80 ± 0.34	0.198
Phenylalanine	3.51 ± 0.47	3.14 ± 0.18	0.516
Glutamic Acid	4715.67 ± 130.46	4025.62 ± 189.39	0.704
Xylitol	0.76 ± 0.03	0.75 ± 0.05	0.756
2-Deoxyribose	0.62 ± 0.02	0.61 ± 0.01	0.785
Fucose	2.06 ± 0.13	2.10 ± 0.14	0.820
Sorbitol	2.68 ± 0.53	2.71 ± 0.21	0.956

Note: $p < 0.05$ and $p < 0.01$ compared with the YX group. GX represents oats grown on saline–alkaline land, and YX represents oats grown in non-saline–alkaline land. The same below. NDF represents neutral detergent fiber, ADF represents acid detergent fiber, DMI represents dry matter intake, DDM represents digestible dry matter, RFV represents relative feeding value. EAA = sum of valine, lysine, phenylalanine, methionine, leucine, isoleucine, tryptophan, histidine, and threonine; NEAA = sum of serine, glycine, arginine, alanine, tyrosine, asparagine, glutamic acid, spermidine, choline, citrulline, glutamine, hydroxyproline, aspartic acid, proline, cysteine, cystine, and taurine; TAAs = total amino acids. ΣSFA: total saturated fatty acids; ΣMUFA: total monounsaturated fatty acids; ΣPUFA: total polyunsaturated fatty acids; Σn-3: total omega-3 PUFA; Σn-6: total omega-6 PUFA; EPA: eicosapentaenoic acid (C20:5n-3); DHA: docosahexaenoic acid (C22:6n-3). Saturated fatty acids include C6:0, C8:0, C10:0, C11:0, C12:0, C13:0, C14:0, C15:0, C16:0, C17:0, C18:0, C20:0, C21:0, C23:0, C24:0. Polyunsaturated fatty acids include C18:2n6, C18:3n3, C20:2n6, C20:3n3, C20:5n3, C22:2n6, C22:5n3, C22:6n3. Monounsaturated fatty acids include C14:1n5, C15:1n5, C16:1n7, C17:1n7, C18:1n9, C24:1n9. Total n-3 fatty acids include C18:3n3, C20:3n3, C20:5n3, C22:5n3, C22:6n3. Total n-6 fatty acids include C18:2n6, C20:2n6, C22:2n6. TFA represents total fatty acids.

3.2.2. Calculation of Oat Quality Indices

As demonstrated in Table 3, the GX group had higher DMI and RFV values ($p < 0.01$) than the YX group, suggesting that the oats in the GX group provided higher feeding values for ruminants. The digestible dry matter (DDM) and dry matter intake (DMI) values were used to evaluate the quality of the oats grown in various regions. They were negatively correlated with ADF and NDF contents.

3.2.3. Targeted Metabolomics Analysis of Oats

Free Amino Acid-Targeted Metabolomics Analysis of Oats

Table 3 shows significant differences in the free amino acid compositions and contents of oats grown in different regions. Overall, oats grown in saline conditions had significantly higher levels of essential free amino acids (EAAs), non-essential free amino acids (NEAAs), and total free amino acids (TAAs) than oats grown in non-saline soils, which implies that saline environments may promote the accumulation of free amino acids in oats. Glycine, serine, and threonine were more abundant in the saline oats. Furthermore, the GX group had higher valine, lysine, arginine, alanine, phenylalanine, and methionine levels than the YX group, indicating significant regional variations.

Fatty-Acid-Targeted Metabolomics Analysis of Oats

As shown in Table 3, a variety of common fatty acids were detected in the oats, with palmitic acid showing the highest levels in the saline oats (259.896 ± 19.93). Similarly, there were significant regional differences in the contents of linoleic acid, palmitoleic acid, arachidic acid, and α-linolenic acid: all were higher in saline than in non-saline areas. In the GX group, the contents of PUFA (302.81 ± 13.24) and SFA (55.37 ± 1.22) were significantly higher than those in the YX group. Similarly, the contents of omega-3 (261.11 ± 17.30) and omega-6 (41.71 ± 4.06) in the GX group were also significantly higher than those in the YX group. The two groups had no significant differences in MUFA, EPA, and DHA contents. However, the TFA content in the GX group (365.91 ± 12.09) was significantly higher than in the YX group.

Monosaccharide-Targeted Metabolomics Analysis in Oats

As shown in Table 3, 25 differential saccharides were detected in the oats from different regions, with sucrose representing the most abundant sugar from the saline–alkaline areas. Moreover, the levels of glucose, mannose, galactose, arabinose, and maltose also differed significantly between the regions, being higher in oats from saline soils than from non-saline land ($p < 0.05$).

3.2.4. Untargeted Metabolomics Analysis of Oats Metabolites

Identification and Analysis of Differential Metabolites in Oats

Figure S2A,B in the Supplementary Materials show that the spectra and the total ion chromatogram (TIC) of the quality control (QC) samples were compared and overlapped. The experimental findings demonstrated slight variance due to instrumental error because the response intensities and retention periods of the chromatographic peaks were identical.

The peaks obtained from the extraction of all experimental and QC samples were analyzed using PCA and are shown in the Supplementary Materials, Figure S3A,B. This suggests that the studies were reproducible because the QC samples were closely clustered in positive and negative ion modes. After the QC samples were eliminated, PLS-DA and OPLS-DA were run to improve group differentiation. The GX and YX groups demonstrated intra-group clustering and inter-group separation in both analysis modes, with the OPLS-

DA results demonstrating more pronounced effects (see Supplementary Materials, Figure S3C,D). To prevent the supervised model from being overfitted during the modeling phase, a replacement test was employed to test the model and guarantee its validity. Figure S3E,F in the Supplementary Materials show the replacement test plots of the OPLS-DA model, in which the replacement retention gradually decreased, and both R^2 and Q^2 of the stochastic model gradually decreased, indicating that the original model did not suffer from overfitting and that the model showed good stability.

Bioinformatics Analysis of Differential Metabolites in Oats

As shown in Table S5, using the predefined criteria for differential metabolites (DFMs) (variable importance in projection [VIP] > 1 and $p < 0.05$), a total of 58 DFMs were detected in both positive and negative ion modes, with 31 in the positive ion mode and 27 in the negative ion mode. The KEGG pathway enrichment analysis shown in Figure 1A illustrates the top 12 pathways, with most related to pyrimidine, free amino acid, purine, and carbohydrate metabolism. A differential enrichment score was constructed to further compare the pathways associated with the DFMs leading to changes in feed metabolites between regions, as shown in Figure 1B.

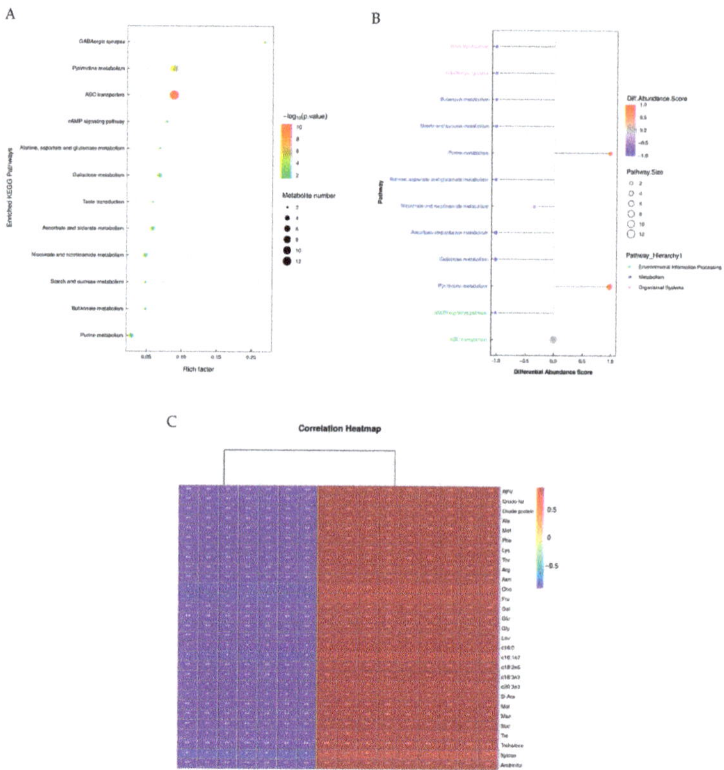

Figure 1. (**A**) Top 12 enriched KEGG pathways of the comparison between YX and GX groups, (**B**) a differential abundance score map of differential metabolic pathways, (**C**) correlation heatmap between oat quality parameters and oat metabolomics analysis. The colors red and blue represent positive and negative correlations, respectively. * $p < 0.05$ and ** $p < 0.01$.

The results in Table 4 show that 14 metabolic pathways were upregulated while 2 were downregulated in the GX group compared with the YX group (DA score > 0.5, $p < 0.05$). Butanoate metabolism; starch and sucrose metabolism; galactose metabolism; alanine, aspartate, and glutamate metabolism; ascorbic acid and aldolate metabolism; the TCA cycle; phenylalanine metabolism; the cAMP signaling pathway; taste transduction; GABAergic synaptic pathways; and fructose and mannose metabolism were among the metabolic pathways that were upregulated. On the other hand, purine and pyrimidine metabolism was downregulated. Succinate, 4-aminobutyric acid, sucrose, trehalose, raffinose, melibiose, D-allulose, D-galacturonic acid, D-arabinose, and L-arabinose-1,4-lactone were among the significant metabolites that were upregulated. In comparison, the downregulated metabolites included His-Lys, Ile-Pro, and His-Ser, as well as deoxyadenosine, deoxyguanosine, 2′-deoxycytidine, thymine, and cytosine.

Table 4. DFMs (differential metabolites) from oats in the key differential enriched KEGG pathways (false discovery rates < 0.05).

Metabolic Pathways (GX vs. YX)	**Metabolites**
Upregulation in the GX group cAMP signaling pathway	Succinate, 4-Aminobutyric acid (GABA)
Taste transduction	Sucrose, Saccharose 4-Aminobutyric acid (GABA)
GABAergic synapse	Succinate, Butanedionic acid 4-Aminobutyric acid (GABA) Ethylenesuccinic acid
Butanoate metabolism	Succinate, Succinic acid Butanedionic acid Ethylenesuccinic acid 4-Aminobutyricacid (GABA) gamma-Aminobutyric acid
Starch and sucrose metabolism	Trehalose, Sucrose 1-α-D-Glucopyranosyl-2-β-D-fructofuranoside
Galactose metabolism	Sucrose, Cane sugar Raffinose, Saccharose, Melibiose 1-α-D-Glucopyranosyl-2-β-D-fructofuranoside
Alanine, aspartate, and glutamate metabolism	Succinate, Butanedionic acid Ethylenesuccinic acid gamma-Aminobutyric acid
Ascorbate and aldarate metabolism	D-Galactarate, D-Mucic acid D-arabinose L-Arabinono-1,4-lactone
Biosynthesis of secondary metabolites	Trehalose
Citrate cycle (TCA cycle)	Succinate, Butanedionic acid Ethylenesuccinic acid
Phenylalanine metabolism	Succinate, Butanedionic acid Ethylenesuccinic acid
Pyruvate metabolism	Succinate, Butanedionic acid Ethylenesuccinic acid

Table 4. *Cont.*

Metabolic Pathways (GX vs. YX)	Metabolites
Amino sugar and nucleotide sugar metabolism	D-arabinose, L-Arabinose, L-Arabinopyranose
Fructose and mannose metabolism	D-Allose
Downregulation in the GX group	
Purine metabolism	Deoxyguanosine, His-Lys Deoxyadenosine
Pyrimidine metabolism	Ile-Pro, His-ser, 5-Methyluracil Thymine, Cytosine, Cytidine 2′-deoxycytidine, Deoxythymidine

Overall, carbohydrate and free amino acid metabolism were upregulated, while pathways associated with nucleotide metabolism were downregulated in the GX group compared with the YX group. This result indicates marked differences in the metabolic pathways in saline- and non-saline-grown oats.

3.2.5. Correlation Between Oats Quality Parameters and Metabolomics

The correlation heatmap presented in Figure 1C illustrates the relationship between oat quality (nutritional quality, free amino acids, fatty acids, and sugars) and untargeted metabolites. Overall, there was a significant correlation between oat metabolites and oat quality. Specifically, RFV, crude protein, Ala, Met, Phe, Lys, Thr, Arg, Asn, Cho, Fru, Gal, Glu, Gly, Lev, c16:0, c16:1n7, c18:2n6, c18:3n3, c20:3n3, D-Ara, Mal, man, suc, tre, trehalose, xylose, and arabinitol were positively correlated with D-galactarate, D-allose, D-arabinose, raffinose, palatinose, sucrose, melibose, erythritol and succinate and negatively correlated with GABA, His-Ser, Ile-Pro, His-Lys, 5′-deoxyadenosine, deoxyguanosine, and palmitic acid.

3.3. *Meat Quality Analysis*

3.3.1. Carcass Traits

Table 5 displays the differences in the carcass quality of Tibetan sheep meat between the YB and B0 groups. The carcass quality of Tibetan lamb meat did not differ significantly between the two groups, except for a significant difference in the area of the eye muscle, which was smaller in the animals belonging to the YB group than in the animals belonging to the B0 group ($p < 0.05$).

Table 5. The effect of different regional oats on carcass quality of Tibetan sheep meat.

Items	Groups		*p*
	YB	B0	
Carcass quality			
Average daily feed intake (kg)	1.45 ± 0.05	1.41 ± 0.08	0.441
Remaining after the feeding (kg)	0.63 ± 0.06	0.71 ± 0.08	0.494
Refuse percentage (%)	30.13 ± 0.02	33.46 ± 0.04	0.046
Initial weight (kg)	15.73 ± 0.88	16.14 ± 0.65	0.316
Final weight (kg)	34.55 ± 0.79	35.84 ± 1.23	0.154
Rib fat thickness (cm)	1.56 ± 0.70	1.53 ± 0.23	0.800
Abdominal fat thickness (cm)	1.79 ± 0.14	1.89 ± 0.17	0.360
Backfat thickness (cm)	1.57 ± 0.12	1.60 ± 0.15	0.620
Eye muscle area (cm^2)	17.20 ± 1.64	26.00 ± 4.90	0.010

Table 5. *Cont.*

Items	Groups		p
	YB	B0	
	Edible quality		
Color	6.89 ± 1.20	6.00 ± 1.59	0.255
Aroma	7.78 ± 0.97	6.11 ± 1.45	0.011
Juiciness	7.89 ± 1.45	6.11 ± 1.30	0.013
Taste	8.22 ± 1.10	6.67 ± 1.41	0.019
Texture	8.44 ± 0.73	6.22 ± 1.56	0.001
General acceptability	8.44 ± 0.88	7.00 ± 1.41	0.019
pH45min	6.16 ± 1.08	6.72 ± 0.28	0.080
pH24h	5.91 ± 0.34	5.82 ± 0.43	0.510
*L**	32.79 ± 0.57	36.50 ± 0.97	0.001
*a**	21.11 ± 0.95	16.89 ± 1.79	0.002
*b**	7.91 ± 0.83	10.57 ± 0.97	0.004
Thawing loss (%)	6.94 ± 2.38	7.80 ± 1.11	0.001
Cooking loss (%)	32.15 ± 4.79	52.25 ± 4.79	0.001
Cooked meat percentage (%)	67.67 ± 1.86	32.94 ± 1.23	0.003
water holding capacity (%)	15.70 ± 1.84	13.78 ± 1.85	0.036
Shear force (N)	130.14 ± 18.66	177.99 ± 16.50	0.020
Hardness (g)	11.00 ± 9.31	38.37 ± 32.58	0.010
Elasticity (mm)	2.62 ± 0.43	3.07 ± 0.54	0.160
Adhesion (g)	5.04 ± 3.90	22.52 ± 20.67	0.024
Chewability (mJ)	17.75 ± 2.93	37.86 ± 6.67	0.015
Cohesion (g)	0.46 ± 0.09	0.51 ± 0.14	0.312
Moisture (%)	74.45 ± 0.01	70.97 ± 0.04	0.348
Protein (%)	22.69 ± 0.36	19.80 ± 0.42	0.023
Fat (%)	0.60 ± 0.07	0.80 ± 0.07	0.001
	Amino acid composition (mg/100 g)		
Arginine	1365.11 ± 15.35	260.31 ± 11.81	<0.001
Threonine	253.24 ± 0.57	774.26 ± 14.94	<0.001
Proline	236.12 ± 3.56	168.11 ± 1.79	<0.001
Asparagine	2.94 ± 0.05	6.03 ± 0.07	<0.001
Histidine	280.19 ± 8.70	1015.98 ± 23.55	0.001
Cystine	27.91 ± 3.13	314.68 ± 14.38	0.001
Phenylalanine	396.48 ± 37.01	551.87 ± 28.18	0.001
Glycine	1061.81 ± 15.46	739.96 ± 5.44	0.001
Hydroxyproline	3.40 ± 0.18	2.02 ± 0.08	0.002
Alanine	315.43 ± 1.14	297.18 ± 1.31	0.002
Leucine	580.62 ± 17.07	250.85 ± 14.56	0.003
Lysine	610.88 ± 36.57	324.15 ± 21.30	0.004
Isoleucine	383.43 ± 13.07	203.85 ± 13.79	0.007
Citrulline	4.86 ± 0.41	2.60 ± 0.39	0.008
Creatinine	2.41 ± 0.24	0.86 ± 0.10	0.008
Taurine	36.16 ± 3.11	14.13 ± 0.78	0.008
Valine	388.26 ± 16.68	216.45 ± 11.02	0.008
Glutamate	153.96 ± 10.37	62.13 ± 8.04	0.010
Creatine	125.90 ± 10.35	60.28 ± 2.06	0.013
Aminoadipic Acid	0.47 ± 0.06	0.21 ± 0.04	0.015
Cysteine	0.06 ± 0.01	0.23 ± 0.05	0.024
Glutamine	19.03 ± 3.11	8.38 ± 0.33	0.029
Methionine	45.60 ± 7.22	92.40 ± 10.33	0.032
Tryptophan	1.92 ± 0.04	3.25 ± 0.44	0.038
Aspartate	101.43 ± 7.93	80.51 ± 0.00	0.042
Serine	270.82 ± 2.33	285.00 ± 7.57	0.060
Tyrosine	336.99 ± 14.85	293.83 ± 10.74	0.087
Ornithine	0.32 ± 0.02	0.28 ± 0.02	0.157
Choline	332.25 ± 16.98	317.66 ± 20.87	0.342
EAA	2940.62 ± 20.23	3433.06 ± 33.55	0.004
NEAA	4397.37 ± 54.82	2914.39 ± 24.00	0.002
TAA	7337.99 ± 64.91	6347.45 ± 55.70	0.001

Table 5. *Cont.*

Items	Groups		p
	YB	B0	
Fatty acid composition (mg/100 g)			
Pentadecanoic acid (c15:0)	1.63 ± 0.08	4.84 ± 0.04	<0.001
Pentadecenoic acid (c15:1n5)	0.40 ± 0.14	26.72 ± 0.30	<0.001
Palmitic acid (c16:0)	240.99 ± 25.20	223.98 ± 14.76	<0.001
Eicosapentaenoic acid (c20:5n3)	4.09 ± 0.26	7.17 ± 0.21	<0.001
Behenic acid (c22:0)	0.43 ± 0.03	28.76 ± 0.89	<0.001
Erucic acid (c22:1n9)	0.43 ± 0.10	3.35 ± 0.06	<0.001
Docosapentaenoic acid (c22:5n3)	6.41 ± 0.29	3.15 ± 1.74	<0.001
Tricosanoic acid (c23:0)	0.06 ± 0.00	19.85 ± 0.52	<0.001
Lignoceric acid (c24:0)	0.05 ± 0.00	77.22 ± 2.04	<0.001
Palmitoleic acid (c16:1n7)	20.43 ± 0.31	9.06 ± 0.30	0.001
Heptadecanoic acid (c17:0)	5.87 ± 0.72	19.85 ± 0.31	0.001
Tridecanoic acid (c13:0)	0.05 ± 0.00	0.21 ± 0.01	0.001
Nervonic acid (c24:1n9)	1.34 ± 0.07	4.85 ± 0.28	0.002
Linoleic acid (c18:2n6)	60.44 ± 1.47	80.84 ± 2.50	0.003
Octanoic acid (c8:0)	0.01 ± 0.00	0.17 ± 0.02	0.004
Decanoic acid (c10:0)	0.15 ± 0.02	1.13 ± 0.12	0.004
Linoleic acid (c18:3n6)	0.90 ± 0.05	1.43 ± 0.02	0.004
Eicosenoic acid (c20:1n9)	1.24 ± 0.24	5.59 ± 0.80	0.005
Eicosatrienoic acid (c20:3n6)	2.60 ± 0.16	7.40 ± 0.51	0.005
Arachidonic acid (c20:4n6)	31.03 ± 3.26	0.59 ± 0.10	0.006
Myristic acid (c14:0)	17.29 ± 1.70	34.18 ± 1.60	0.007
Myristoleic acid (c14:1n5)	0.89 ± 0.35	4.54 ± 0.73	0.009
Linolenic acid (c18:3n3)	4.74 ± 0.59	7.61 ± 0.02	0.011
Docosadienoic acid (c22:2n6)	0.33 ± 0.18	1.16 ± 0.03	0.011
Lauric acid (c12:0)	0.87 ± 0.08	1.79 ± 0.18	0.015
Adrenic acid (c22:4n6)	2.44 ± 0.25	1.17 ± 0.06	0.015
Oleic acid (c18:1n9)	475.35 ± 39.93	303.06 ± 22.72	0.028
Elaidic acid (c18:1tn9)	3.73 ± 0.27	4.61 ± 0.23	0.033
Heptadecenoic acid (c17:1n7)	5.26 ± 1.09	30.72 ± 10.41	0.035
Stearic acid (c18:0)	189.17 ± 12.95	148.05 ± 3.45	0.036
Undecanoic acid (c11:0)	0.01 ± 0.00	0.01 ± 0.00	0.085
Arachidic acid (c20:0)	1.00 ± 0.10	1.17 ± 0.08	0.131
Heneicosanoic acid (c21:0)	0.14 ± 0.01	0.50 ± 0.21	0.138
ΣPUFA	112.98 ± 4.62	110.52 ± 2.13	0.338
ΣSFA	457.72 ± 31.29	561.70 ± 20.37	0.059
ΣPUFA: ΣSFA	0.25 ± 0.01	0.20 ± 0.01	0.050
ΣMUFA	509.06 ± 40.10	392.51 ± 20.49	0.067
Σn-3	15.24 ± 0.20	17.93 ± 1.69	0.111
Σn-6	97.74 ± 4.45	92.60 ± 2.39	0.168
EPA	4.09 ± 0.26	7.17 ± 0.21	<0.001
TFA	1079.75 ± 76.01	1064.73 ± 42.99	0.441

Note: $p < 0.05$ and $p < 0.01$ compared with the B0 group. L^* represents Lightness, a^* represents Redness, b^* represents Yellowness; EAA = sum of valine, lysine, phenylalanine, methionine, histidine, leucine, tryptophan, leucine, and threonine; NEAA = sum of arginine, proline, asparagine, cystine, glycine, hydroxyproline, alanine, citrulline, creatinine, taurine, glutamic acid, creatine, aminoadipic acid, cysteine, glutamine, aspartic acid, serine, tyrosine, ornithine, and choline; TAAs = total amino acids. ΣSFA: total saturated fatty acids; ΣMUFA: total monounsaturated fatty acids; ΣPUFA: total polyunsaturated fatty acids; Σn-3: total omega-3 PUFA; Σn-6: total omega-6 PUFA; EPA: eicosapentaenoic acid (C20:5n-3). Saturated fatty acids include C8:0, C10:0, C11:0, C12:0, C13:0, C14:0, C15:0, C16:0, C17:0, C18:0, C20:0, C21:0, C22:0, C23:0, C24:0. Polyunsaturated fatty acids include C18:2n6, C18:3n3, C18:3n6, C20:3n6, C20:4n6, C20:5n3, C22:2n6, C22:5n3, C22:4n6. Monounsaturated fatty acids include C14:1n5, C15:1n5, C16:1n7, C17:1n7, C18:1n9, C18:1tn9, C20:1n9, C22:1n9, C24:1n9. Total n-3 fatty acids include C18:3n3, C20:5n3, C22:5n3. Total n-6 fatty acids include C18:2n6, C18:3n6, C20:3n6, C20:4n6, C22:2n6, C22:4n6. TFA represents total fatty acid.

3.3.2. Meat Sensory Evaluation Analysis

The results of the sensory scores of Tibetan sheep meat in the YB and B0 groups are shown in Table 5 and Figure S4. There were significant differences in the Tibetan sheep meat in the two groups ($p < 0.05$), apart from a lack of difference in meat color. The meat of animals belonging to the YB group scored higher than the B0 group in terms of aroma, juiciness, texture, and overall acceptability, indicating that meat from Tibetan sheep fed on saline oats had better sensory quality.

3.3.3. Meat-Eating Quality Analysis

The results of the analysis of the eating quality of meat in groups YB and B0 are shown in Table 5. There were significant differences ($p < 0.05$) in the eating quality of the meat between the two groups in terms of thawing loss, cooking loss, cooked meat percentage, shear force, hardness, adhesion, and chewiness, with lower values for cooking loss, hardness, and chewiness in animals belonging to the YB group together with greater cooked meat percentage values relative to the animals belonging to the B0 group. No significant differences in the remaining indices, namely, pH, elasticity, and cohesion, were observed between the two groups.

3.3.4. Meat Nutritional Quality Analysis

Table 5 indicates that the nutritional quality of the LL muscles of the Tibetan sheep in the two groups differed significantly ($p < 0.05$) in terms of protein and fat contents, with the meat of the animals belonging to the YB group having a greater protein content and lower fat content. However, there was no discernible variation in the groups' moisture levels.

3.3.5. Targeted Metabolomics Analysis of Meat

Free Amino Acid Analysis of Meat

As shown in Table 5, the free amino acid composition of the LL muscle of housed Tibetan sheep from saline and non-saline areas showed significant differences in the levels of most free amino acids ($p < 0.05$), with arginine, proline, creatine, glycine, leucine, isoleucine, creatinine, and valine being higher in the animals belonging to the YB group than those in the animals belonging to the B0 group. On the other hand, animals belonging to the YB group had lower levels of threonine, asparagine, and phenylalanine than animals belonging to the B0 group. Still, animals belonging to the YB group had significantly higher levels of NEAAs and TAAs than the B0 group ($p < 0.05$).

Fatty Acid Analysis of Meat

As shown in Table 5, the contents of most fatty acids in the LL muscles differed significantly between the two groups, with animals belonging to the YB group having lower levels of c15:0, c15:1n5, c20:5n3, c22:0, c22:1n9, c10:0, c18:3n6, c20:1n9, c20:3n6, and c20:4n6 than animals belonging to the B0 group. The levels of c16:0, c22:5n3, c16:1n7, c20:4n6, and c22:4n6 were higher than those in animals belonging to the B0 group ($p < 0.05$). In animals belonging to the YB group, the content of EPA was significantly lower than that in animals belonging to the B0 group ($p < 0.05$). There were no significant differences in PUFA, SFA, omega-3, omega-6, or total fatty acids.

3.3.6. Untargeted Metabolomics Analysis of Tibetan Sheep Meat

Quality Control Analysis

The comparison of the spectrum overlap of the TICs for the QC samples is presented in the Supplementary Materials, Figure S5A,B. The peaks' response intensities and retention periods significantly overlapped, suggesting negligible fluctuation attributable to instrumental error during the experiment. The peak regions from all the experimental and QC samples were then evaluated using PCA, as shown in the Supplementary Materials,

Figure S6A,B. As seen in the figures, the QC samples were closely clustered in positive and negative ion modes, indicating the good reproducibility of the experiment. After excluding the QC samples, further evaluations using PLS-DA and OPLS-DA were performed to differentiate the samples further. Groups YB and B0 displayed intra-group clustering and inter-group separation in both analyses, with OPLS-DA demonstrating a more significant effect, as shown in the Supplementary Materials, Figure S6C,D. The replacement test was used to verify the supervised model's validity to prevent overfitting during the modeling phase. Figure S6E,F in the Supplementary Materials show the replacement test plots of the OPLS-DA model, with a gradual decrease in the replacement retention and R^2 and Q^2 of the stochastic model, indicating an absence of overfitting in the original model with good stability.

Bioinformatics Analysis of Differential Metabolites in Meat

Overall, after database matching and secondary mass spectrometry verification, a total of 1133 effectively annotatable metabolites were obtained in the metabolomics experiments of this study, which were found in animals belonging to the YB group and the B0 group through a combination of positive and negative ion modes, with 632 identified in the positive ion mode and 501 in the negative ion mode. As shown in Table S6, differential metabolites were identified using the criteria of VIP > 1 and $p < 0.05$, finding 73 differential metabolites between groups Y and B0, of which 35 metabolites were identified in the positive ion mode and 38 in the negative ion mode; of these, 11 were enriched in KEGG pathways. The most significantly enriched pathways for free amino acid and lipid metabolism, as well as protein digestion and absorption, are presented in the KEGG pathway enrichment analysis in Figure 2A. To further compare the differential metabolite-associated pathways that may cause variations in feed metabolites across regions, various enrichment score plots were developed, as seen in Figure 2B.

The results in Table 6 show a marked upregulation of all metabolic pathways in animals belonging to the YB group compared with the B0 group (DA score > 0.5, $p < 0.05$). These pathways involved fructose and mannose metabolism, carbohydrate digestion and absorption, aminoacyl-tRNA biosynthesis, the mTOR signaling pathway, the phosphotransferase system, and lipid and free amino acid metabolism. The key upregulated metabolites included D-mannose-6-phosphate, D-glucose-6-phosphate, 2,6-diaminohexanoic acid, arginine, isoleucine, lysine, cysteine, and choline phosphate.

Table 6. DFMs (differential metabolites) from *longissimus Lumborum* of Tibetan sheep in the key differential enriched KEGG pathways.

Metabolic Pathway (YB VS B0)	Metabolites
Upregulation in the YB group	
Fructose and mannose metabolism	D-mannose 6-phosphate
Carbohydrate digestion and absorption	D-glucose 6-phosphate, Robison ester
Starch and sucrose metabolism	D-glucose 6-phosphate Robison ester
Aminoacyl–tRNA biosynthesis	Arginine, Isoleucine L-Lysine, DL-isoleucine
mTOR signaling pathway	Arginine, (S)-2-Amino-5-guanidinovaleric acid, L-Arg
Zeatin biosynthesis	Adenine, Thiomethyladenosine S-methyl-5′-thioadenosine 6-Aminopurine

Table 6. *Cont.*

Metabolic Pathway (YB VS B0)	Metabolites
ABC transporters	Arginine, Isoleucine, L-Lysine Deoxyadenosine, L-Arg S-methyl-l-cysteine (S)-2-Amino-5-guanidinovaleric acid 2-Amino-3-methylvaleric acid 2,6-Diaminohexanoic acid
Phosphotransferase system (PTS)	D-glucosamine 6-phosphate Vitamin C, Ascorbate, Ascorbic acid D-mannose 6-phosphate D-glucosaminic acid D-glucose 6-phosphate
Pentose phosphate pathway (PPP)	D-glucosaminic acid D-Glucosaminate 2-Amino-2-deoxy-D-gluconate
Glycerophospholipid metabolism	1,2-dihexadecanoyl-sn-glycero-3-phosphocholine Lecithin
2-Oxocarboxylic acid metabolism	Isoleucine, N-.alpha.-acetyl-l-ornithine L-Lysine, 2-Amino-3-methylvaleric acid N-Acetylornithine, 2,6-Diaminohexanoic acid
Biosynthesis of free amino acids	Arginine, Isoleucine, L-homoserine N-.alpha.-acetyl-l-ornithine S-Adenosyl-L-homocysteine Lysine acid, 2,6-Diaminohexanoic acid 2-Amino-3-methylvaleric acid N-Acetylornithine, L-Homoserine 2-Amino-4-hydroxybutyric acid
Protein digestion and absorption	Arginine, Isoleucine, L-Lysine (S)-2-Amino-5-guanidinovaleric acid
Biofilm formation–Vibrio cholerae	D-glucose 6-phosphate, Robison ester
Biosynthesis of various secondary metabolites—part 3	Arginine, L-Lysine (S)-2-Amino-5-guanidinovaleric acid
Lysine biosynthesis	L-homoserine, L-Lysine 2,6-Diaminohexanoic acid
Arginine biosynthesis	Arginine N-.alpha.-acetyl-l-ornithine (S)-2-Amino-5-guanidinovaleric acid N-Acetylornithine
Cysteine and methionine metabolism	L-homoserine S-methyl-5′-thioadenosine S-Adenosyl-L-homocysteine 2-Amino-4-hydroxybutyric acid
Ascorbate and aldarate metabolism	Ascorbate, Vitamin C
HIF-1 signaling pathway	Vitamin C, Ascorbate
Vitamin digestion and absorption	Vitamin C, L-Ascorbic acid
Bile secretion	Taurochenodeoxycholate, Chenodeoxycholoyltaurine
Lipoic acid metabolism	Octanoic acid, Octanoate
Cholesterol metabolism	Taurochenodeoxycholate Chenodeoxycholoyltaurine

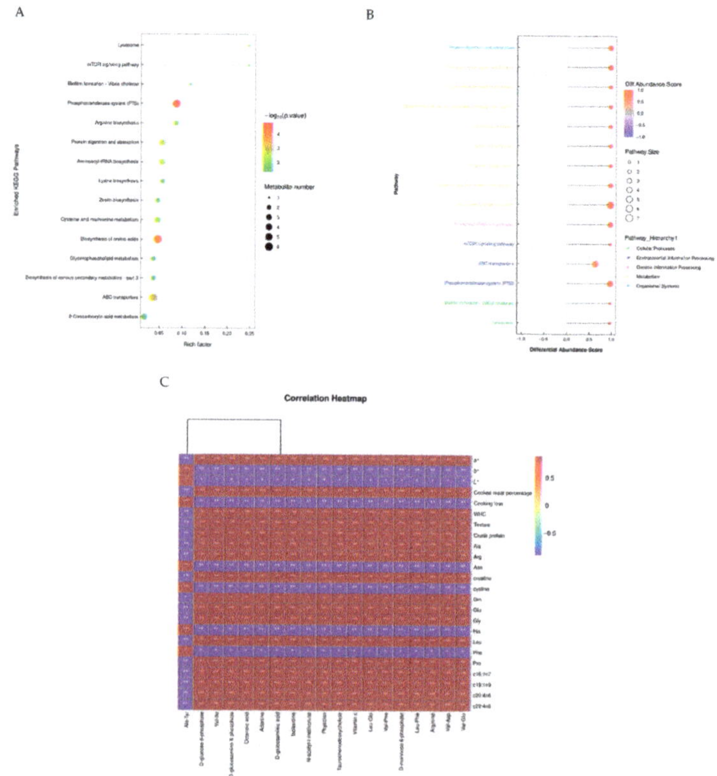

Figure 2. (**A**) Top 12 enriched KEGG pathways of the comparison between B0 and YB groups, (**B**) a differential abundance score map of differential metabolic pathways, (**C**) correlation heatmap between meat quality parameters and meat metabolomics analysis. The colors red and blue represent positive and negative correlations, respectively. * $p < 0.05$ and ** $p < 0.01$.

3.3.7. Correlation Analysis

Correlations between the meat phenotypic data and the untargeted metabolomics results were established to investigate the relationship between muscle metabolism and meat quality in Tibetan sheep meat under various feeding conditions, as illustrated in Figure 2C.

The metabolites in the meat included D-glucose 6-phosphate, Val-Ile, D-glucosamine 6-phosphate, octanoic acid, adenine, D-glucosaminic acid, isoleucine, N-acetyl-l-methionine, physcion, taurochenodeoxycholate, vitamin C, Leu-Glu, Val-Phe, D-mannose 6-phosphate, Leu-Phe, arginine, Val-Asp, and Val-Glu, which were positively correlated with *a**; the cooked meat percentage; WHC; texture; crude protein; and Ala, Arg, Gln, Glu, and Gly and negatively correlated with *b**, *L**, cooking loss, Asn, cystine, His, and Phe.

3.4. *Relationship Between Quality and Metabolites of Oats Grown in Saline Soil and Quality and Metabolites of Tibetan Sheep Meat*

Figure 3 presents a correlation clustering heatmap that depicts the relationship between oat quality, metabolites, and the quality of Tibetan sheep meat and its metabolites. The figure illustrates a significant correlation between the two groups. The horizontal axes denote meat quality and associated metabolites, while the vertical axes represent oat quality and its metabolites.

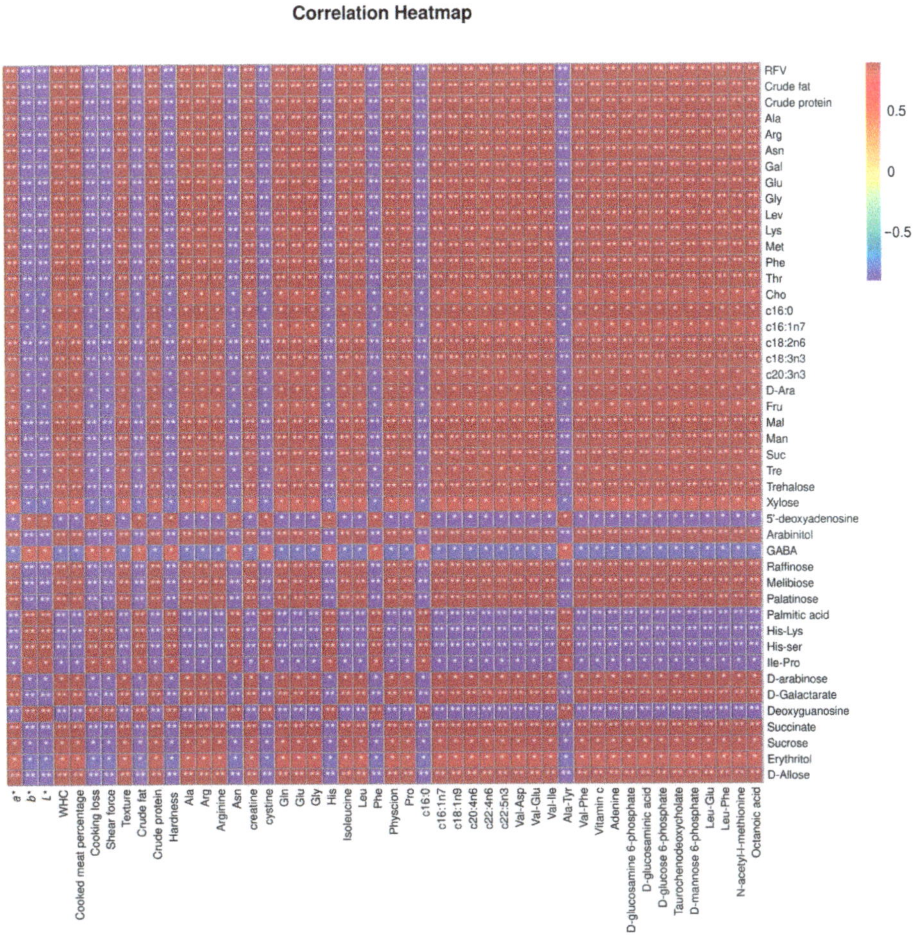

Figure 3. Clustering heatmap of correlation between metabolites and quality of oats and meat. * $p < 0.05$ and ** $p < 0.01$.

Positive correlations were observed between crude protein, crude fat, Ala, Arg, Asn, Gal, Glu, Gly, Lev, Lys, Met, Phe, Thr, Man, Suc, Tre, trehalose, xylose, raffinose, melibiose, palatinose, succinate, sucrose, erythritol, and D-allose in oats and *a**, WHC, cooked meat percentage, crude protein, Ala, Arg, Asn, vitamin C, adenine, D-glucosamine 6-phosphate, D-glucosaminic acid, D-glucose 6-phosphate, taurochenodeoxycholate, D-mannose 6-phosphate, and octanoic acid in meat. At the same time, these were negatively correlated with *b**, *L**, cooking loss, and shear force. These associations offer insights into the correlation between oat composition and muscle characteristics.

4. Discussion

In this study, it was found that the relatively high contents of mineral elements such as Ca, K, Mg, and Na in saline–alkaline soils may enhance the stress resistance of oats, enabling them to adapt to saline–alkaline environments and thus accumulate more osmotic regulatory substances to improve their quality. This may be because when salt concentration increases, ion imbalance and osmotic stress in plants can affect their morphology, biomass, and biochemical processes. Therefore, plants need to maintain a concentration of mineral elements within a specific range to achieve optimal physiological functions [25]. Moreover, changes in the availability of mineral contents in the soil can affect plant growth and quality. Under saline–alkaline stress, plants will eliminate excessive Na^+ and Cl^- ions in vacuoles or older parts to minimize damage caused by excessive salt ions. Meanwhile, plants will biosynthesize osmotic substances, such as soluble sugars and amino acids, and activate enzymatic and non-enzymatic antioxidant defense systems to remove excess reactive oxygen species (ROS), protecting plant cells from oxidative damage [26]. Salt-tolerant plants may selectively absorb elements such as Ca, K, and Mg from the soil under saline–alkaline stress, increasing their mineral content and thus enhancing their nutritional value. The accumulation of Mg and Ca is beneficial for maintaining ion balance and regulating plant growth in saline–alkaline environments. Moreover, K can enhance plant stress resistance, strengthen cellular responses to adverse environmental conditions, and improve plant tolerance to challenging conditions such as saline–alkaline stress and drought [27]. Jin et al. [28] confirmed that the addition of potassium fulvate (PF), an organic fertilizer, to soil can alleviate nutritional antagonism and osmotic stress in oats under saline–alkaline conditions; increase the concentrations of total nitrogen, total potassium, and their available forms (ammonium nitrogen and nitrate nitrogen) in the soil; enhance the absorption of water and essential mineral nutrients by plants; and promote the growth of oats.

The use of male sheep as experimental subjects usually arises from the necessity to regulate variables, streamline the experimental framework, or investigate male-specific issues. This strategy requires careful evaluation of specific study aims, with careful consideration of the potential influence of gender variations on the credibility of scientific conclusions. This experiment exclusively employed male sheep, which presents limitations. In forthcoming trials, we aim to thoroughly examine the variations in meat quality among female sheep, male sheep, and castrated rams.

In this experiment, Tibetan sheep were raised in Gonghe and Haiyan. Geographical differences may potentially introduce the influence of environmental variables, such as temperature, humidity, and altitude. However, both regions belong to the alpine area of the Qinghai–Tibet Plateau, with similar climatic and altitude characteristics, and the purpose of this study is to clarify the differences between forage and meat quality. Therefore, in our study design, artificially adjustable variables were controlled, and group differences were solely established through roughage (oats from different regions), with the concentrate composition being identical between the two groups. Oats were harvested in the same growing season to avoid the impact of seasonality on nutrient accumulation. All meat quality analyses and metabolomics detections were performed in the same laboratory, eliminating inter-laboratory systematic errors. These measures reduced the influence of natural environmental differences. Future studies will further exclude the interference of environmental variables through cross-location feeding experiments to verify the robustness of the conclusions of this study.

In this study, the fat content of oats grown in saline conditions (GX group) was significantly higher than that of the non-saline oats (YX group). Li et al. [29] conducted a lipidomic analysis of salt-treated alfalfa varieties. They found that the lipid levels in the plants changed under salt stress, indicating that salt treatment affected the plants' fat and fatty acid contents. This may have been due to changes in the structural integrity and fluidity of cell membranes in plant leaves during salt stress. Ge et al. [30] found that the content of galactolipids (DGDG and MGDG) and major phospholipids (PC and PE) increased in the leaves of sorghum seedlings under salt stress; this is likely due to salt stress-induced changes in the glycerolipid pathway between the cytoplasm and plastid, facilitating the conversion of PC to PA and providing precursors for galactolipid synthesis. According to the study, herbage on saline–alkaline land had a higher concentration of fatty acids (including SFA, PUFA, n-3, and n-6) than non-saline–alkaline land. This finding may be directly related to saline–alkaline stress.

Free amino acids are stress-response metabolites that accumulate under saline and alkaline stress; this accumulation may result from either de novo synthesis or protein degradation, accelerating post-stress recovery and osmoprotection [31]. Qian et al. [32] demonstrated that under saline–alkaline stress, the amino acid content increases significantly, and amino acid biosynthesis exhibits a positive response to saline–alkaline stress. The current study's findings indicated that serine, valine, glycine, arginine, and alanine concentrations were elevated in the GX group compared with the YX group, potentially linked to salinity stress. Untargeted metabolomics results further demonstrated upregulation in alanine, aspartate, glutamate, and phenylalanine metabolism, aligning with the targeted metabolism outcomes for free amino acids. Furthermore, correlation analysis suggested that the accumulation of glutamate and succinate may have enhanced the salinity tolerance of the oat grasses in the YX group. Proteomic analysis of saline alfalfa by Gao et al. [33] revealed that key differentially expressed proteins were primarily enriched in the antioxidant system and starch and sucrose metabolism, as well as in secondary metabolism, suggesting that saline and alkaline stress increases antioxidant functions in the plants together with the production of secondary metabolites, such as sucrose, maltose, glucose, and trehalose, and the promotion of osmotic homeostasis. This study's untargeted metabolomics and subsequent KEGG analysis indicated that starch and sucrose metabolism were upregulated, resulting in increased carbohydrate buildup. Starch is the primary carbohydrate in oats and a vital element, significantly contributing to the structural integrity of oat grains. As described by Rostamabadi et al. [34], the resistant starch in oat starch can bind to bile acids, preventing their reabsorption, promoting the conversion of cholesterol into other bile acids, and maintaining the balance of bile acid excretion. Consequently, this process reduces the cholesterol level in the blood, which benefits human health.

Overall, in the nutritional quality and metabolomics analysis of saline-grown and non-saline-grown oats, the GX saline oats group showed significant advantages as ruminant feed.

Despite significant differences in meat quality and composition among animals in different dietary groups, there were no statistical differences in growth performance, such as growth rate and final weight. It is hypothesized that the dietary formulation in the experiment generated differences between groups solely through roughage (oats from different regions), with the concentrate composition being identical between the two groups, all of which met the basic nutritional requirements for animal growth. Therefore, no differences were observed in the overall growth indicators, while dietary differences may mainly affect meat quality by regulating muscle metabolic pathways. Moreover, the sheep involved in the experiment were all of the same breed, and the consistency in their genetic background and growth stage contributed to the stability of growth characteristics and some carcass indicators [35].

The tenderness of meat determines consumer acceptability and satisfaction. The shear force value of the meat in group YB was less than that of group B0, indicating more significant tenderness. The sensory evaluations suggested that the sheep muscles from animals belonging to the YB group showed improved texture. Correlation analysis revealed a positive relationship between muscle texture and water-holding capacity (WHC) with D-glucose-6-phosphate. The upregulation of D-glucose-6-phosphate in animals belonging to the YB group may enhance glycolysis by modulating starch and sucrose metabolic pathways, resulting in ATP production and degradation of glycogen. This process likely leads to a rapid decrease in pH, facilitating the hydrolysis of fibrous muscle proteins and improving muscle tenderness, which may account for the increased tenderness observed in animals belonging to the YB group. Muscle tenderness is influenced by factors including collagen content, knob length, and protein degradation [36]. The present study found that the contents of free glutamic acid and glycine in the muscles of Tibetan sheep fed with oats from saline soils were significantly increased, accompanied by a decrease in shear force value, improvement in tenderness, and an increase in aroma. This may be because glutamic acid, as a key substrate for muscle protein synthesis, can promote the degradation of collagen between muscle fibers through its accumulation, thereby reducing the degree of cross-linking between muscle fibers, decreasing shear force, and improving tenderness. Meanwhile, as synergistic activators of umami receptors, glutamic acid and glycine can directly enhance taste signal transmission and improve the umami perception of meat quality with their increased contents [37]. Tibetan sheep fed with saline oats had increased levels of linoleic acid in their muscles, along with an improved water-holding capacity score. This may be due to the fact that linoleic acid, as an n-6 polyunsaturated fatty acid, can lower the melting point of muscle fat, making it easier for fat to melt during oral mastication and release flavor substances, thereby enhancing the perception of juiciness. Meanwhile, linolenic acid can reduce the production of fat oxidation products by inhibiting the activity of muscle lipoxygenase, avoiding the dry and hard taste caused by fat oxidation, and indirectly maintaining juiciness. The positive correlation between these two fatty acids and juiciness indicators further verifies the regulatory effect of fatty acid composition on meat juiciness [38].

Generally, higher a^* and lower L^* and b^* values indicate better muscle color within a specific range. In this study, animals belonging to the YB group had higher a^* values, with the meat showing a brighter red color than that in the B0 group, which may be related to the production of oxygenated myoglobin. It has been reported that L^* values are positively associated with muscle WHC; water exudation causes changes in the refractive index of the muscle surface, resulting in a higher L^* value [33]. In the present study, animals belonging to the YB group showed better WHC values, indicating less water exudation from the muscle and, thus, a lower L^* value. The lower cooking loss and higher cooked meat rate in animals belonging to the YB group in this study may have been due to the changes in the structure of the muscle fibers, leading to the improved water retention capacity of the muscle proteins [39]. The results of the correlation study showed that D-glucose-6-phosphate was negatively connected with cooking loss and the L^* value and significantly and positively correlated with the a^* value, cooked meat rate, WHC, and texture. This suggests that glycolysis controls the quality of meat. Collectively, saline-region Tibetan lamb showed improved eating and sensory characteristics.

In the present study, it was found that Tibetan sheep meat in the YB group had higher protein and lower fat contents, which is similar to the findings of Friha et al. [40], who reported that rearing lambs on saline soils resulted in better growth performance, leaner carcasses, and higher meat quality. This result may be related to the composition and proportions of free amino acids and fatty acids. Free amino acids are the building blocks of muscle proteins, and their contents and compositions affect meat texture, flavor, nutritional value, and antioxidant capacity [41]. Leucine, isoleucine, and valine, collectively known as branched-chain free amino acids (BCAAs) due to their characteristic side-chain structures and specific metabolic pathways, promote protein synthesis and tissue growth by stimulating mRNA translation through mTORC1 signaling [42]. In addition to encouraging the release of prolactin, growth hormone, and insulin from the corresponding endocrine organs, arginine enhances the expression of essential proteins and enzymes involved in substrate oxidation and mitochondrial biosynthesis, controlling muscle oxidation and lowering excess body fat in animals [43]. Proline acts as a substrate and is involved in the synthesis of pyruvate and glucose, which is associated with the collagen content of muscles. At the same time, threonine increases the meat's sweetness, making it more palatable to consumers [44]. In this study, animals belonging to the YB group were associated with increased proline, serine, and lysine deposition in the LL muscles through upregulation of free amino acid metabolism. Compared with animals belonging to the YB group, animals in the B0 group were associated with reduced deposition of branched-chain free amino acids in the muscle by downregulating leucine, isoleucine, and valine biosynthesis. Furthermore, by upregulating the metabolism of creatinine and arginine, animals belonging to the YB group were able to collect higher quantities of arginine. Therefore, it is anticipated that the meat from animals belonging to the YB group raised in saline conditions has free amino acid concentrations that are more in accordance with what is required for human health.

Glycogen is the primary source of energy for the production of glycolytic substrates in postmortem muscle when the electron transport chain is terminated due to hypoxia; pyruvate is unable to enter the mitochondria, and the muscle accumulates lactic acid and hydrogen ions to maintain homeostasis, resulting in the denaturation of muscle proteins and a decrease in pH [45]. The phosphotransferase system (PTS) re-reduces glucose phosphorylation through phosphorylation and carbon source transfer, with associated enzymes hydrolyzing intracellular cAMP and lowering its concentration. This lowers glycogen metabolism by blocking the cAMP-dependent protein kinase (APK) signaling pathway and phosphorylase activity. AMPK, the primary regulator of lipid and glucose metabolism in the cell, is activated when the cell is depleted of energy [46]. Lamberigts et al. [47] found that intraperitoneal injection of α-lipoic acid inhibited AMPK activity in the hypothalamus, reducing food intake and energy expenditure. In the present study, PTS and α-lipoic acid metabolism were significantly upregulated in the animals belonging to the YB group compared with the B0 group, suggesting that glycogen metabolism was inhibited in this group. Hexokinase (HK), the first rate-limiting enzyme in the glycolytic pathway, phosphorylates glucose to generate D-glucose-6 phosphate (G6P); moreover, negative feedback involving intracellular G6P, a central regulator of glycogen synthesis in skeletal muscle, regulates HK [48]. Glucose-6-phosphate dehydrogenase (G6PD) catalyzes the conversion of G6P into glucose-6-phosphate-δ-lactone, which engages glucose in the pentose phosphate pathway to form NADPH. NADPH is a reducing agent that generates reduced glutathione, which helps to counteract oxidative stress and maintain cellular homeostasis and normal metabolic activities [49]. In the present study, the results of the untargeted metabolomics indicated that G6P levels were higher in animals belonging to the YB group than in the B0 group, accompanied by upregulated glutathione metabolism. The current study proposes that animals belonging to the YB group might show diminished glycolysis due to reduced HK activity and elevated glucose-6-phosphate dehydrogenase activity in the LL muscles of Tibetan sheep, thus sustaining a higher pH and enhancing water retention and inter-fiber bonding. This would yield meat with a denser structure, demonstrating enhanced tenderness and texture. At the same time, reduced oxygen diffusion from the muscle surface would enable greater absorption and lower light reflection, leading to a decrease in the muscle L^*.

In high-altitude hypoxic regions, glucose metabolism produces NADPH via the pentose phosphate pathway (PPP), stabilizing the hypoxia-inducible factor HIF-1α protein [50].

Therefore, it was hypothesized that both the PP and HIF-1α pathways were significantly upregulated in animals belonging to the YB group. The PPP is a fundamental component of cellular metabolism. It is essential for maintaining carbon homeostasis, providing nucleotide and free amino acid biosynthesis precursors, supplying reducing molecules for antioxidant activities, and resisting oxidative stress [51]. HIF-1α regulates the activities of antioxidant enzymes such as glutathione peroxidase and heme oxygenase. Further, elevated expression of HIF-1α may also increase the expression of most antioxidant proteins in muscle tissues by enhancing the expression of Nrf2, increasing the antioxidant capacity of the organism [52]. Vitamin C, ascorbic acid, is a non-enzymatic antioxidant that effectively scavenges ROS [53]. Initially isolated from bovine bile, taurine is a sulfur-containing free amino acid that modulates the intracellular enzymatic antioxidant defense system by activating various signaling pathways and enhancing catalase activity in response to stress [54]. Compared with the B0 group, animals belonging to the YB group were associated with increased levels of vitamin C and taurine, resulting from the upregulation of bile secretion and vitamin digestion and absorption, leading to improved antioxidant activities in the muscle and thus improving the meat quality and nutritional value.

In summary, as shown in Figure 4, it is hypothesized that the saline oat YX group accumulated a large number of carbohydrates during adaptation to saline stress. As a result, the levels of carbohydrates such as D-glucose-6-phosphate and D-mannose-6-phosphate in the meat were significantly increased. The upregulation of these compounds resulted in enhanced activation of the PPP and PTS pathways in the LL muscles of the sheep. The abundance of G6P specifically blocks hexokinase activity in the upregulated PTS pathway, therefore postponing a decline in pH. This postponement in pH lowering enhances water retention in the muscle, diminishes cooking losses, elevates the percentage of cooked meat and water-holding capacity, and preserves meat quality.

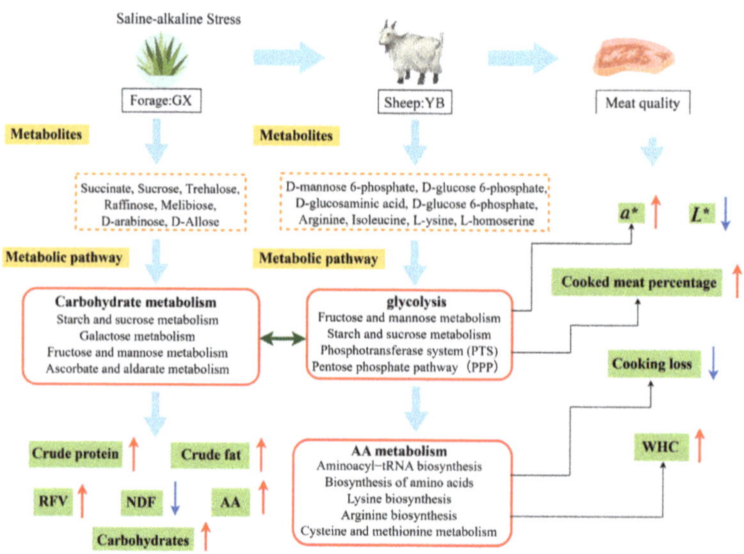

Figure 4. Hypothesized scheme pathways and potential mechanisms related to changes in oat quality, the muscle metabolome, and meat quality. Blue and red colors indicate significantly downregulated and upregulated metabolic pathways in each comparison, respectively. *L** represents Lightness, *a** represents Redness.

Furthermore, the transfer of G6P to the PPP was improved by reducing hypoxia and the requirement to raise cellular antioxidant capacity by increasing NADPH production. The resulting NADPH further increased the muscle's antioxidant capacity, thus lessening the harm that oxidative stress caused to the animals. Meanwhile, Tibetan lamb in saline areas had higher-quality meat because the upregulation of free amino acid pathways in saline oats resulted in the upregulation of free amino acid pathways in muscle, as well as the deposition of free amino acids in muscle.

5. Conclusions

Oats cultivated in saline–alkaline land affect the meat quality of Tibetan sheep. Specifically, Tibetan sheep in the YB group fed with saline oats exhibited better meat color, a higher cooked meat rate and water-holding capacity, and lower cooking loss. This resulted from the higher protein, monosaccharide, and amino acid contents in saline oats, which promoted the upregulation of various amino acids such as arginine and lysine in muscles; the digestion and absorption of carbohydrates such as D-glucose-6-phosphate and D-mannose-6-phosphate; and the accumulation of antioxidant components, including ascorbic acid and taurine. In conclusion, the increase in compatible solutes in saline oats had a significant effect on maintaining the WHC of Tibetan sheep muscles, improving tenderness, and enhancing sensory quality. Therefore, this finding provides a promising theoretical basis for optimizing meat sheep production and feeding management.

Supplementary Materials: The following supporting information can be downloaded at https://www.mdpi.com/article/10.3390/foods14173044/s1: Figure S1: (Dataset from oats) (A) The distribution of RSD of free amino acids in QC samples of oats. (B) The distribution of RSD of fatty acids in QC samples of oats. Figure S2: (Dataset from oats) (A) The total ion chromatograms of quality control samples in positive ion modes. (B) The total ion chromatograms of quality control samples in negative ion modes. Figure S3: (Dataset from oats) Quality control of oat samples. (A) PCA analysis of all the samples based on peaks detected in positive ion modes. (B) PCA analysis of all the samples based on peaks detected in negative ion modes. Multivariate statistical analysis of oats in different regions: (C) PLS-DA and (D) OPLS-DA scores of the overall sample in the positive ion mode and permutations test of (E) PLS-DA and (F) OPLS-DA in the negative ion detection mode. Figure S4: (Dataset from sheep) Radar plot of sensory evaluation of Tibetan sheep in YB and B0 groups. Figure S5: (Dataset from sheep) (A) Total ion chromatograms of quality control samples in positive ion modes. (B) Total ion chromatograms of quality control samples in negative ion modes. Figure S6: (Dataset from sheep) Quality control of meat samples. (A) PCA analysis of all the samples based on peaks detected in positive ion modes. (B) PCA analysis of all the samples based on peaks detected in negative ion modes. Multivariate statistical analysis of oat in different regions: (C) PLS-DA and (D) OPLS-DA scores of the overall sample in the positive ion mode and permutations test of (E) PLS-DA and (F) OPLS-DA in the positive ion detection mode. Table S1: Standard curve of free amino acids and fatty acids in oat. Table S2: Standard curve of carbohydrates in oats. Table S3: Parameters of GC-MS. Table S4: Scoring criteria of sensory evaluation of meat. Table S5: Detailed results of differential metabolites in the oats in the positive and negative ion detection modes (OPLS-DA VIP > 1 and p-value < 0.05) (YX vs. GX). Table S6: Detailed results of differential metabolites in the longissimus lumborum in the positive and negative ion detection modes (OPLS-DA VIP > 1 and p-value < 0.05) (YB vs. B0). Formula (S1): Calculation of the content of free amino acids in oats. Formula (S2): Calculation of the content of fatty acids in oats. Formula (S3): Calculation of the content of carbohydrates in oats.

Author Contributions: Conceptualization, X.X., L.H. and S.H.; Data curation, X.X. and Z.Y.; Formal analysis, X.X., L.H. and L.G.; Funding acquisition, L.H.; Investigation, S.H., Z.W. and B.Y.; Methodology, X.X. and L.H.; Project administration, L.H.; Resources, S.H.; Software, X.X., L.G. and S.S.; Supervision, L.H.; Validation, C.Y. and S.H.; Visualization, X.X. and L.H.; Writing—original draft,

X.X. and L.H.; Writing—review and editing, L.H. All authors have read and agreed to the published version of the manuscript.

Funding: The authors are thankful for the support of Qinghai University for the use of laboratory facilities. The current work was funded by Qinghai Provincial Department of Science and Technology "Construction of Standardized Production System for Improving quality and efficiency of Tibetan sheep industry (2022-NK-169)" and "Research and Integration of Production Technology for Special Tibetan Sheep in High Saline and Alkaline Habits (2022-NK-169-3)".

Institutional Review Board Statement: The study was approved by the Qinghai University Ethics Review Committee (QUA- 2022- 0515 2022-05-15).

Informed Consent Statement: Informed consent was obtained from all subjects involved in the study.

Data Availability Statement: The original contributions presented in this study are included in the article. Further inquiries can be directed to the corresponding author.

Acknowledgments: The authors are thankful for the support of Qinghai University for the use of laboratory facilities. The authors acknowledge Xingang Lv (Shanghai Applied Protein Technology Co., Ltd., China) for the technical support.

Conflicts of Interest: The authors declare that the research was conducted in the absence of any commercial or financial relationships that could be construed as potential conflicts of interest.

References

1. Liu, Y.; Li, R.; Ying, Y.; Zhang, Y.; Huang, Y.; Wu, H.; Lin, K. Non-Genetic Factors Affecting the Meat Quality and Flavor of Inner Mongolian Lambs: A Review. *Front. Vet. Sci.* **2022**, *9*, 1067880. [CrossRef]
2. Liang, M. Microbial Characteristics and Functions and Their Influencing Factors of Rhizosphere Soil of Four Plants in the Caka Salt Lake area of Qinghai, China. Master's Thesis, Forestry Northwest A & F University, Xi'an, China, 2023.
3. Morsy, S.; Elbasyoni, I.S.; Baenziger, S.; Abdallah, A.M. Gypsum Amendment Influences Performance and Mineral Absorption in Wheat Cultivars Grown in Normal and Saline-sodic Soils. *J. Agron. Crop Sci.* **2022**, *208*, 675–692. [CrossRef]
4. Du, Y.; Zhang, Y.; Zhang, Z.; Li, X.; Li, L.; Chai, X.; Zeng, F. Soil Improvement Effect of Planting Seven Gramineous Plants in Saline-Alkali Soil in the Northern Xinjiang. *Prataculturral Sci.* **2023**, *40*, 1210–1219.
5. Wu, W.; Ge, R.; Wang, J.; Wei, X.; Zhao, Y.; Pu, X.; Xu, C. Comparative Study on Production Performance of Different Oat (*Avena sativa*) Varieties and Soil Physicochemical Properties in Qaidam Basin. *Plants* **2025**, *14*, 1978. [CrossRef] [PubMed]
6. Liu, J.; Peng, D.; Kong, X.; Zhang, Q.; Wang, X.; Liu, H.; Yang, H. Evaluation of Production Performance and Nutritional Value of Five Forage Oats Varieties in Saline-Alkali Soil of Songnen Plain. *Feed. Res.* **2023**, *46*, 132–136.
7. Bai, J.; Liu, J.; Zhang, N.; Yang, J.; Sa, R.; Wu, L. Effect of Alkali Stress on Soluble Sugar, Antioxidant Enzymes and Yield of Oat. *J. Integr. Agric.* **2013**, *12*, 1441–1449. [CrossRef]
8. Yang, S.; Chu, N.; Zhou, H.; Li, J.; Feng, N.; Su, J.; Deng, Z.; Shen, X.; Zheng, D. Integrated Analysis of Transcriptome and Metabolome Reveals the Regulation of Chitooligosaccharide on Drought Tolerance in Sugarcane (*Saccharum* spp. Hybrid) under Drought Stress. *Int. J. Mol. Sci.* **2022**, *23*, 9737. [CrossRef]
9. Ge, S.; Han, L.; Hou, S.; Yuan, Z.; Gui, L.; Sun, S.; Yang, C.; Wang, Z.; Yang, B. Influence of Cooking Methods on Flavor Parameters and Sensory Quality of Tibetan Sheep Meat Examined Using an Electronic Nose, an Electronic Tongue, GC-IMS, and GC-MS. *Foods* **2025**, *14*, 2181. [CrossRef]
10. Zhang, X.; Han, L.; Hou, S.; Raza, S.H.A.; Wang, Z.; Yang, B.; Sun, S.; Ding, B.; Gui, L.; Simal-Gandara, J.; et al. Effects of Different Feeding Regimes on Muscle Metabolism and Its Association with Meat Quality of Tibetan Sheep. *Food Chem.* **2022**, *374*, 131611. [CrossRef] [PubMed]
11. Liang, J. Effects of *Salicornia europaea* L. Extract on Growth Performance, Digestibility, Slaughter Performance and Meat Quality of Sheep. Master's Thesis, Xinjiang Agricultural University, Urumqi, China, 2022.
12. Qiu, J.; Huang, M.; Wang, Z.; Wang, X.; Feng, Q.; Qian, X.; Gao, Y.; Zhang, Z.; Li, H.; Guo, H. Effects of Alfalfa Grown in Saline Alkali Soil on Growth and Slaughter Performance, Meat Quality, Organ Development and Serum Biochemical Indexes of Mutton Sheep. *J. Econ. Anim.* **2022**, *27*, 84–89.
13. Moreno, G.M.; Borba, H.; Araújo, G.G.; Sañudo, C.; Sobrinho, A.G.S.; Buzanskas, M.E.; Júnior, D.M.L.; de Almeida, V.V.; Neto, O.B. Oscar Boaventura. Meat Quality of Lambs Fed Different Saltbush Hay (*Atriplex Nummularia*) Levels. *Ital. J. Anim. Sci.* **2015**, *14*, 3302. [CrossRef]

14. Pearce, K.L.; Norman, H.C.; Hopkins, D.L. The Role of Saltbush-Based Pasture Systems for the Production of High Quality Sheep and Goat Meat. *Small Rumin. Res.* **2009**, *91*, 29–38. [CrossRef]
15. Liu, H.; Xu, T.; Xu, S.; Ma, L.; Han, X.; Wang, X.; Zhang, X.; Hu, L.; Zhao, N.; Chen, Y.; et al. Effect of Dietary Concentrate to Forage Ratio on Growth Performance, Rumen Fermentation and Bacterial Diversity of Tibetan Sheep under Barn Feeding on the Qinghai-Tibetan Plateau. *PeerJ* **2019**, *7*, e7462. [CrossRef]
16. Ma, T.; Deng, K.-D.; Tu, Y.; Jiang, C.-G.; Zhang, N.-F.; Li, Y.-L.; Si, B.-W.; Lou, C.; Diao, Q.-Y. Effect of Dietary Concentrate:Forage Ratios and Undegraded Dietary Protein on Nitrogen Balance and Urinary Excretion of Purine Derivatives in Dorper × thin-Tailed Han Crossbred Lambs. *Asian-Australas. J. Anim. Sci.* **2014**, *27*, 161–168. [CrossRef]
17. Song, X.; Yang, N.; Su, Y.; Lu, X.; Liu, J.; Liu, Y.; Zhang, Z.; Tang, Z. Suaeda Glauca and Suaeda Salsa Employ Different Adaptive Strategies to Cope with Saline–Alkali Environments. *Agronomy* **2022**, *12*, 2496. [CrossRef]
18. Han, F.; Fu, G.; Yu, C.; Wang, S. Modeling Nutrition Quality and Storage of Forage Using Climate Data and Normalized-Difference Vegetation Index in Alpine Grasslands. *Remote Sens.* **2022**, *14*, 3410. [CrossRef]
19. Shahi, M.; Azarnivand, H.; Esfahan, E.Z.; Jafari, M. Exploring the Potential Forage Quality of Some Halophytic Species at Different Phenological Stages. *BMC Plant Biol.* **2025**, *25*, 794. [CrossRef]
20. Zhang, X.; Li, S.; Liu, Z.; Chen, L.; Wu, J. Nutritional Quality of Common Forage Species and Their Influences across Natural Alpine Rangelands, Northern Tibetan Plateau. *Acta Agrestia Sin.* **2024**, *32*, 3797–3806.
21. Gao, Y.; Cheng, T.; Zhang, C.; Yan, Y.; Zhang, L.; Liu, Q.; Liu, Y.; Qiao, Q. Analysis of Leaf Forage Value and Screening of Different Populations of *Pteroceltis Tatarinowii*, A Rare and Endemic Species in China. *Front. Plant Sci.* **2023**, *14*, 1164451. [CrossRef] [PubMed]
22. Ma, Y.; Han, L.; Zhang, S.; Zhang, X.; Hou, S.; Gui, L.; Sun, S.; Yuan, Z.; Wang, Z.; Yang, B. Insight into the Differences of Meat Quality Between Qinghai White Tibetan Sheep and Black Tibetan Sheep from the Perspective of Metabolomics and Rumen Microbiota. *Food Chem. X* **2023**, *19*, 100843. [CrossRef] [PubMed]
23. Zhang, X.; Zhao, L.; Tian, H.; Ma, Z.; Zhang, Q.; Pu, M.; Cao, P.; Zhang, D.; Zhang, Y.; Zhao, Y.; et al. Integrative Analysis of Transcriptomics and Metabolomics Provides Insights into Meat Quality Differences in Hu Sheep with Different Carcass Performance. *Foods* **2025**, *14*, 2477. [CrossRef] [PubMed]
24. AOAC. *Official Methods of Analysis of the Association of Official Analytical Chemists*; AOAC International: Gaithersburg, MD, USA, 2005.
25. Song, X.; Su, Y.; Zheng, J.; Zhang, Z.; Liang, Z.; Tang, Z. Study on the Effects of Salt Tolerance Type, Soil Salinity and Soil Characteristics on the Element Composition of Chenopodiaceae Halophytes. *Plants* **2022**, *11*, 1288. [CrossRef]
26. Ullah, A.; Tariq, A.; Sardans, J.; Peñuelas, J.; Zeng, F.; Graciano, C.; Asghar, M.A.; Raza, A.; Xiong, Y.-C.; Chai, X.; et al. Alhagi Sparsifolia Acclimatizes to Saline Stress by Regulating Its Osmotic, Antioxidant, and Nitrogen Assimilation Potential. *BMC Plant Biol.* **2022**, *22*, 453. [CrossRef] [PubMed]
27. Liu, H.; Li, Y.; Li, S. Cu and Na Contents Regulate N Uptake of Leymus Chinensis Growing in Soda Saline-Alkali Soil. *PLoS ONE* **2020**, *15*, e0243172. [CrossRef] [PubMed]
28. Jin, X.; Wang, J.; Liu, X.; Chang, J.; Li, C.; Lu, G. Potassium Fulvate Alleviates Salt–Alkali Stress and Promotes Comprehensive Growth of Oats in Saline–Alkali Soils of the Qaidam Basin. *Plants* **2025**, *14*, 1982. [CrossRef]
29. Li, Z. Evaluation of Salt Tolerance and Nutritional Quality for Wild Barley New Cultivar (*Hordeum brevisubulatum* cv. Hexi). Master's Thesis, Lanzhou University, Lanzhou, China, 2021.
30. Ge, S.; Liu, D.; Chu, M.; Liu, X.; Wei, Y.; Che, X.; Zhu, L.; He, L.; Xu, J. Dynamic and adaptive membrane lipid remodeling in leaves of sorghum under salt stress. *Crop J.* **2022**, *10*, 1557–1569. [CrossRef]
31. Singer, S.D.; Lehmann, M.; Zhang, Z.; Subedi, U.; Burton Hughes, K.; Lim, N.Z.; Ortega Polo, R.; Chen, G.; Acharya, S.; Hannoufa, A.; et al. Elucidation of Physiological, Transcriptomic and Metabolomic Salinity Response Mechanisms in *Medicago sativa*. *Plants* **2023**, *12*, 2059. [CrossRef]
32. Qian, G.; Wang, M.; Zhou, J.; Wang, X.; Zhang, Y.; Liu, Y.; Zhu, P.; Han, L.; Li, X.; Liu, C.; et al. Analysis of Widely Targeted Metabolites of Quinoa Sprouts (*Chenopodium quinoa* Willd.) under Saline-Alkali Stress Provides New Insights into Nutritional Value. *Food Chem.* **2024**, *448*, 138575. [CrossRef]
33. Gao, Y.; Long, R.; Kang, J.; Wang, Z.; Zhang, T.; Sun, H.; Li, X.; Yang, Q. Comparative Proteomic Analysis Reveals That Antioxidant System and Soluble Sugar Metabolism Contribute to Salt Tolerance in Alfalfa (*Medicago sativa* L.) Leaves. *J. Proteome Res.* **2019**, *18*, 191–203. [CrossRef]
34. Rostamabadi, H.; Karaca, A.C.; Deng, L.; Colussi, R.; Narita, I.M.P.; Kaur, K.; Aaliya, B.; Sunooj, K.V.; Falsafi, S.R. Oat Starch—How Physical and Chemical Modifications Affect the Physicochemical Attributes and Digestibility? *Carbohydr. Polym.* **2022**, *296*, 119931. [CrossRef]
35. Yagoubi, Y.; Hajji, H.; Smeti, S.; Mahouachi, M.; Kamoun, M.; Atti, N. Growth Performance, Carcass and Noncarcass Traits and Meat Quality of Barbarine Lambs Fed Rosemary Distillation Residues. *Animal* **2018**, *12*, 2407–2414. [CrossRef] [PubMed]

36. Bertrand, A.; Gatzke, C.; Bipfubusa, M.; Lévesque, V.; Chalifour, F.P.; Claessens, A.; Rocher, S.; Tremblay, G.F.; Beauchamp, C.J. Physiological and Biochemical Responses to Salt Stress of Alfalfa Populations Selected for Salinity Tolerance and Grown in Symbiosis with Salt-Tolerant Rhizobium. *Agronomy* **2020**, *10*, 569. [CrossRef]
37. Lee, S.; Ko, K.; Kim, G.; Park, J.; Ryu, Y. Comparison of Meat Quality, Including Fatty Acid Content and Amino Acid Profile, and Transcriptome Profile among Hanwoo, Korea Black Cattle, and Jeju Black Cattle. *Food Sci. Anim. Resour.* **2025**, *45*, 553–572. [CrossRef]
38. Torrecilhas, J.A.; Pereira, G.L.; Vito, E.S.; Fiorentini, G.; Ramirez-Zamudio, G.D.; Fonseca, L.S.; Torres, R.d.N.S.; Simioni, T.A.; Duarte, J.M.; Machado Neto, O.R.; et al. Changes in the Lipid Metabolism of the *Longissimus thoracis* Muscle in Bulls When Using Different Feeding Strategies during the Growing and Finishing Phases. *Metabolites* **2023**, *13*, 1042. [CrossRef]
39. Yi, J.; Chen, F.; Li, H.; Yu, Q.; Sun, C.; Wen, R. Oxidative-Textural Coupling Mechanisms in Chicken Breast Stored at Low-Temperature Gradients: Super-Chilling as a Quality Preserving Strategy. *Food Res. Int.* **2025**, *217*, 116740. [CrossRef]
40. Friha, M.; Hamdi, H.; Ayeb, N.; Hajlaoui, A.; Durand, D.; Majdoub-Mathlouthi, L. Potential Use of Natural Saline Pasture for Grazing Lambs: Effect on Digestibility, Growth Performances, Carcass and Meat Quality. *Small Rumin. Res.* **2022**, *210*, 106669. [CrossRef]
41. Zhang, J.; Zhang, Y.; Wang, J.; Jin, H.; Qian, S.; Chen, P.; Wang, M.; Chen, N.; Ding, L. Comparison of Antioxidant Capacity and Muscle Amino Acid and Fatty Acid Composition of Nervous and Calm Hu Sheep. *Antioxidants* **2023**, *12*, 459. [CrossRef] [PubMed]
42. Kamei, Y.; Hatazawa, Y.; Uchitomi, R.; Yoshimura, R.; Miura, S. Regulation of Skeletal Muscle Function by Amino Acids. *Nutrients* **2020**, *12*, 261. [CrossRef]
43. Kang, L.; Li, X.; Zhao, X.; Liu, T.; Jin, Y.; Duan, Y. Effects of L-Arginine Supplementation on Fat Deposition and Meat Quality in Growing Lambs: Interactions with Gut Microbiota and Metabolic Signalling Pathways. *Food Chem.* **2025**, *479*, 143677. [CrossRef]
44. Zhang, L.; Qin, Q.; Li, Q.; Yu, Y.; Song, Z.; He, L.; Sun, Y.; Ye, L.; Wang, G.; Xu, J. Molecular Adaptations and Quality Enhancements in a Hybrid (*Erythroculter ilishaeformis* ♀× *Ancherythroculter nigrocauda* ♂) Cultured in Saline-Alkali Water. *Biology* **2025**, *14*, 718. [CrossRef]
45. Zhang, M.; Dunshea, F.R.; Warner, R.D.; DiGiacomo, K.; Osei-Amponsah, R.; Chauhan, S.S. Impacts of Heat Stress on Meat Quality and Strategies for Amelioration: A Review. *Int. J. Biometeorol.* **2020**, *64*, 1613–1628. [CrossRef] [PubMed]
46. Zhang, X.; Han, L.; Gui, L.; Raza, S.H.A.; Hou, S.; Yang, B.; Wang, Z.; Ma, Y.; Makhlof, R.T.M.; Alhuwaymil, Z.; et al. Metabolome and Microbiome Analysis Revealed the Effect Mechanism of Different Feeding Modes on the Meat Quality of Black Tibetan Sheep. *Front. Microbiol.* **2022**, *13*, 1076675. [CrossRef]
47. Lamberigts, C.; Wang, Y.; Dierckx, T.; Buys, N.; Everaert, N.; Buyse, J. The Influence of Thyroid State on Hypothalamic AMP-Activated Protein Kinase Pathways in Broilers. *Gen. Comp. Endocrinol.* **2021**, *311*, 113838. [CrossRef]
48. Liang, G.; Ma, Y.; Deng, P.; Li, S.; He, C.; He, H.; Liu, H.; Fan, Y.; Li, Z. Hexokinases in Gastrointestinal Cancers: From Molecular Insights to Therapeutic Opportunities. *Semin. Oncol.* **2025**, *52*, 152351. [CrossRef]
49. Li, B.; Ming, H.; Qin, S.; Nice, E.C.; Dong, J.; Du, Z.; Huang, C. Redox Regulation: Mechanisms, Biology and Therapeutic Targets in Diseases. *Signal Transduct. Target. Ther.* **2025**, *10*, 72. [CrossRef]
50. Vetrovoy, O.; Stratilov, V.; Potapova, S.; Tyulkova, E. Oxidative Stress Accompanies HIF1-Dependent Impairment of Glucose Metabolism in the Hippocampus of Adult Rats That Survived Prenatal Severe Hypoxia. *Dev. Neurosci.* **2024**, *46*, 297–307. [CrossRef] [PubMed]
51. Bayramoglu Akkoyun, M.; Temel, Y.; Bengü, A.Ş.; Akkoyun, H.T. Ameliorative Effects of Astaxanthin against Copper(II) Ion-Induced Alteration of Pentose Phosphate Pathway and Antioxidant System Enzymes in Rats. *Environ. Sci. Pollut. Res. Int.* **2021**, *28*, 62919–62926. [CrossRef]
52. Tripathi, S.; Parmar, D.; Raval, S.; Mishra, R.; Singh, G. Attenuation of Chromium (VI) and Arsenic (III)-Induced Oxidative Stress and Hepatic Apoptosis by Phloretin, Biochanin-a, and Coenzyme Q10 via Activation of SIRT1/Nrf2/HO-1/NQO1 Signaling. *J. Biochem. Mol. Toxicol.* **2024**, *38*, e23817. [CrossRef] [PubMed]
53. Shojadoost, B.; Yitbarek, A.; Alizadeh, M.; Kulkarni, R.R.; Astill, J.; Boodhoo, N.; Sharif, S. Centennial Review: Effects of Vitamins A, D, E, and C on the Chicken Immune System. *Poult. Sci.* **2021**, *100*, 100930. [CrossRef] [PubMed]
54. Liu, S.; Zhong, K.; Zhong, J.; Fan, X.; Qin, X. Mitigating Effect of Taurine Combined with Corona Dormancy on Oxidative Stress in Trachinotus Ovatus under Low-Temperature Stress. *Int. J. Mol. Sci.* **2025**, *26*, 2927. [CrossRef]

foods

MDPI

Article

Effect of Fruit Powders as Natural Alternatives to Sodium Nitrite on Lipid Oxidation in Clean-Label Salami

Adriana-Ioana Moraru Manea, Ileana Cocan, Delia-Gabriela Dumbrava and Mariana-Atena Poiana *

Faculty of Food Engineering, University of Life Sciences "King Mihai I" from Timisoara, Calea Aradului No. 119, 300645 Timisoara, Romania; adriana.manea@usvt.ro (A.-I.M.M.); ileanacocan@usvt.ro (I.C.); deliadumbrava@usvt.ro (D.-G.D.)

* Correspondence: marianapoiana@usvt.ro; Tel.: +40-726239838

Abstract: Public concerns about the health risks of synthetic antioxidants have prompted the meat industry to look for natural alternatives rich in phenols with strong antioxidant properties. This study investigates the use of blackcurrant (BCP), lingonberry (LP), and sour cherry (SCP) powders as natural substitutes for synthetic nitrites in reformulating two clean-label salami types, smoked and cooked and smoked and scalded, with a focus on their effects on oxidative stability during processing and refrigerated storage (4 °C). Nitrite-free formulations were prepared with each fruit powder at three inclusion levels to provide total phenolic contents of 90, 200, and 300 mg gallic acid equivalents (GAE)/kg of processed meat. A nitrite-containing control (90 mg/kg) and an additive-free control were included for comparison. The phytochemical profiles of powders were characterized by total phenolic, flavonoid, monomeric anthocyanin contents, and L-ascorbic acid levels. Antioxidant activity was assessed via 1,1-diphenyl-2-picrylhydrazyl (DPPH) radical scavenging and ferric reducing antioxidant power (FRAP) assays. Salami samples were analyzed for proximate composition, and lipid oxidation was monitored at 0, 15, and 30 days of storage using peroxide value, inhibition of oxidation, p-anisidine value, TOTOX, and thiobarbituric acid value. Fruit powders demonstrated dose- and type-dependent inhibition of primary and secondary lipid oxidation, enhancing oxidative stability during processing and storage. After 30 days of storage, oxidation markers in fruit-enriched salami remained below recommended thresholds, confirming effective control of lipid oxidation. The inhibitory potential followed the order BCP > LP > SCP, consistent with antioxidant profiles as reflected by DPPH and FRAP values. BCP at 300 mg GAE/kg showed a stronger lipid oxidation inhibition than sodium nitrite. Promising improvements in lipid oxidation resistance were also observed with LP at 300 mg GAE/kg and BCP at 200 mg GAE/kg. These findings highlight the potential of fruit-derived antioxidants to support the development of more sustainable, value-added meat products without compromising quality.

Keywords: fruit powders; natural antioxidants; clean-label salami; nitrite substitutes; lipid oxidation

1. Introduction

Meat products are vital for human nutrition, providing excellent sources of proteins, fats, essential amino acids, vitamins, minerals, and other critical nutrients. However, their complex composition makes them susceptible to significant chemical transformations during processing and storage, potentially forming harmful substances like nitrosamines and lipid oxidation products [1]. Oxidation of both lipids and proteins not only diminishes shelf life but can also generate detrimental end-products [2]. Specifically, lipid oxidation is a primary driver of quality degradation in meat products during storage and processing. This complex process initiates with the peroxidation of unsaturated fatty acids in phospholipid membranes, yielding hydroperoxides as primary oxidation products. Subsequently, these compounds decompose into secondary compounds such as aldehydes, ketones, alkenes, and alcohols, negatively impacting both the sensory qualities and nutritional value of meat and meat products [3,4]. Currently, synthetic preservatives are widely employed to protect foods from quality deterioration. They mitigate the damaging effects of free radicals in meat systems, thereby minimizing oxidative degradation during processing and storage and consequently extending product shelf life [5,6].

Synthetic nitrates and nitrites (as potassium or sodium salts) are widely used in processed meat for their vital role in quality, particularly color stability, and their ability to retard bacterial spoilage and lipid oxidation [7]. In cured meats, nitrites exert a protective antioxidative effect as nitric oxide (NO) binds to heme iron, preventing lipid peroxidation and enhancing storage stability. Despite these benefits, nitrite remains one of the most controversial additives due to the toxicity associated with the formation of nitrosamines through nitrosation reactions between secondary amines and nitrosating agents such as nitrates, nitrites, or nitrogen oxides during meat processing. This issue highlights the need to shift towards clean-label strategies that address current sustainability and food safety concerns [8,9]. Consumer health concerns have become increasingly prominent in recent years, reflected in a willingness to pay more for nitrite-free meat products [10,11]. This trend has led to the reformulation of meat products, specifically by reducing or substituting nitrites through the incorporation of plant materials rich in phytochemicals. To address this challenge, natural extracts and ingredients offer a valuable alternative to conventional synthetic additives, with the potential to effectively slow product quality deterioration [12–15]. Common phenolic compounds are key plant components, exhibiting strong antioxidant capabilities by scavenging reactive nitrogen/oxygen species and free radicals, inhibiting free radical-forming enzymes, binding metals, and activating antioxidant enzymes. In meat products, polyphenols effectively inhibit oxidation, preventing discoloration and quality deterioration [16]. Numerous studies attest to the efficacy of natural antioxidants in meat products [17,18]. Replacing synthetic nitrates and antioxidants with plant-based alternatives (powders, extracts, and other natural derivatives) has been shown to strongly inhibit protein and lipid oxidation, slow degradation processes, retard off-flavor development (rancidity), improve color stability, enhance microbiological quality, and extend the shelf life of fresh and processed meat products. Crucially, these natural alternatives achieve comparable effectiveness to synthetic preservatives without compromising sensory or nutritional properties [19]. Incorporating plant-derived bioactive compounds with antioxidant function into meat products can preserve their composition and quality while offering health benefits to consumers [20]. Fruit powders, in particular, with their significant content of high-value phenolic compounds and other bioactive ingredients, can effectively prevent the initiation or propagation of lipid oxidation reactions. Multiple research efforts confirm the efficacy of these plant-derived compounds. For instance, mulberry fruit powder significantly extended the shelf life of minced beef due to its antioxidant properties [21]. Similarly, fruit powders derived from red grapes, gooseberry, and tomato have been shown to preserve the quality of restructured chicken slices for up to 20 days under refrigerated storage conditions [22]. Plant-derived powders, rich in complex phytochemical compounds, offer a viable alternative to individual synthetic antioxidants for extending the shelf life of meat products and may also have associated nutritional and health benefits. Certain vegetable powders have been shown to significantly enhance the oxidative stability of turkey meat patties by 20–30% under accelerated oxidation conditions [23]. It has also been reported that fortifying beef burgers with blueberries resulted in improved sensory quality and increased primary and secondary lipid oxidation stability, demonstrating its high potential as a natural functional ingredient with preservative capacity [24]. In general, powdered plant materials have notable antimicrobial and antioxidant properties in minced meat products [25]. The results of the study conducted by Martínez-Zamora et al. [26] revealed the high potential of spices, fruits, and vegetables to limit the oxidative deterioration of lipids and proteins, making them a suitable alternative to synthetic counterparts.

Our previous research highlighted the potential of tomato processing byproducts [27] and bell pepper processing byproducts [28] as promising substitutes for sodium nitrite. These byproducts can ensure lipid oxidative stability during the cold storage of sausages for 20 days, contributing to the development of value-added meat products. The application of natural extracts and spices, fruits, and vegetables in formulating clean-label meat products as substitutes for synthetic nitrites/nitrates and antioxidants is a growing trend. However, limited studies document their effect on the oxidative stability of meat products during prolonged cold storage. Addressing the challenges previously outlined, this study aimed to evaluate the efficacy of blackcurrant, lingonberry, and sour cherry powders as natural alternatives to nitrites in the reformulation of two distinct salami types intended to be clean-label, smoked and cooked, and smoked and scalded, focusing on their effects on oxidative stability during processing and refrigerated storage (4 °C). Our central hypothesis is that the bioactive compounds inherent in these fruit powders could serve as a viable replacement for synthetic additives commonly used in meat products, thereby not only addressing potential health concerns but also significantly improving the functional profile, sustainability, and overall quality of the final product. Specifically, this research focused on: (1) determining the phytochemical content and antioxidant activities of the fruit powders, and (2) evaluating their ability to effectively inhibit lipid oxidation throughout both the processing and cold storage (up to 30 days) of the salami samples. Nitrite-free salami formulations were developed by incorporating fruit powders at doses calculated to provide total phenolic contents of 90, 200, and 300 mg gallic acid equivalents (GAE)/kg of processed meat. The salami samples were analyzed for their proximate composition, and the progression of lipid oxidation was systematically evaluated using specific chemical indices at 0, 15, and 30 days of storage at 4 °C.

2. Materials and Methods

2.1. Materials

Fresh meat and pork fat for the salami recipe were sourced from Comtim Romania SRL (Timis, Romania). Other ingredients were procured from specific suppliers: salt from Salrom-SNS SA (Bucharest, Romania), nitrite salt from Daz Activ Trade SRL (Botoșani, Bucharest, Romania), and spices from Profood Rom SRL (Miercurea Ciuc, Harghita County, Romania). The spice blend included granulated garlic, ground thyme, ground white and black pepper, ground allspice, sweet paprika, hot paprika, ground nutmeg, ground coriander, and ground cumin.

Frozen fruits were procured from Romanian producers. Blackcurrants (*Ribes nigrum* L.) came from SC Forelit (Salard, Bihor County, Romania), lingonberries (*Vaccinium vitis-idaea* L.) from SC Vladalex Impex (Targu Mures, Mures County, Romania), and sour cherries (*Prunus cerasus* L.) from SC Gradina Padurii (Oradea, Bihor County, Romania).

For chemical analyses, analytical grade reagents were supplied by Chemical Company (Iasi, Romania), Sigma-Aldrich (St. Louis, MO, USA), Chimreactiv (Bucharest, Romania), and Adra Chim (Bucharest, Romania).

2.2. Obtaining the Fruit Powder

Prior to processing, the frozen fruit samples were allowed to thaw at room temperature for approximately two hours. Subsequently, the samples were subjected to convective drying in a BINDER drying chamber (Binder GmbH, Tuttlingen, Germany) at 60 °C for a total of 15 h, divided into three 5 h sessions conducted over three consecutive days. Drying at this temperature is known to limit enzymatic degradation, thereby improving the preservation of polyphenolic compounds and maintaining the functional quality of the dried plant matrix [29]. The dried fruits reached a final moisture content below 5%, with final values of 4.59% for sour cherry, 4.93% for blackcurrant, and 4.71% for lingonberry. Achieving such low moisture levels is essential for microbial stability, as water activity values in the range of 0.2–0.3, usually associated with a moisture content below 5%, are considered microbiologically safe [30]. After drying, the samples were cooled to 20 °C and ground to a fine powder using a Grindomix GM 2000 laboratory mill (Retsch GmbH, Germany), then passed through a 60-mesh sieve to ensure uniform particle size. The powders obtained from blackcurrants (BCP), lingonberry (LP), and sour cherries (SCP) were vacuum-packed in polypropylene bags and stored at room temperature in the dark until further analysis or application.

2.3. Manufacture of Salami Formulas

This study focused on developing two salami types formulated as clean-label products by replacing sodium nitrite with sour cherry (SCP), blackcurrant (BCP), or lingonberry (LP) powders. The two types, smoked and cooked (SI) and smoked and scalded (SII), differ in their specific recipes and processing parameters. Both product types undergo the same overall processing time, but SI is subjected to a final heat treatment at 72 °C with 5% relative air humidity, while SII is treated at the same temperature under approximately 80% relative humidity, which allows a comparative evaluation of the antioxidative efficacy of fruit powders in mitigating oxidative degradation under typical industrial meat production conditions. All salami formulations were produced under standard conditions at the S.C. Cavarantana meat processing unit in the village of Cavaran, Caraș-Severin County, Romania, following the traditional manufacturing procedures employed at this facility. For each salami type (SI and SII), 11 distinct experimental variants were prepared. The design included a positive control containing 90 mg sodium nitrite per kilogram of processed meat and a negative control without nitrite. The positive control samples used a salt mixture containing 0.5% (g/g) sodium nitrite, while the negative control samples used salt without added sodium nitrite. In the remaining variants, sodium nitrite was entirely replaced with fruit powders incorporated at concentrations calculated to provide target total phenolic contents (TPC) of 90, 200, or 300 mg gallic acid equivalents (GAE) per kilogram of processed meat. The specific amount of each fruit powder (SCP, BCP, and LP) incorporated into the recipes was determined by its individual total phenolic compound content. The minimum total phenolic content (TPC) level of 90 mg GAE/kg in the processed meat, supplied by the fruit powders, was selected to correspond to the sodium nitrite concentration used in the traditional SI and SII recipes (90 mg/kg). The selection of these fruit powder doses for the sausage formulation was based on our previous studies, where supplementation with tomato byproducts [27] and yellow and red bell pepper [28], providing total phenolic content levels of 90, 180, and 270 mg GAE/kg of processed meat, yielded promising results in limiting oxidative degradation during processing and 21 days of cold storage. All salami variants were prepared in a cold room maintained at a temperature below 10 °C. The codes for the salami formulations are defined as follows:

SI-C: Nitrite-free smoked and cooked salami (negative control sample);
SI-CN: Smoked and cooked salami with added sodium nitrite (positive control);
SI-SCP90, SI-SCP200, SI-SCP300: Smoked and cooked nitrite-free salami with sour cherry powder added to provide a TPC of 90, 200, and 300 mg GAE/kg of processed meat;
SI-BCP90, SI-BCP200, SI-BCP300: Smoked and cooked nitrite-free salami with blackcurrant powder added to provide a TPC of 90, 200, and 300 mg GAE/kg of processed meat;
SI-LP90, SI-LP200, SI-LP300: Smoked and cooked nitrite-free salami with lingonberry powder added to provide a TPC of 90, 200, and 300 mg GAE/kg of processed meat;
SII-C: Nitrite-free smoked and scalded salami (negative control sample);
SII-CN: Smoked and scalded salami with added sodium nitrite (positive control);
SII-SCP90, SII-SCP200, SII-SCP300: Smoked and scalded nitrite-free salami with sour cherry powder added to provide a TPC of 90, 200, and 300 mg GAE/kg of processed meat;
SII-BCP90, SII-BCP200, SII-BCP300: Smoked and scalded nitrite-free salami with blackcurrant powder added to provide a TPC of 90, 200, and 300 mg GAE/kg of processed meat;
SII-LP90, SII-LP200, SII-LP300: Smoked and scalded nitrite-free salami with lingonberry powder added to provide a TPC of 90, 200, and 300 mg GAE/kg of processed meat.

Ingredient quantities for SI salami (per kg of pork/fat mixture) are presented in Table 1.

Table 1. Ingredients for SI salami manufacturing (control and fruit powder supplemented).

Sample	Pork Meat (g)	Pork Fat (g)	Salt (g)	Salt + 0.5% (*w*/*w*) Sodium Nitrite (g)	Spice Mixture (g)	SCP (g)	BCP (g)	LP (g)
SI-C	780	220	18	-	15	-	-	-
SI-CN	780	220	-	18	15	-	-	-
SI-SCP90	780	220	18	-	15	9.42	-	-
SI-SCP200	780	220	18	-	15	20.94	-	-
SI-SCP300	780	220	18	-	15	31.41	-	-
SI-BCP90	780	220	18	-	15	-	6.13	-
SI-BCP200	780	220	18	-	15	-	13.62	-
SI-BCP300	780	220	18	-	15	-	20.43	-
SI-LP90	780	220	18	-	15	-	-	7.83
SI-LP200	780	220	18	-	15	-	-	17.40
SI-LP300	780	220	18	-	15	-	-	26.10

The spice mixture used in the SI formulation contained 7 g of garlic granules, 3 g of ground thyme, 3 g of ground white pepper, and 2 g of ground allspice.

Table 2 details the ingredients per kg of pork/fat mixture in SII salami.

Table 2. Ingredients for SII salami manufacturing (control and fruit powder supplemented).

Sample	Pork meat (g)	Pork Fat (g)	Salt (g)	Salt + 0.5% (*w*/*w*) Sodium Nitrite (g)	Spice Mixture (g)	SCP (g)	BCP (g)	LP (g)
SII-C	800	200	18	-	28	-	-	-
SII-CN	800	200	-	18	28	-	-	-
SII-SCP90	800	200	18	-	28	9.42	-	-
SII-SCP200	800	200	18	-	28	20.94	-	-
SII-SCP300	800	200	18	-	28	31.41	-	-
SII-BCP90	800	200	18	-	28	-	6.13	-
SII-BCP200	800	200	18	-	28	-	13.62	-
SII-BCP300	800	200	18	-	28	-	20.43	-
SII-LP90	800	200	18	-	28	-		7.83
SII-LP200	800	200	18	-	28	-	-	17.40
SII-LP300	800	200	18	-	28	-	-	26.10

The spice mixture used in SII formulations contained 9 g of garlic granules, 10 g of sweet paprika, 1 g of hot paprika, 3 g of ground black pepper, 2 g of ground nutmeg, 1 g of ground coriander, and 2 g of ground cumin.

The technological flow followed for SI production involves cutting pork meat and fat into pieces of approximately 50 g, mincing them using a meat grinder with a 5 mm sieve (Luohe, YGM-100, Henan, China), adding the other ingredients in the recipe, and homogenizing with a mixer (Inotec IM-150E, Reutlingen, Germany) for 10 min. The resulting mixture is stuffed into collagen membranes with a diameter of 45 mm using a filling machine (Albert Handtmann, VF-608, 88400 Biberach an der Riss, Germany), and the resulting salami sticks are sealed at both ends with metal clips using a clip sealing machine (Tipper Tie Technopack, KDCVT 400, 21509 Glinde, Germany). The salamis are left to mature in a cold room for 24 h at a temperature of 4 °C. The heat treatment of the raw salami formulations was carried out in a smoke oven (Doleschal, SC6001, A-4400 Steyr, Austria), undergoing the following phases: preheating at 60 °C for 15 min, drying at 70 °C for 30 min, smoking at 70 °C for 30 min, and cooking at 72 °C and 5% relative air humidity until an internal temperature of 70 °C is reached in the salami.

SII followed the same manufacturing process as SI, except for the final heat treatment stage, which was conducted at 72 °C in an environment with approximately 80% relative air humidity until the internal temperature of the salami reached 70 °C. After heat treatment, all salami formulas are cooled, packaged in oxygen-permeable bags, and stored in a cold room at 4 °C and a relative air humidity of 70–80%, after being divided according to each experimental time (0, 15, and 30 days). All formulations were replicated independently twice. For each replicate, six pieces of salami were produced per treatment. The salami has a recommended shelf life of 30 days.

Figure 1 displays the salami samples after this 30-day cold storage period.

The salami samples from day 0 were tested for proximate chemical composition and lipid oxidation, while samples taken after 15 and 30 days of storage were tested to assess the progression of lipid oxidation based on specific chemical indices, such as the peroxide value (PV), para-anisidine value (pAV), and thiobarbituric acid value (TBA). The total oxidation value (TOTOX) and oxidation inhibition (IO, %) were also calculated.

2.4. *Phytochemical Profile and Antioxidant Activity of Fruit Powder*

2.4.1. Obtaining the Extract for Evaluating the Total Phenolic Content, Total Flavonoid Content, and Antioxidant Activity

Alcoholic extracts were prepared by adding 10 mL of 70% (v/v) ethanol to 0.5 g of fruit powders (SCP, BCP, and LP), following the method of Litwinek et al. [31] with minor modifications. Extraction was carried out for 2 h at ambient temperature under continuous stirring with a magnetic stirrer (IDL GmbH & Co KG, Nidderau, Germany), after which the mixtures were centrifuged for 10 min at 10,000 rpm (Hettich EBA 21, Andreas Hettich GmbH & Co. KG., Tuttlingen, Germany). The supernatant was collected, and the remaining residue was re-extracted with a new portion of 70% ethanol (v/v) for another 60 min under the same conditions. The collected supernatants were combined, and the volume of the mixture was adjusted to 20 mL with 70% ethanol (v/v) and stored at −20 °C in the dark until analysis of total phenolic content, total flavonoid content, and antioxidant activity. The extraction procedure was carried out in triplicate for each fruit powder.

2.4.2. Evaluation of Total Phenolic Content

The total phenolic content (TPC) was quantified according to the Folin–Ciocalteau method [32]. Before analysis, the alcoholic extracts from the fruit powder were diluted in a ratio of 1:10 (v/v) with distilled water. Subsequently, 0.5 mL of diluted extracts was combined with 2.5 mL of Folin-Ciocalteu reagent, previously diluted in a ratio of 1:10 v/v with distilled water, and 2 mL of 7.5% Na_2CO_3 solution. The resulting mixture was incubated at 50 °C for 30 min, followed by measurement of the absorbance at 750 nm using the SPECORD 205 UV-Vis spectrophotometer (Analytik Jena Inc., Jena, Germany) against a control sample prepared under the same conditions. A gallic acid calibration curve was generated by plotting values of standards' absorbances versus their concentrations in the range 0.1–1.0 µM GAE/mL. The TPC in extracts was calculated from the regression equation, and the results were converted to mg gallic acid equivalent (GAE)/g dry weight (d.w.) of fruit powder.

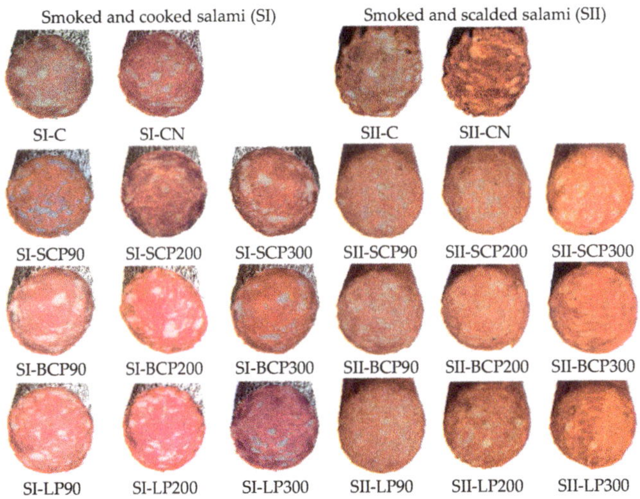

Figure 1. Salami formulas after 30 days of cold storage. SI-C, SII-C: Nitrite-free SI and SII salami; SI-CN, SII-CN: SI and SII salami with added sodium nitrite; SI-SCP90, SI-SCP200, SI-SCP300, SII-SCP90, SII-SCP200, SII-SCP300: Nitrite-free SI and SII salami with sour cherry powder added to provide a TPC of 90, 200 and 300 mg GAE/kg of meat; SI-BCP90, SI-BCP200, SI-BCP300, SII-BCP90, SII-BCP200, SII-BCP300: Nitrite-free SI and SII salami with blackcurrant powder added to provide a TPC of 90, 200 and 300 mg GAE/kg of meat; SI-LP90, SI-LP200, SI-LP300, SII-LP90, SII-LP200, SII-LP300: Nitrite-free SI and SII salami with lingonberry powder added to provide a TPC of 90, 200 and 300 mg GAE/kg of meat.

2.4.3. Evaluation of Total Flavonoid Content (TFC)

Total flavonoid content in the fruit powder extracts was quantified using an adapted colorimetric assay [33]. In brief, a 3.0 mL portion of the ethanol extract, previously obtained, was introduced to a mixture containing 4.5 mL of deionized water and 1.0 mL of a 0.3% (w/v) aqueous solution of sodium nitrite. This initial mixture underwent a 6 min incubation period at a controlled temperature of 20 °C. Subsequently, 1.0 mL of 10% (w/v) aluminum nitrate solution was incorporated, followed by an additional incubation period of 6 min. The procedure continued with the addition of 10.0 mL of 4% (w/w) sodium hydroxide solution, after which the final volume of the reaction mixture was adjusted to 25.0 mL using a 70% ethanol solution (v/v) and allowed to stand at room temperature for 15 min to complete the reaction. The absorbance was measured at 510 nm against a 70% (v/v) ethanol blank using a UV-Vis spectrophotometer. A calibration curve was generated employing a series of quercetin (QE) standard solutions, covering a concentration range of 0.5 to

50 μg/mL. The total flavonoid content of fruit powder was ultimately expressed as milligrams of quercetin equivalents per 100 g of dry weight (mg QE/100 g d.w.).

2.4.4. Evaluation of Monomeric Anthocyanin Content

The anthocyanin content in fruit powder extract was determined spectrophotometrically as monomeric anthocyanin by the pH-differential method [34]. Fruit powder extracts were prepared by mixing 0.25 g of fruit powder with 5 mL of a 0.1 N HCl and 96% (*v*/*v*) ethanol mixture (1:9, *v*/*v*). The extraction proceeded for one hour under continuous stirring. The suspension was then centrifuged for 5 min at 5000 rpm, and the residue obtained after removing the supernatant was re-extracted twice under the same extraction conditions. The combined supernatants were adjusted to a final volume of 15 mL. The extraction was performed in triplicate for each fruit powder. The alcoholic extracts obtained from SCP, BCP, and LP were diluted 1:5, *v*/*v* in 0.025 M potassium chloride buffer (pH 1.0) and 0.4 M sodium acetate buffer (pH 4.5) and the absorbance of each was measured at both 520 and 700 nm with a UV-Vis spectrophotometer Specord 205 (Analytik Jena Inc., Jena, Germany) against distilled water as a blank in a glass cuvette with an optical path length of 10 mm. TMA content was calculated according to Darniadi et al. [34] and expressed as mg cyanidin-3-glucoside equivalents (C3G) per 100 g d.w. using a molar extinction coefficient of 26 900 L/cm·mol and molecular weight of 449.2 g/mol for C3G.

2.4.5. Determination of L-Ascorbic Acid

L-Ascorbic acid (AsAc) or vitamin C content was measured by titrimetric method with 2,6-dichlorophenol-indophenol sodium salt solution. AsAc was extracted by homogenizing 1 g of the fruit powder with 20 mL of 2% (*w*/*v*) oxalic acid solution. The mixture was allowed to stand for about 1 h at 20 °C and filtered through Whatman filter paper No 2 to remove any remaining plant material. The filtrate was mixed with the kaolin decolorizer and re-filtered [35]. Next, 10 mL of the obtained clear filtrate was titrated with 0.025% (*m*/*v*) 2,6-dichloroindophenol solution until a pink color was obtained. The final AsAc content was calculated according to Žlabur [36] and expressed as mg/100 g d.w.

2.4.6. Evaluation of the Antioxidant Activity of Fruit Powder via 1,1-Diphenyl-2-picrylhydrazyl (DPPH) Method

The antioxidant activity of the fruit powder extracts, specifically their ability to scavenge free radicals, was quantified via the 1,1-diphenyl-2-picrylhydrazyl (DPPH) assay [37] with a 0.1 mM DPPH solution in 70% (*v*/*v*) ethanol. The previously obtained ethanolic extracts of each fruit powder were diluted with 70% (*v*/*v*) ethanol at a volumetric ratio of 1:50. Subsequently, a 1.0 mL aliquot of the diluted extracts was combined with 2.5 mL of the 0.1 mM DPPH solution in 70% (*v*/*v*) ethanol. The resulting mixtures were homogenized using a hot plate stirrer (IDL GmbH & Co KG, Nidderau, Germany) and then subjected to a 30 min incubation period in darkness at 20 °C. The absorbance of each incubated mixture was measured at a wavelength of 517 nm against 70% (*v*/*v*) ethanol as the reference blank. Under identical operational conditions, a control sample, consisting of 1.0 mL of 70% (*v*/*v*) ethanol mixed with 2.5 mL of the 0.1 mM DPPH solution in 70% (*v*/*v*) ethanol, was also prepared and analyzed. The DPPH radical scavenging activity was calculated as shown in Equation (1), where $A_{control}$ and A_{sample} represent the absorbance values of the control and sample, respectively.

$$\text{DPPH Scavenging Activity}(\%) = \frac{A_c - A_s}{A_c} \times 100 \quad (1)$$

A calibration curve correlating DPPH scavenging activity (%) with Trolox concentration (μg/mL) was generated using standard solutions of Trolox within the concentration range of 1.0 to 25 μg Trolox/mL [38]. The antioxidant activity of fruit powder was calculated and expressed as μM Trolox equivalents (TE) per gram of dry weight (d.w.).

2.4.7. Assessment of Antioxidant Activity by Ferric Reducing Antioxidant Power (FRAP) Assay

The ferric reducing antioxidant power (FRAP) assay was used to determine the total antioxidant potential of the samples. This method assesses the ability of antioxidant constituents in ethanol extracts to reduce ferric ions (Fe^{3+}) from a colorless tripyridyltriazine complex to ferrous ions (Fe^{2+}) in an acidic medium. This redox reaction, driven by electron donation from antioxidant species, results in the formation of an intense blue-colored complex with tripyridyltriazine (TPTZ), which exhibits maximum absorption at 593 nm [39]. The FRAP working solution was prepared by mixing 100 mL of acetate buffer (pH 3.6), 10 mL of a 10 mM TPTZ solution in 40 mM HCl, and 10 mL of a 20 mM $FeCl_3 \cdot 6H_2O$ solution. Prior to measurement, the initial alcoholic extracts of the fruit powders were diluted with distilled water at a 1:50 (*v*/*v*) ratio. For the assay, 0.5 mL of the diluted extracts was allowed to react with 2.5 mL of the FRAP working solution at 37 °C for 30 min. The resulting absorbance was measured at 593 nm, against a blank solution prepared without the sample under identical conditions. The antioxidant capacity was quantified as μM Fe^{2+} equivalents per gram of dry weight, based on a calibration curve generated with $FeSO_4 \cdot 7H_2O$ standard solutions in the concentration range of 0.05 to 0.5 μM Fe^{2+} equivalents/mL. Each analysis was conducted in three independent replicates.

2.5. *Proximate Composition and Energy Value Evaluation of Salami Formulations*

The proximate analysis of the salami formulations was performed following appropriate methods recommended by AOAC [40], such as AOAC 950.46 for moisture, AOAC 973.48 for protein, AOAC 960.39 for fat, AOAC 937.09 for NaCl, and AOAC 999.11 for ash, respectively. The carbohydrate content was calculated by subtracting from 100 the sum of the protein, ash, lipid, NaCl, and moisture content. The energy values for each salami formula, expressed in kcal/100 g, were calculated taking into account their carbohydrate, fat, and protein content and using the specific energy factors for protein (4 calories per gram), for fat (9 calories per gram), and carbohydrates (4 calories per gram) [41]. Each analysis was conducted in three independent replicates.

2.6. *Assessing the Progression of Lipid Oxidation*

Lipid oxidation in salami samples, stored under refrigerated conditions (4 °C) for 0, 15, and 30 days, was assessed by measuring specific chemical indices. These included peroxide value (PV), para-anisidine value (pAV), and thiobarbituric acid (TBA) value. Additionally, inhibition of oxidation (IO) and total oxidation (TOTOX) values were calculated. All analyses were conducted in three independent replicates.

2.6.1. Peroxid Value (PV)

To assess the extent of primary lipid oxidation in salami samples, peroxide value (PV) was determined. This indicator quantifies the concentration of hydroperoxides generated during the initial oxidation stage. The analysis followed the iodometric method [42] and was performed on the lipids extracted from salami samples with a mixture of chloroform and methanol (2:1, *v*/*v*) following the procedure described by Seo et al. [43]. PV was reported as milliequivalents of active oxygen per kilogram of lipid.

2.6.2. Para-Anisidine Value (pAV)

To assess the extent of secondary lipid oxidation, the quantification of aldehydic compounds, as secondary oxidation products, was performed by means of the para-anisidine value (pAV). This analysis was performed following the methodology detailed in the International Organization for Standardization (ISO) protocol 6885:2008 [44]. The basic principle of this test involves the nucleophilic addition of the amine group of para-anisidine to the electrophilic carbonyl group of aldehydes, resulting in a Schiff base that exhibits maximum absorption at a wavelength of 350 nm. For the analysis, a precisely weighed 2-g aliquot of extracted fat according to Seo et al. [43] was dissolved in 25 mL of isooctane, and its baseline absorbance was spectrophotometrically determined at 350 nm using isooctane as the reference (Abs_I). Subsequently, a 5 mL portion of this initial solution was reacted with 1 mL of a para-anisidine reagent (0.25% w/v in glacial acetic acid) in a separate vessel. Following a 10 min reaction period, the absorbance of this mixture was measured at 350 nm against a control comprising 5 mL of isooctane and 1 mL of the anisidine reagent, (Abs_{II}). The pAV was then computed using the relationship presented in Equation (2):

$$\mathrm{pAV} = 25 \times \frac{1.2 \times \mathrm{Abs_{II}} - \mathrm{Abs_I}}{\mathrm{m}} \tag{2}$$

where Abs_I and Abs_{II} represent the absorbance of the fat sample in isooctane and in isooctane with para-anisidine solution, respectively; m—mass of the fat sample (g).

2.6.3. Inhibition of Oxidation (IO)

The inhibition of lipid oxidation (IO), achieved by adding either sodium nitrite or fruit powder, was quantified following the method by Mariod et al. [45]. This assessment relied on comparing the peroxide value (PV) increase in samples after 15 and 30 days of storage, relative to day 0, against the PV increase observed in the control sample over the same storage period. The calculation was performed using the relationship detailed in Equation (3).

$$\mathrm{IO}(\%) = (1 - \frac{\text{increase in PV of sample}}{\text{increase in PV of control}}) \times 100 \tag{3}$$

2.6.4. TOTOX Value

Based on the findings of a previous study [2], the simultaneous assessment of both primary and secondary markers of lipid oxidation is considered more effective for a comprehensive understanding of oxidative deterioration. This dual approach provides insight into the overall extent of oxidation by including the accumulation of secondary oxidation products alongside the oxidative status indicated by primary species. The TOTOX value was calculated by combining the peroxide value (PV) and the para-anisidine value (pAV), according to Equation (4) [46].

$$\text{TOTOX value} = \mathrm{pAV} + 2 \times \mathrm{PV} \tag{4}$$

2.6.5. Thiobarbituric Acid (TBA) Value

To quantify the secondary products of lipid oxidation, particularly malondialdehyde (MDA), the thiobarbituric acid (TBA) test was performed [47]. This colorimetric method is based on the principle that MDA reacts with TBA under heated acidic conditions to produce a colored complex, the concentration of which is proportional to the MDA content. To this end, 5 g of the minced salami sample was homogenized by shaking for 5 min in 20 mL of 5% trichloroacetic acid solution (w/v). The resulting homogenate was then centrifuged at 12,000 rpm for 10 min to obtain a clear supernatant. Subsequently, a 4 mL aliquot of the supernatant was combined with an equal volume (4 mL) of a 0.02 M aqueous TBA solution. The resulting mixture was then heated for 60 min at 100 °C in a water bath to facilitate color development. Following incubation and cooling to ambient temperature, the absorbance of the resulting solution was measured at 532 nm using a Specord 205 spectrophotometer (Analytik Jena Inc., Jena, Germany). A reagent blank, devoid of the fat sample, was used as the reference for these spectrophotometric determinations. A fat-free sample prepared under the same operating conditions was used as a reference. The MDA concentration in the samples was determined by interpolating the absorbance values against a standard calibration curve prepared with MDA solutions ranging from 10 to 50 μg/mL. TBA values were expressed as mg MDA per kg of sample and represent the average of three independent measurements.

2.7. Statistical Analysis

The data obtained are reported as mean values, followed by the standard deviation (SD). A one-way analysis of variance (ANOVA) was used to evaluate the statistical significance of the differences between the fruit powders, as well as between the salami samples, in response to the storage period and the supplementation variant. Levene's test was utilized to confirm homogeneity of variances. Following this, post hoc analysis was conducted via Tukey's test, and all differences were considered statistically significant at $p < 0.05$.

3. Results and Discussion

3.1. Phytochemical Profile and Antioxidant Activity of Fruit Powder

Table 3 presents the phytochemical profiles of the investigated fruit powders, blackcurrant (BCP), lingonberry (LP), and sour cherry (SCP), including total phenolic, flavonoid, and monomeric anthocyanin contents, ascorbic acid, and antioxidant activity measured via ferric reducing antioxidant power (FRAP) and DPPH radical scavenging assays.

Table 3. Phytochemical profile and antioxidant activity of fruit powder.

Fruit Powder	TPC (mg GAE/g d.w.)	TFC (mg QE/g d.w.)	TMA (mg C3G/100 g d.w.)	AsAc (mg/100 g d.w.)	FRAP (μM Fe^{2+}/g d.w.)	DPPH (μM TE/g d.w.)
BCP	15.45 ± 0.51 [a]	6.98 ± 0.19 [a]	859.04 ± 1.89 [a]	621.09 ± 1.73 [a]	419.41 ± 1.39 [a]	307.44 ± 1.57 [a]
LP	12.07 ± 0.38 [b]	6.11 ± 0.14 [b]	312.58 ± 1.75 [b]	64.52 ± 0.58 [b]	370.47 ± 1.27 [b]	278.65 ± 1.32 [b]
SCP	10.01 ± 0.24 [c]	4.85 ± 0.11 [c]	509.73 ± 1.43 [c]	45.91 ± 0.47 [c]	294.59 ± 1.18 [c]	220.71 ± 1.24 [c]

BCP: blackcurrant powder; LP: lingonberry powder; SCP: sour cherry powder; TPC: total phenolic content, mg gallic acid equivalents (GAE)/g d.w.; TFC: total flavonoid content, mg quercetin equivalents (QE)/g d.w.; TMA: total monomeric anthocyanins, cyanidin-3-glucoside (C3G)/100 g d.w.; AsAc: ascorbic acid, mg/100 g d.w.; FRAP: ferric reducing antioxidant power, μM Fe^{2+}/g d.w.; DPPH: 1,1-diphenyl-2-picrylhydrazyl, μM Trolox equivalents (TE)/g d.w. Results are expressed as the mean value of three independent analyses ± standard deviation (SD). Means in the same column bearing different superscripts are significantly different (one-way ANOVA, $p < 0.05$).

3.1.1. Total Phenolic Content (TPC)

BCP exhibited the highest TPC, with values 54.35% and 28.00% greater than SCP and LP, respectively. This is consistent with Kim [48], who reported a high TPC of approximately 710.33 mg GAE/100 g fresh weight (f.w.) in fresh blackcurrants, as well as Sadowska et al. [49], who found blackcurrant powders ranging from 613.7 to 1577.2 mg GAE/100 g d.w., influenced by drying methods applied. It is worth noting that the TPC value determined for BCP in this study aligns with a previously reported range [49]. A slightly higher TPC value for blackcurrant powder was also documented, reaching 19.84 mg GAE/g d.w. [50]. For LP, the TPC values align with Daukšienė et al. [32], who reported a wide range (6.75–27.19 mg/g d.w.) in dried fruit depending on the preparation method, highlighting the impact of drying techniques on phenolic compounds retention.

Meanwhile, freeze-dried sour cherry powders exhibited the lowest total phenolic content (TPC), ranging from 1.87 to 6.80 mg GAE/g [51], with broader variability reported among cultivars (96.56–268.98 mg GAE/100 g f.w.) [52]. Additionally, depending on the extraction parameters applied, ultrasound-assisted extraction methods have yielded TPC values ranging from 8.45 to 17.50 mg GAE/g d.w. [53].

3.1.2. Total Flavonoid Content (TFC)

As presented in Table 3, BCP demonstrated the highest flavonoid content, exceeding that of SCP by 43.92% and LP by 14.24%. Previous research conducted by Yaman [54] on various sour cherry genotypes reported total flavonoid content (TFC) values ranging from 0.68 to 1.35 mg QE/g f.w. When adjusted for moisture content, our results are in good agreement with these previously reported ranges. Borowiec et al. [50] documented a TFC of 8.28 mg QE/g d.w. in blackcurrant powder, which aligns well with the findings of the present study. In contrast, Blejan et al. [55] reported a considerably higher value (24.43 mg QE/g d.w.) in lyophilized blackcurrant pomace. This discrepancy can be attributed to the higher concentration of polyphenols in the skins and seeds retained in pomace. Moreover, freeze-drying, as applied to the pomace, is known to better preserve flavonoid compounds compared with conventional techniques such as air- or oven-drying.

3.1.3. Total Monomeric Anthocyanin Content (TMA)

The TMA in BCP was significantly higher (859.04 mg C3G/100 g d.w.) than in SCP (509.73 mg) and LP (312.58 mg), consistent with Lee et al. [56], who reported similar values for blackcurrants (813.60 mg/100 g d.w.). Borowiec et al. [50] documented lower TMA in blackcurrant powders (520 mg/100 g d.w.), possibly due to differences in source material or drying methods. For SCP, the TMA values align well with those reported by Sokół-Łętowska et al. [52], 83.24 mg C3G/100 g f.w., when adjusted to dry weight equivalents.

Lingonberry TMA levels (306–396 mg C3G/100 g dry weight) reported by Ozola and Kampuse [57], as well as lower fresh weight levels documented by Dróżdż et al. [58] and Koponen et al. [59], support the variability observed and emphasize the influence of origin and processing. Consistent with Ozola and Kampuse [57], TMA content in dried lingonberry byproducts ranged from 306 to 396 mg C3G/100 g d.w., varying by drying method. Dróżdż et al. [58] reported 33–47 mg/100 g f.w. in Polish wild lingonberries, while Koponen et al. [59] found 77.5 mg/100 g f.w. in berries grown in Finland. Adjusted to dry weight, these values align with our findings.

3.1.4. L-Ascorbic Acid Content (AsAc)

BCP demonstrated remarkably high ascorbic acid content (621.09 mg/100 g dry weight), being approximately 9.63 and 13.53 times richer than LP (64.52 mg) and SCP (45.91 mg), respectively. These findings align with Sadovska et al. [49], who documented ascorbic acid contents from 445.2 to 861.2 mg/100 g d.w. in blackcurrant powders depending on drying conditions. The value for the convection drying method; (62050; mg/100, g, dw), was comparable to our data. Urbonaviciene et al. [60] reported ascorbic acid levels in lingonberries ranging from 8.8 to 9.6 mg/100 g f.w., while Wojdyło et al. [61] found values between 5.5 and 22.1 mg/100 g f.w. in sour cherries; both are consistent with our results after moisture adjustment.

3.1.5. Antioxidant Activities (FRAP and DPPH Assays)

Antioxidant capacity, evaluated via ferric reducing antioxidant power (FRAP) and DPPH radical scavenging assays (Table 3), revealed the highest electron-donating and radical-quenching activities in BCP. FRAP values for BCP were higher than those of SCP by 42.37% and LP by 13.21%, while DPPH scavenging followed a similar pattern, with BCP significantly ($p < 0.05$) exceeding SCP and LP by 39.30% and 10.33%, respectively. BCP consistently demonstrated the highest antioxidant activity in both assays, which can be attributed to its elevated levels of total phenolics and flavonoids. These compounds not only act as electron or hydrogen donors but also contribute to the formation of stable radical intermediates, an essential mechanism in mitigating lipid oxidation in fat-rich food matrices.

Kim [48] observed lower DPPH free radical scavenging activity in blackcurrants (265.58 μM TE/g d.w.). Similarly, Sokół-Łętowska et al. [52] reported FRAP antioxidant levels in various sour cherry cultivars that closely match our findings. Daukšienė et al. [32] documented FRAP values for dried lingonberries between 164.5 and 550.67 μmol/g d.w., aligning well with our data. These findings reinforce that the antioxidant effects of fruit powder polyphenols arise from their molecular structure, enabling efficient electron donation and the stabilization of free radicals. These strong antioxidant properties underscore the potential of fruit powders as natural preservatives, positioning them as promising alternatives to synthetic additives in meat products.

3.2. The Proximate Composition of Salami Formulations

The proximate composition and energy value for both SI and SII salami formulations are shown in Table 4.

Table 4. Proximate composition and energy value of developed salami formulations (day 0).

Sample	Proximate Composition						Energy Value (kcal/100 g)
	Moisture (g/100 g)	Protein (g/100 g)	Lipids (g/100 g)	Ash (g/100 g)	NaCl (g/100 g)	CRB (g/100 g)	
SI-C	61.24 ± 0.06 [a]	11.52 ± 0.06 [a]	22.02 ± 0.06 [a]	1.85 ± 0.02 [e]	2.19 ± 0.03 [a]	1.18	248.98
SI-CN	61.21 ± 0.04 [a]	11.50 ± 0.05 [a]	22.01 ± 0.07 [a]	1.86 ± 0.01 [e]	2.21 ± 0.02 [a]	1.21	248.92
SI-SCP90	60.87 ± 0.03 [d]	11.44 ± 0.05 [a]	21.84 ± 0.04 [b]	1.98 ± 0.01 [d]	2.21 ± 0.01 [a]	1.66	248.95
SI-SCP200	60.55 ± 0.04 [e]	11.37 ± 0.03 [a]	21.62 ± 0.06 [c]	2.11 ± 0.02 [b]	2.23 ± 0.02 [a]	2.12	248.54
SI-SCP300	60.27 ± 0.05 [g]	11.31 ± 0.05 [a]	21.43 ± 0.07 [c]	2.19 ± 0.03 [a]	2.22 ± 0.01 [a]	2.58	248.42
SI-BCP90	61.06 ± 0.05 [b]	11.45 ± 0.04 [a]	21.90 ± 0.05 [b]	1.93 ± 0.02 [d]	2.21 ± 0.03 [a]	1.45	248.70
SI-BCP200	60.82 ± 0.04 [d]	11.43 ± 0.05 [a]	21.76 ± 0.06 [b]	2.01 ± 0.01 [c]	2.21 ± 0.01 [a]	1.78	248.63
SI-BCP300	60.61 ± 0.05 [e]	11.38 ± 0.03 [a]	21.63 ± 0.05 [c]	2.08 ± 0.02 [b]	2.22 ± 0.03 [a]	2.08	248.50
SI-LP90	60.95 ± 0.03 [c]	11.44 ± 0.04 [a]	21.87 ± 0.06 [b]	1.90 ± 0.02 [d]	2.22 ± 0.01 [a]	1.61	249.08
SI-LP200	60.69 ± 0.04 [e]	11.40 ± 0.05 [a]	21.69 ± 0.06 [c]	1.97 ± 0.03 [d]	2.23 ± 0.02 [a]	2.01	248.91
SI-LP300	60.46 ± 0.03 [f]	11.37 ± 0.04 [a]	21.53 ± 0.05 [c]	2.03 ± 0.02 [c]	2.24 ± 0.03 [a]	2.36	248.75
SII-C	63.27 ± 0.05 [a]	12.63 ± 0.04 [a]	19.51 ± 0.03 [a]	1.92 ± 0.03 [d]	2.10 ± 0.02 [a]	0.57	228.39
SII-CN	63.28 ± 0.07 [a]	12.65 ± 0.06 [a]	19.53 ± 0.04 [a]	1.93 ± 0.02 [d]	2.13 ± 0.03 [a]	0.48	228.27
SII-SCP90	62.88 ± 0.03 [d]	12.57 ± 0.05 [a]	19.37 ± 0.05 [b]	2.07 ± 0.02 [b]	2.12 ± 0.01 [a]	0.99	228.59
SII-SCP200	62.56 ± 0.04 [f]	12.50 ± 0.04 [a]	19.21 ± 0.06 [b]	2.18 ± 0.03 [b]	2.13 ± 0.02 [a]	1.42	228.59
SII-SCP300	62.31 ± 0.05 [g]	12.43 ± 0.05 [a]	19.02 ± 0.05 [b]	2.28 ± 0.04 [a]	2.15 ± 0.03 [a]	1.81	228.13
SII-BCP90	63.05 ± 0.04 [b]	12.59 ± 0.03 [a]	19.43 ± 0.03 [b]	2.02 ± 0.02 [c]	2.10 ± 0.03 [a]	0.81	228.45
SII-BCP200	62.82 ± 0.05 [d]	12.55 ± 0.06 [a]	19.30 ± 0.04 [b]	2.11 ± 0.02 [b]	2.11 ± 0.02 [a]	1.11	228.35
SII-BCP300	62.61 ± 0.04 [f]	12.51 ± 0.05 [a]	19.19 ± 0.03 [b]	2.19 ± 0.01 [b]	2.12 ± 0.01 [a]	1.38	228.28
SII-LP90	62.95 ± 0.03 [c]	12.59 ± 0.03 [a]	19.40 ± 0.05 [b]	1.99 ± 0.02 [c]	2.10 ± 0.03 [a]	0.97	228.85
SII-LP200	62.71 ± 0.04 [e]	12.54 ± 0.05 [a]	19.25 ± 0.07 [b]	2.08 ± 0.02 [b]	2.12 ± 0.01 [a]	1.31	228.60
SII-LP300	62.48 ± 0.05 [f]	12.48 ± 0.04 [a]	19.11 ± 0.06 [b]	2.15 ± 0.03 [b]	2.13 ± 0.02 [a]	1.65	228.52

CRB: Carbohydrates. SI-C, SII-C: Nitrite-free SI and SII salami; SI-CN, SII-CN: SI and SII salami with added sodium nitrite; SI-SCP90, SI-SCP200, SI-SCP300, SII-SCP90, SII-SCP200, SII-SCP300: Nitrite-free SI and SII salami with sour cherry powder added to provide a TPC of 90, 200 and 300 mg GAE/kg of meat; SI-BCP90, SI-BCP200, SI-BCP300, SII-BCP90, SII-BCP200, SII-BCP300: Nitrite-free SI and SII salami with blackcurrant powder added to provide a TPC of 90, 200 and 300 mg GAE/kg of meat; SI-LP90, SI-LP200, SI-LP300, SII-LP90, SII-LP200, SII-LP300: Nitrite-free SI and SII salami with lingonberry powder added to provide a TPC of 90, 200 and 300 mg GAE/kg of meat. Results are expressed as the mean ± standard deviation (SD) of three independent analyses. Means in the same column bearing different superscripts are significantly different (one-way ANOVA, $p < 0.05$).

The values for the proximate composition fall within the ranges established for cooked and scalded salami by Order 210/2006 [62]. The moisture content of both SI and SII salami formulations was affected by the type and dose of fruit powder added, as well as the ingredients and the heat treatment applied (cooking or scalding). The moisture content of SI samples ranged from 60.27 g/100 g to 61.24 g/100 g, while that of SII samples varied between 62.31 g/100 g and 63.27 g/100 g. For both salami types, the highest moisture content values were recorded in the control samples: SI-C (61.24 g/100 g) and SI-CN (61.21 g/100 g), SII-C (63.27 g/100 g), and SII-CN (63.28 g/100 g). The addition of fruit powders (SCP, BCP, and LP) resulted in a gradual and statistically significant decrease ($p < 0.05$) in moisture content for both salami types. The reduction in moisture content is mainly due to the low moisture content of the fruit powder (below 5%). This observation aligns with Mahapatra et al. [63], who reported reduced moisture content in heat-treated meatballs fortified with fruit and fruit byproduct powders compared with controls. Fu et al. [64] reported that increasing levels of cherry powder in sausages led to greater moisture loss during 30 days of refrigerated storage. This suggests that higher concentrations of cherry powder may adversely impact the product's water-holding capacity. One possible reason is that the cherry powder interferes with protein–water interactions. Polyphenols present in the powder may bind to muscle proteins, displacing water molecules that would otherwise form hydrogen bonds and contribute to water retention. This substitution diminishes the ability of proteins to retain water in the meat matrix.

The protein contribution from the fruit powder in the salami formulations doesn't significantly impact the overall protein values in SI and SII salami types. The protein content in the SI salami formulation exhibited a slight reduction, decreasing from 11.52 g/100 g in the SI-CN sample to 11.31 g/100 g in the SI-SCP300 formulas. Similarly, the SII salami formulation also showed a decrease in protein content, falling from 12.65 g/100 g in the SII-CN formula to 12.43 g/100 g in the SII-SCP300 sample. Compared to the protein content in the control samples, both SI and SII salami formulations exhibited a minor reduction in protein content as the concentration of added fruit powder increased. Our observations are consistent with the findings reported by Cocan et al. [28], who documented a reduction in protein content following the incorporation of bell pepper byproduct powder into pork sausages.

Supplementation of salami with varying doses of fruit powder significantly ($p < 0.05$) affected lipid content. As the amount of added powder increased, a gradual decrease in lipid content was observed. In SI salami, the lipid content declined from 22.02 g/100 g in the SI-C sample to 21.43 g/100 g in the SI-SCP300 formulation. Similarly, in SII, it decreased from 19.53 g/100 g in the SII-CN sample to 19.02 g/100 g in the SII-SCP300 formulation. The SI samples consistently exhibited lipid content approximately 2.5 g/100 g higher than that of SII, a difference attributed to the specific manufacturing recipe, which incorporated a greater amount of fat in the SI formulations. A gradual reduction in lipid content with increased fruit powder supplementation was also reported by Mahapatra et al. [63], who observed a decrease in lipid levels in meatballs fortified with starfruit and guava powders. Similar results were obtained by Zaini et al. [65], where the low lipid content of banana peel powder contributed to reducing the lipid levels in chicken sausages.

The ash content increased proportionally with the quantity of SCP, BCP, and LP integrated into the salami formulas. This can be attributed to the minerals and vitamins supplied by the fruit powders. Ash content in the smoked and cooked (SI) salami ranged from 1.85 to 2.19 g/100 g, while the smoked and scalded (SII) salami showed values between 1.92 and 2.28 g/100 g. The highest levels were recorded in SI-SCP300 (2.19 g/100 g) and SII-SCP300 (2.28 g/100 g). A similar trend in ash content was observed in the investigation conducted by Zaini et al. [65], who noted a progressive increase in the ash content of analyzed chicken sausage samples (3.04–5.77%) with incremental additions of banana peel powder. Furthermore, Lopez-Vargas et al. [66] reported a comparable pattern, indicating a 2.5–5% increase in the ash content of the formulated pork burger samples following the inclusion of passion fruit peel powder.

The NaCl content in the analyzed SI and SII salami formulations showed no significant differences ($p > 0.05$) in response to fruit powder incorporation. Values ranged from 2.19–2.24 g/100 g for SI and 2.10–2.15 g/100 g for SII. Our findings align with the values documented in the literature, particularly those by Zanardi et al. [67], who reported a content of 1–2.5% for Italian salami.

Carbohydrate content gradually increased in both types of salami with incremental SCP, BCP, and LP supplementation. In SI samples, values ranged from 1.19 to 2.58 g/100 g, while in SII formulations, they ranged between 0.48 and 1.81 g/100 g.

The inclusion of fruit powder in the salami formulations did not significantly impact their energy value. However, a modest reduction in energy value was evident across both salami types, correlating with an increase in the level of fruit powder incorporation. A similar trend has been documented in other investigations, with an increase in the proportion of tomato processing byproducts incorporated into pork sausages [27].

3.3. *Progression of Lipid Oxidation*

Several lipid oxidation indices were evaluated to monitor the progression of oxidation in salami samples stored for 0, 15, and 30 days under refrigeration (4 °C). The peroxide value indicated primary oxidation, while the para-anisidine value and thiobarbituric acid reactive substances reflected secondary oxidation. Inhibition of lipid oxidation and the TOTOX value complemented the assessment of the oxidative status of the investigated products.

3.3.1. Peroxide Value

Chemical spoilage in meat products during processing and storage primarily results from lipid degradation by autooxidation. The initial stage of lipid oxidation leads to the formation of hydroperoxide or the loss of polyunsaturated fatty acids. Measurement of lipid hydroperoxide formation, commonly expressed as the peroxide value, has long served as a key indicator of primary oxidation compounds in meat and meat products [2].

Figure 2 illustrates the effect of supplementing salami formulations with sodium nitrite and fruit powder on peroxide value (PV) during cold storage for 0, 15, and 30 days.

Immediately following processing (day 0), PV measured were low, ranging from 0.31 to 0.59 meq O_2/kg in SI samples and from 0.43 to 0.71 meq O_2/kg in SII samples. A low PV value is an indicator of freshness and quality, indicating that the product has been manufactured, stored, and handled properly [68].

The results indicate that incorporating fruit powders into salami significantly enhances oxidative stability during processing. This beneficial effect is both dose-dependent and species-dependent, as higher concentrations of fruit powder more effectively inhibit oxidative processes and improve lipid stability. This is evidenced by the lower peroxide values observed in samples enriched with higher doses (e.g., BCP300, LP300, SCP300), as well as by the superior oxidative protection provided by BCP compared with LP and SCP.

The differences in PV on day 0 for SI and SII are attributed to their distinct processing conditions. The SII salami, heat-treated at 72 °C under 80% relative humidity, exhibited enhanced oxidative stability and lower PV compared with the SI samples, which underwent heat treatment at the same temperature but under low relative humidity (5%). Under such conditions, lipid oxidation is primarily driven by elevated temperature and increased oxygen availability. For SI samples, low relative humidity increases lipid exposure to oxygen by limiting the protective moisture layer, thereby accelerating oxidation. Additionally, surface drying can concentrate pro-oxidants such as metal ions, further promoting peroxidation [69]. In contrast, the higher relative humidity used for SII creates a water barrier that reduces oxygen diffusion to lipid surfaces, thus slowing the oxidation rate [70]. These findings underscore the importance of precisely controlling relative humidity and temperature during meat processing to optimize lipid stability and overall product quality.

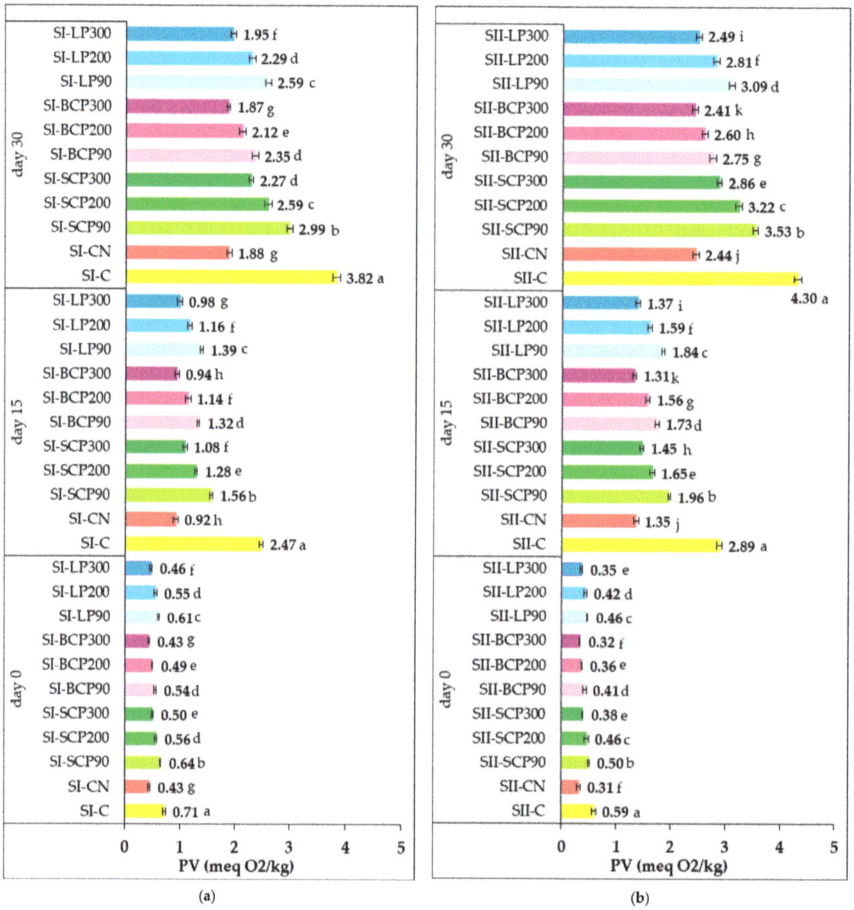

Figure 2. Changes in peroxide value (PV) during cold storage of smoked and cooked salami (SI) (**a**) and smoked and scalded salami (SII) (**b**), following supplementation with sodium nitrite and fruit powder. SI-C, SII-C: Nitrite-free SI and SII salami; SI-CN, SII-CN: SI and SII salami with added sodium nitrite; SI-SCP90, SI-SCP200, SI-SCP300, SII-SCP90, SII-SCP200, SII-SCP300: Nitrite-free SI and SII salami with sour cherry powder added to provide a TPC of 90, 200 and 300 mg GAE/kg of meat; SI-BCP90, SI-BCP200, SI-BCP300, SII-BCP90, SII-BCP200, SII-BCP300: Nitrite-free SI and SII salami with blackcurrant powder added to provide a TPC of 90, 200 and 300 mg GAE/kg of meat; SI-LP90, SI-LP200, SI-LP300, SII-LP90, SII-LP200, SII-LP300: Nitrite-free SI and SII salami with lingonberry powder added to provide a TPC of 90, 200 and 300 mg GAE/kg of meat. Results are expressed as the mean ± standard deviation (SD) of three independent analyses. Means labeled with different letters within the same storage period are significantly different (one-way ANOVA, $p < 0.05$).

During cold storage, a general increase in PV was observed, indicating progressive lipid oxidation. Extending the storage period from 15 to 30 days significantly intensified primary oxidation processes in the samples. Significant differences ($p < 0.05$) in PV were recorded between the control and the samples supplemented with nitrite or fruit powders in both SI and SII formulations. These variations are linked to the antioxidant effects of SCP, BCP, and LP powders added at three dosage levels. Samples supplemented with higher levels of fruit-derived total phenolics showed lower PV at both 15 and 30 days of storage, reflecting enhanced oxidative stability. When normalized for equivalent total phenolic content (TPC), the powders ranked in effectiveness as follows: BCP > LP > SCP.

After 30 days, SI samples (60.27–61.24% moisture) exhibited lower PV values compared to SII samples (62.31–63.27%). The greater lipid peroxidation in SII, despite initially lower oxidation, suggests that its higher moisture content facilitated oxidative degradation over extended storage.

At the end of storage, salami samples supplemented with fruit powders exhibited PVs ranging from 1.87 to 2.99 meq O_2/kg for SI and from 2.41 to 3.53 meq O_2/kg for SII. In contrast, the negative control samples (SI-C and SII-C) showed higher PVs of 3.82 and 4.30 meq O_2/kg, respectively. Samples containing nitrite (SI-CN and SII-CN) demonstrated a strong antioxidative effect, with PVs of 1.88 and 2.44 meq O_2/kg, respectively. Among the fruit powders tested, BCP was the most effective in limiting hydroperoxide formation.

At all storage points, BCP added to both SI and SII salami formulations at 300 mg GAE/kg of processed meat exhibited the strongest inhibition of primary lipid oxidation, as evidenced by consistently low PV values. No significant differences in PV were observed between SI-BCP300 and SI-CN, or between SII-BCP300 and SII-CN samples, indicating that BCP's inhibitory effect was comparable to that of nitrite.

After 30 days of cold storage, the peroxide values of lipids extracted from the fruit-powder-enriched salami samples remained below 5 meq O_2/kg. This level is characteristic of non-rancid fats, which typically show low PV, often under 5 meq O_2/kg [71]. In general, peroxide values between 0 and 6 suggest that oxidation has not significantly progressed, while values from 7 to 10 may indicate the initial stages of rancidity. A PV exceeding 10 is generally associated with advanced lipid oxidation, rendering the fat unsuitable for consumption [71]. Previous research by Yi et al. [68] suggests that the upper limit for acceptable PVs in oils and fats lies between 5 and 10 meq O_2/kg. In line with these findings, the PVs observed in this study remained within the maximum allowable limit of 10 milliequivalents of active oxygen per kilogram of fat, as defined by the Codex Standard for Named Animal Fats (CODEX STAN 211–1999) [72]. The inhibition of primary oxidation in salami through the incorporation of fruit powders aligns with existing literature, which demonstrates that adding various fruit types and fruit-derived products to meat effectively prevents lipid oxidation and significantly extends shelf life [73,74]. Similarly, other studies have reported a decrease in the PV of soybean oil with increasing levels of natural antioxidants [75]. These results support the use of fruit powders as functional ingredients with preservative properties during meat processing and storage.

3.3.2. Inhibition of Oxidation

Table 5 shows the changes in inhibition of lipid oxidation (IO) during cold storage of SI and SII formulations supplemented with sodium nitrite and fruit powder.

Table 5. Changes in the inhibition of oxidation (IO) during cold storage of smoked and cooked salami (SI) and smoked and scalded salami (SII), supplemented with sodium nitrite and fruit powder.

Sample	IO (%)	
	Day 15	Day 30
SI-CN	72.72 ± 0.24 [a]	53.59 ± 0.23 [a]
SI-SCP90	47.75 ± 0.21 [i]	24.68 ± 0.11 [i]
SI-SCP200	58.83 ± 0.16 [g]	34.85 ± 0.15 [h]
SI-SCP300	66.85 ± 0.19 [d]	42.93± 0.18 [e]
SI-BCP90	55.73 ± 0.12 [h]	41.90 ± 0.20 [f]
SI-BCP200	63.16 ± 0.20 [f]	47.72 ± 0.24 [c]
SI-BCP300	71.10 ± 0.22 [b]	53.82 ± 0.26 [a]
SI-LP90	55.77 ± 0.17 [h]	36.33 ± 0.16 [g]
SI-LP200	65.28 ± 0.19 [e]	44.16 ± 0.19 [d]
SI-LP300	70.45 ± 0.26 [c]	52.22 ± 0.21 [b]
SII-CN	54.83 ± 0.16 [c]	42.43 ± 0.15 [b]
SII-SCP90	36.35 ± 0.23 [i]	18.22 ± 0.10 [i]
SII-SCP200	48.20 ± 0.27 [f]	25.46 ± 0.12 [h]
SII-SCP300	53.42 ± 0.19 [d]	33.09 ± 0.16 [f]
SII-BCP90	42.71 ± 0.15 [g]	37.08 ± 0.20 [d]
SII-BCP200	47.82 ± 0.25 f	39.58 ± 0.22 [c]
SII-BCP300	57.08 ± 0.26 [a]	43.66 ± 0.26 [a]
SII-LP90	39.99 ± 0.17 [h]	29.21 ± 0.19 [g]
SII-LP200	49.04 ± 0.20 [e]	35.66 ± 0.21 [e]
SII-LP300	55.70 ± 0.22 [b]	42.38 ± 0.14 [b]

SI-C, SII-C: Nitrite-free SI and SII salami; SI-CN, SII-CN: SI and SII salami with added sodium nitrite; SI-SCP90, SI-SCP200, SI-SCP300, SII-SCP90, SII-SCP200, SII-SCP300: Nitrite-free SI and SII salami with sour cherry powder added to provide a TPC of 90, 200 and 300 mg GAE/kg of meat; SI-BCP90, SI-BCP200, SI-BCP300, SII-BCP90, SII-BCP200, SII-BCP300: Nitrite-free SI and SII salami with blackcurrant powder added to provide a TPC of 90, 200 and 300 mg GAE/kg of meat; SI-LP90, SI-LP200, SI-LP300, SII-LP90, SII-LP200, SII-LP300: Nitrite-free SI and SII salami with lingonberry powder added to provide a TPC of 90, 200 and 300 mg GAE/kg of meat. Results are expressed as the mean ± standard deviation (SD) of three independent analyses. Means labeled with different letters within the same storage period are significantly different (one-way ANOVA, $p < 0.05$).

The IO (%) was calculated from PV to assess early lipid oxidation and reflects the anti-peroxidative effect of fruit powders in SI and SII formulations [76]. In all salami samples enriched with fruit powders, a clear inhibitory effect on lipid oxidation was observed during refrigerated storage at both 15 and 30 days. The inhibition of lipid oxidation was dose-dependent across both SI and SII formulations. Among the tested powders, BCP showed the strongest inhibitory effect, followed by LP and SCP, sustained through to day 30. On day 15, the IO of BCP300 in SI (71.10%) was slightly below that of nitrite (72.77%), while in SII, BCP300 outperformed nitrite (57.08% vs. 54.83%). By day 30, the inhibitory effect of both nitrite and fruit powders declined. In SI, BCP300 matched nitrite's IO (53.82% vs. 53.59%), while in SII it remained slightly higher (43.66% vs. 42.43%). LP300 also demonstrated comparable effectiveness, exhibiting a slightly lower IO than nitrite in SI (52.22%) and an almost identical value in SII (42.38%). SCP exhibited a weaker inhibitory effect compared to BCP and LP, with IO values for SCP300 approximately 20–22% lower than nitrite after 30 days. Both BCP and LP maintained their antioxidant activity throughout prolonged storage, especially at doses to provide 300 mg GAE/kg, making them promising natural alternatives to nitrite in processed meat

3.3.3. Para-Anisidine Value (pAV)

Prolonged cold storage of meat products can lead to lipid peroxidation, compromising quality. While peroxide value (PV) reflects early oxidation, it does not account for later-stage changes [77]. Para-anisidine value (pAV) is a reliable indicator of secondary oxidation, measuring aldehydes like 2-alkenals and 2,4-alkadienals formed from peroxide breakdown during advanced oxidative stages [2]. Figure 3 illustrates the pAV values measured on days 0, 15, and 30 of refrigerated storage for the various salami formulations.

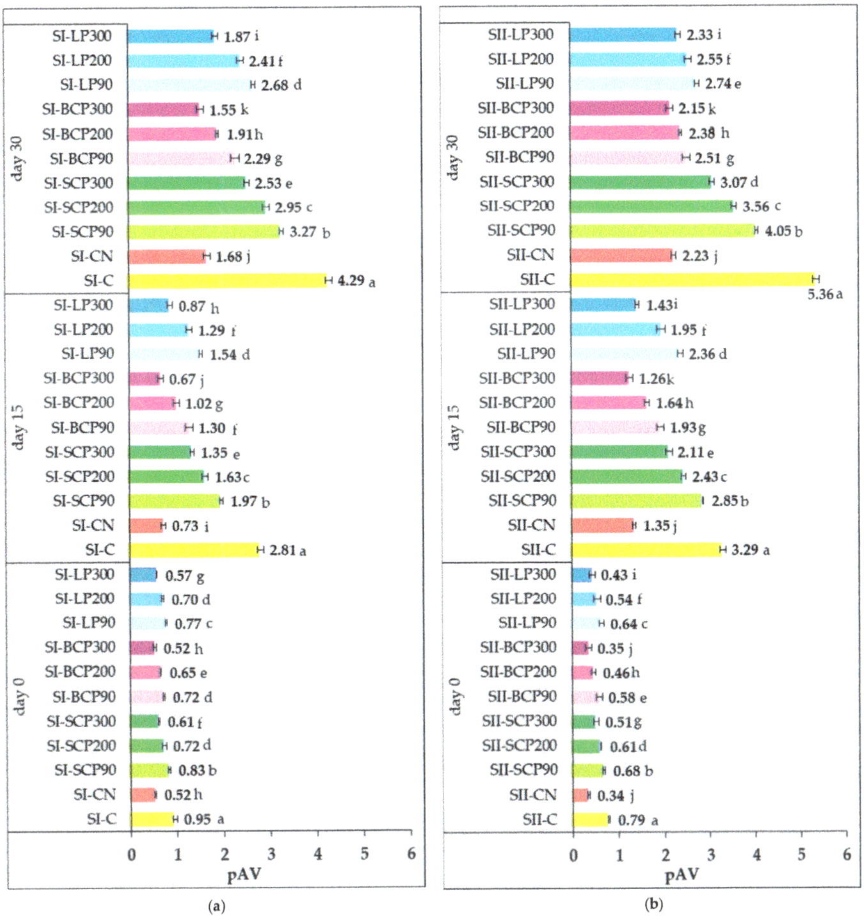

Figure 3. Impact of sodium nitrite and fruit powder supplementation on para-anisidine value (pAV) during cold storage of smoked and cooked salami (SI) (**a**) and smoked and scalded salami (SII) (**b**). SI-C, SII-C: Nitrite-free SI and SII salami; SI-CN, SII-CN: SI and SII salami with added sodium nitrite; SI-SCP90, SI-SCP200, SI-SCP300, SII-SCP90, SII-SCP200, SII-SCP300: Nitrite-free SI and SII salami with sour cherry powder added to provide a TPC of 90, 200 and 300 mg GAE/kg of meat; SI-BCP90, SI-BCP200, SI-BCP300, SII-BCP90, SII-BCP200, SII-BCP300: Nitrite-free SI and SII salami with blackcurrant powder added to provide a TPC of 90, 200 and 300 mg GAE/kg of meat; SI-LP90, SI-LP200, SI-LP300, SII-LP90, SII-LP200, SII-LP300: Nitrite-free SI and SII salami with lingonberry powder added to provide a TPC of 90, 200 and 300 mg GAE/kg of meat. Results are expressed as the mean ± standard deviation (SD) of three independent analyses. Means labeled with different letters within the same storage period are significantly different (one-way ANOVA, $p < 0.05$).

The changes in pAV reflect the replacement of sodium nitrite with fruit powders, added at levels corresponding to total phenolic contents of 90, 200, and 300 mg GAE/kg of processed meat. The data show that SCP, BCP, and LP effectively suppressed secondary lipid oxidation in both SI and SII samples throughout storage when compared with the positive and negative controls. On day 0, pAV levels were low across all formulations containing either sodium nitrite or fruit powder, although they increased over time. No significant differences ($p > 0.05$) were observed on day 0 between SI-BCP300 and SII-BCP300 and the respective nitrite-containing controls (SI-CN and SII-CN), indicating that BCP at a level providing 300 mg GAE/kg of processed meat showed a comparable protective effect against secondary oxidation during processing.

Secondary lipid oxidation followed a similar trend to primary oxidation, with higher oxidation rates observed under the dry conditions (5% relative humidity) characteristic of the SI processing. This was probably due to the absence of a protective water barrier. Conversely, the SII samples exhibited lower levels of secondary oxidation, which may be attributed to heat treatment under high humidity (80%). This condition has probably favored the formation of a physical barrier or surface film around the lipid molecules, thereby reducing oxygen diffusion, diluting pro-oxidant catalysts, and limiting the movement of water-soluble radicals into the lipid phase [69,70].

Throughout cold storage, the addition of fruit powders at all tested levels significantly reduced pAV compared with the SI-C and SII-C controls. The inhibitory effect on secondary lipid oxidation intensified with increasing TPC levels supplied by the fruit powders. Significant differences ($p < 0.05$) were observed in the pAV among salami samples enriched with different fruit powders and at varying inclusion levels. Among these, BCP supplementation demonstrated the strongest inhibition of secondary lipid oxidation during processing and after 15 and 30 days of storage. Accordingly, the lowest pAV was recorded in BCP-enriched samples, followed by those containing LP and SCP.

After 30 days of storage, the pAV of the fruit-powder-enriched salami formulas ranged from 1.55 to 3.27 for SI and 2.15 to 4.05 for SII. In contrast, the pAV of the negative control samples (SI-C and SII-C) was higher, at 4.29 and 5.36, respectively. In the positive control samples (SI-CN and SII-CN), a strong inhibitory effect was observed due to the addition of nitrite, quantified by low pAV values of 1.68 for SI-CN and 2.23 for SII-CN.

The results obtained after cold storage suggest that the incorporated fruit powders exhibited a specific inhibitory effect on secondary lipid oxidation processes, dependent on the type and dose, thereby having a positive impact on oxidative stability. Throughout cold storage, BCP300 showed the strongest inhibitory potential on secondary oxidation processes, both in SI and II, higher than that of sodium nitrite. At the end of storage, pAV of 1.55 was recorded for the SI-BCP300 and 2.15 for the SII-BCP300 sample. These values were lower than those observed in the positive control samples, SI-CN (1.68) and SII-CN (2.23).

The decomposition of hydroperoxides into secondary lipid oxidation products after 15 and 30 days of storage was more pronounced in the SII formulations compared with the SI samples, as indicated by the higher pAV. The higher moisture content in the SII samples probably accelerated secondary oxidation during prolonged storage by increasing the mobility of reactive species and facilitating degradation reactions over time.

Overall, both during processing and storage, pAV decreased progressively with increasing TPC supplied by the incorporated fruit powders. This trend aligns with findings from previous studies on smoked and scalded sausages supplemented with powders derived from vegetable processing byproducts [27,28]. The low pAV values indicate the effectiveness of fruit powders in slowing the formation of secondary oxidation products, suggesting that the fats are less prone to rancidity.

Although no official regulatory limits for p-anisidine value (pAV) in edible oils and fats currently exist, a threshold of 10 is commonly referenced, with values below 4 recommended to ensure high quality [78]. In addition, the GOED Voluntary Monograph [79] considers a pAV below 20 as the acceptable limit for product quality.

3.3.4. TOTOX Value

Lipid oxidation is a gradual process, with an increase in PV being followed by an increase in pAV. Evaluating a singular class of oxidation products is generally insufficient for a comprehensive and accurate assessment of a sample's oxidative status. A more robust approach involves the concurrent analysis of both primary and secondary lipid oxidation markers [2]. This dual measurement strategy provides insights into the complete oxidative trajectory. Consequently, despite its classification as a derivative indicator rather than a direct analytical technique, TOTOX has been proposed as a tool for quantifying the overall oxidation of fat samples. Figure 4 shows the effect of salami formulas supplementation with fruit powder and nitrite on TOTOX value during cold storage for 15 and 30 days.

Generally, fresh meat products are characterized by low TOTOX values. As shown, the TOTOX for salami samples on day 0 of storage showed low values in the range 1.38–2.37 for SI and 0.96–1.96 for SII. The low TOTOX value could be an indication that the salami samples were properly manufactured and that they would not easily go rancid when properly stored in refrigeration conditions.

The results demonstrate oxidative changes occurred during 15 and 30 days of storage, as evidenced by increases in TOTOX values. A closer look at the results reveals that control samples (SI-C and SII-C) had significantly higher TOTOX values ($p < 0.05$) after both 15 and 30 days of storage, in comparison with both the positive control samples and the fruit-powder-enriched salami samples.

After 30 days of storage, the SI-C and SII-C control samples reached TOTOX values of 11.94 and 13.95, respectively. In contrast, incorporating sodium nitrite resulted in lower TOTOX values of 5.43 for the SI-CN sample and 7.11 for the SII-CN sample. All incorporated fruit powders significantly ($p < 0.05$) reduced lipid oxidative deterioration. Specifically, salami samples supplemented with SCP, BCP, and LP recorded significantly lower ($p < 0.05$) TOTOX values than those measured for the SI-C and SII-C on both day 15 and day 30 of storage. This beneficial effect can be attributed to the phenolic compounds provided by the incorporated fruit powders. Natural antioxidants generally delay lipid oxidation by donating hydrogen atoms or chelating metal ions. Berries, in particular, are rich in polyphenols—such as phenolic acids and flavonoids—which can effectively scavenge free radicals, thereby inhibiting lipid oxidation. Various plant-derived sources, including blackcurrants, blueberries, apples, aronia, and elderberry powders, have been documented as effective antioxidants in pork patties, demonstrating differing capacities to neutralize free radicals and reduce both primary and secondary lipid oxidation [80].

In our study, BCP showed a higher radical scavenging ability than that of LP and SCP, which was proven to converge with their high ability to inhibit lipid oxidation.

It can be stated that the addition of sodium nitrite and fruit powder, respectively, significantly increased the oxidative stability of the salami formulations both during processing and storage, compared with the control ($p < 0.05$). It was observed that TOTOX values decreased with increasing fruit powder dose, with the lowest values recorded in salami samples (both SI and SII formulations) enriched with a dose of fruit powder ensuring a TPC of 90 mg GAE/kg of processed meat.

The fruit species from which the fruit powder was derived also had an effect on the TOTOX values obtained. The lowest TOTOX values were recorded for salami with added SCP, followed by samples fortified with LP and BCP.

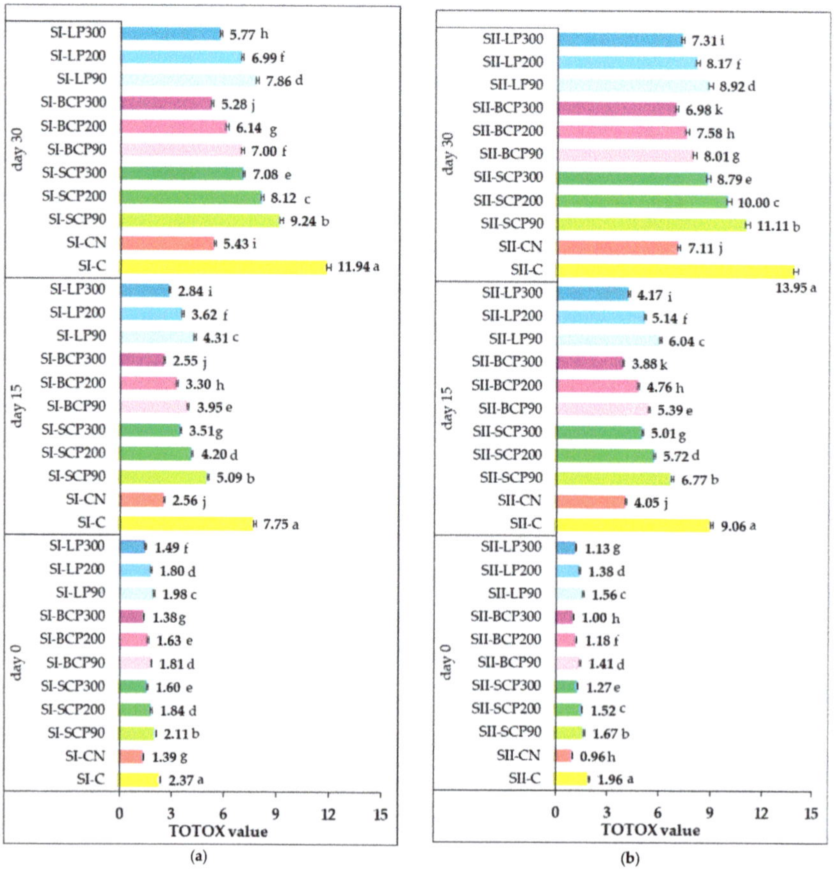

Figure 4. Effect of sodium nitrite and fruit powder supplementation on TOTOX value during cold storage of smoked and cooked salami (SI) (**a**) and smoked and scalded salami (SII) (**b**). SI-C, SII-C: Nitrite-free SI and SI salami; SI-CN, SII-CN: SI and SII salami with added sodium nitrite; SI-SCP90, SI-SCP200, SI-SCP300, SII-SCP90, SII-SCP200, SII-SCP300: Nitrite-free SI and SII salami with sour cherry powder added to provide a TPC of 90, 200 and 300 mg GAE/kg of meat; SI-BCP90, SI-BCP200, SI-BCP300, SII-BCP90, SII-BCP200, SII-BCP300: Nitrite-free SI and SII salami with blackcurrant powder added to provide a TPC of 90, 200 and 300 mg GAE/kg of meat; SI-LP90, SI-LP200, SI-LP300, SII-LP90, SII-LP200, SII-LP300: Nitrite-free SI and SII salami with lingonberry powder added to provide a TPC of 90, 200 and 300 mg GAE/kg of meat. Results are expressed as the mean ± standard deviation (SD) of three independent analyses. Means labeled with different letters within the same storage period are significantly different (one-way ANOVA, $p < 0.05$).

The 15 and 30 days of storage resulted in lower TOTOX values for cooked salami enriched with fruit powder (SI) compared with the scalded samples (SII), despite its higher initial lipid oxidation status. The efficacy of the fruit powder antioxidants (rich in phenolic compounds) is significantly influenced by the salami's moisture content during cold storage. In the drier SI salami, the limited water content could concentrate the antioxidants within the lipid phase or at the lipid–water interface. This concentration, combined with restricted diffusion of pro-oxidants like oxygen, might allow the antioxidants to more effectively scavenge free radicals and chelate metal ions in their immediate vicinity, thereby sustaining their protective action for longer periods and limiting the extent of lipid oxidation. Conversely, in the wetter SII salami, the higher moisture content, while initially protective during heat treatment, might have allowed for greater mobility of oxygen and water-soluble pro-oxidants during cold storage. Although the fruit powders' phenolic compounds still exert antioxidant activity (e.g., radical scavenging, metal chelation), their effectiveness might be diluted or overwhelmed by the faster rate of oxidative reactions facilitated by the higher moisture. This could lead to a quicker depletion of the antioxidants' capacity, allowing lipid oxidation to advance more rapidly and resulting in higher TOTOX values over extended storage.

Regarding the inhibition potential against lipid oxidation during storage, BCP at a dose designed to provide a level of TPC of 300 mg GAE/kg of processed meat demonstrated the ability to effectively replace sodium nitrite in both SI and SII salami. Promising results for improving lipid oxidation resistance were also obtained by supplementing salami formulations with LP300. In SI-LP300 and SII-LP300, an inhibitory effect on oxidative degradation processes, slightly lower than that of sodium nitrite, was recorded at both 15 and 30 days of storage. This was reflected by TOTOX values of 2.84 (day 15) and 5.77 (day 30) for SI-LP300, compared with 2.56 (day 15) and 5.43 (day 30) for SI-CN. Similarly, SII-LP300 recorded TOTOX values of 4.17 (day 15) and 7.31 (day 30), compared with 4.05 (day 15) and 7.11 (day 30) for SII-CN.

Although there is currently no official TOTOX limit set by regulatory authorities for edible oils and fats, the voluntary GOED monograph [79] considers values below 26 to be acceptable. It has been suggested that TOTOX values below 10 can be considered indicative of superior lipid matrix quality [2].

By incorporating fruit powders, all SI formulations and most SII samples maintained TOTOX values below 10 even after 30 days of storage. The only exceptions were SII-SCP90 and SII-SCP-200, which recorded TOTOX values of 10.00 and 11.11, respectively. Our data demonstrated that replacing sodium nitrite with fruit powders in the salami recipe is a suitable application for improving oxidative quality both during processing and 30 days of cold storage. These findings align with other studies reporting a limitation in the increase in TOTOX values in stored meat products due to natural antioxidant addition, thereby demonstrating their potential to significantly slow down lipid oxidation [81].

3.3.5. TBA Value

The TBA assay is a widely used method for evaluating the extent of lipid oxidation in food products. This test quantifies substances that react with thiobarbituric acid, typically aldehydes and ketones formed during the secondary phase of auto-oxidation when lipid peroxides break down [2]. Malondialdehyde (MDA), or 1,3-propanedial, is a key product of secondary oxidation of polyunsaturated fatty acids. In meat products, even small concentrations of MDA significantly contribute to rancid odors, making it a widely recognized indicator of lipid oxidation [2]. Currently, there's no legally established limit for MDA concentrations in meat and meat products. However, different studies have proposed various thresholds for acceptability. Several authors suggest that malondialdehyde (MDA) levels between 2.0 and 2.5 mg/kg represent the upper threshold below which meat and meat products are generally not perceived as rancid [2,80]. Others consider levels above 1.0 mg/kg potentially unacceptable [82,83]. Campo et al. [84] proposed a limit of approximately 2.0 mg MDA/kg for the sensory acceptability of oxidized beef, while Hughes et al. [85] reported that meat products with TBA values between 2.60 and 3.11 mg MDA/kg were still deemed acceptable by consumers.

Figure 5 illustrates the effect of fruit powder incorporation into salami formulations on TBA values during cold storage (0, 15, and 30 days), in comparison with control samples without nitrite and those containing sodium nitrite.

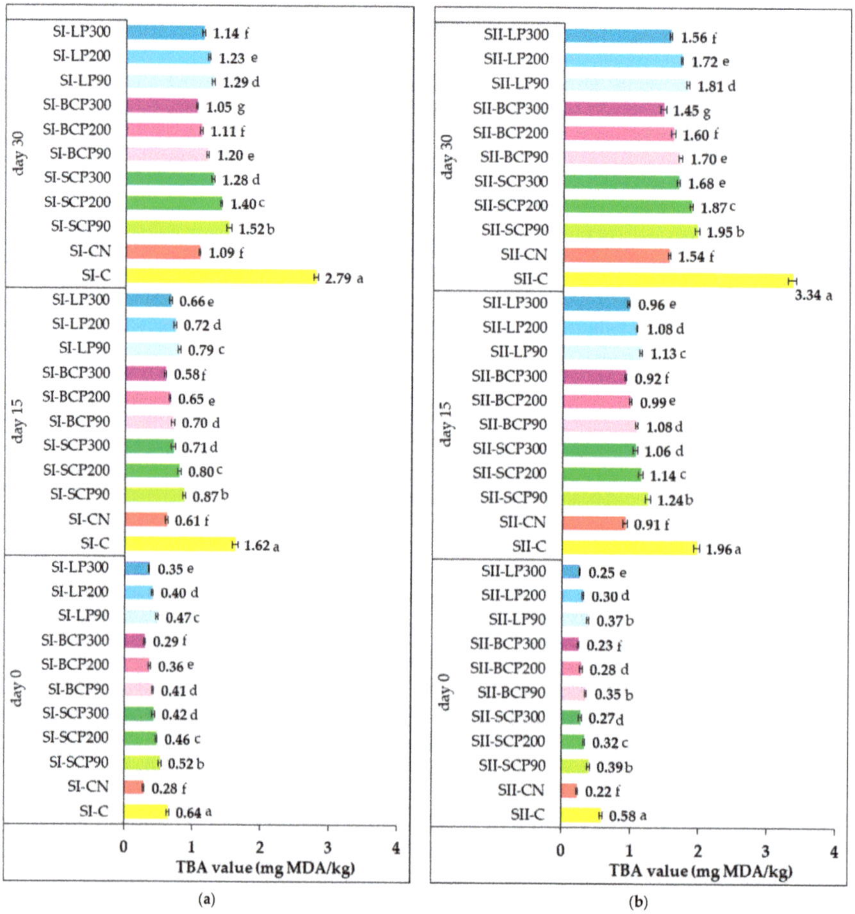

Figure 5. Impact of sodium nitrite and fruit powder supplementation on TBA value during cold storage of smoked and cooked salami (SI) (**a**) and smoked and scalded salami (SII) (**b**). SI-C, SII-C: Nitrite-free SI and SII salami; SI-CN, SII-CN: SI and SII salami with added sodium nitrite; SI-SCP90, SI-SCP200, SI-SCP300, SII-SCP90, SII-SCP200, SII-SCP300: Nitrite-free SI and SII salami with sour cherry powder added to provide a TPC of 90, 200 and 300 mg GAE/kg of meat; SI-BCP90, SI-BCP200, SI-BCP300, SII-BCP90, SII-BCP200, SII-BCP300: Nitrite-free SI and SII salami with blackcurrant powder added to provide a TPC of 90, 200 and 300 mg GAE/kg of meat; SI-LP90, SI-LP200, SI-LP300, SII-LP90, SII-LP200, SII-LP300: Nitrite-free SI and SII salami with lingonberry powder added to provide a TPC of 90, 200 and 300 mg GAE/kg of meat. Results are expressed as the mean ± standard deviation (SD) of three independent analyses. Means labeled with different letters within the same storage period are significantly different (one-way ANOVA, $p < 0.05$).

TBA values revealed variations in lipid oxidation among salami formulas, attributable to thermal treatments during processing, fruit powder type and concentration, and cold storage duration. The initial TBA values in the supplemented salami were generally low. Specifically, SI-CN recorded 0.28 mg MDA/kg, and SII-CN recorded 0.22 mg MDA/kg. Values ranged from 0.29 to 0.52 mg MDA/kg in SI fruit-powder-enriched samples and from 0.23 to 0.39 mg MDA/kg in SII fruit-powder-enriched samples.

During the 15- and 30-day storage periods, the control samples (SI-C and SII-C) exhibited significantly higher TBA values compared with those containing sodium nitrite (SI-CN and SII-CN). The incorporation of fruit powder in different amounts also impacted the TBA values during storage (Figure 5).

After 30 days of storage, TBA values ranged from 1.05 to 1.52 mg MDA/kg for SI and 1.45 to 1.95 mg MDA/kg for SII. TBA values generally decreased with increasing fruit powder dose. At the end of the 30-day storage period, BCP-supplemented salami exhibited the lowest TBA values for both SI (1.05–1.20 mg MDA/kg) and SII (1.45–1.70 mg MDA/kg). These were followed by samples with LP (SI: 1.14–1.29 mg MDA/kg; SII: 1.56–1.81 mg MDA/kg), and then those enriched with SCP (SI: 1.28–1.52 mg MDA/kg; SII: 1.68–1.95 mg MDA/kg).

A closer look at the TBA values of positive control samples (SI-CN and SII-CN) and BCP300-enriched samples revealed no significant differences ($p > 0.05$) after processing (day 0) and after 15 days of cold storage. At the end of storage, no significant differences ($p > 0.05$) were observed in the TBA values of samples with nitrite and those supplemented with BCP200 and LP300. Incorporating BCP300 demonstrated a significantly greater inhibitory effect on secondary lipid oxidation compared with sodium nitrite added at 90 mg/kg of processed meat.

Across SI formulations, TBA values remained below 1 mg MDA/kg after 15 days of storage. Extending storage to 30 days increased TBA values, but they remained below 2 mg MDA/kg, suggesting these samples weren't significantly affected by oxidation or rancidity. Even after 30 days, incorporating BCP200, BCP300, and LP300 into SI salami effectively inhibited secondary lipid oxidation, maintaining TBA values close to 1 mg MDA/kg. For SII samples, TBA values were higher than for SI samples, but they remained below the 2 mg MDA/kg threshold after 30 days of storage, even with fruit powders at doses providing a TPC of 90 mg GAE/kg in the processed meat. Incorporation of BCP300, BCP200, and LP300 into SII led to the lowest TBA values, ranging from 1.45 to 1.60 mg MDA/kg.

These results demonstrate the effectiveness of fruit powders in controlling the extent of lipid oxidation throughout the 30-day cold storage period recommended for the salami types investigated.

4. Conclusions

This study provides clear evidence that fruit-derived powders, particularly blackcurrant (BCP), lingonberry (LP), and sour cherry (SCP), can be used as effective natural antioxidants in salami production, offering a promising alternative to synthetic nitrites. Their efficacy is largely attributed to their rich phytochemical profiles, including phenolics, flavonoids, anthocyanins, and ascorbic acid, which significantly reduced lipid oxidation during both processing and cold storage. Among the powders tested, BCP, at a dose corresponding to 300 mg GAE/kg of meat, consistently provided the highest oxidative stability, comparable to or superior to that of sodium nitrite, as reflected in the lowest values of key oxidation markers (PV, pAV, TBA, and TOTOX), thus confirming its strong potential for use in clean-label formulations. LP and SCP also demonstrated notable antioxidant properties, suggesting their suitability for tailored applications based on specific product characteristics. The antioxidant effect was both dose-dependent and fruit powder-specific, with improved efficacy at higher phenolic levels. In addition, the processing method significantly influenced the oxidative performance: smoked and cooked salami (SI), with a lower moisture content, showed greater stability during storage compared to smoked and scalded salami (SII), where the higher moisture content facilitated more intense oxidative degradation over time. These findings highlight the importance of aligning antioxidant strategies not only with ingredient functionality but also with technological parameters. From an applied perspective, fortification with fruit powders, especially BCP, has been shown to be effective in maintaining oxidative stability below critical thresholds during 30 days of refrigerated storage, supporting their potential use in clean-label meat products. However, to fully validate the technological feasibility of replacing nitrites with fruit-derived antioxidants, further comprehensive studies are needed to assess the microbiological safety of these formulations. Future research is planned to evaluate the efficacy of these powders in controlling microbial growth during storage, particularly their potential to inhibit spoilage microorganisms and relevant foodborne pathogens in nitrite-free formulations. Such investigations will provide essential information on the microbiological safety and technological reliability of these natural additives, thereby supporting their effective integration into clean labeling strategies for industrial meat production.

Author Contributions: Conceptualization, A.-I.M.M. and M.-A.P.; methodology, A.-I.M.M. and M.-A.P.; software, A.-I.M.M., I.C. and M.-A.P.; formal analysis, A.-I.M.M., I.C., D.-G.D. and M.-A.P.; investigation, A.-I.M.M., I.C., D.-G.D. and M.-A.P.; data curation, A.-I.M.M., I.C. and D.-G.D.; writing—original draft preparation, A.-I.M.M., I.C., D.-G.D. and M.-A.P.; writing—review and editing, A.-I.M.M., I.C., D.-G.D. and M.-A.P.; supervision, I.C. and M.-A.P. All authors have read and agreed to the published version of the manuscript.

Funding: The publication of the present paper was supported by the University of Life Sciences "King Mihai I" from Timisoara, Romania, and the results will be included in the PhD Thesis of Adriana-Ioana Moraru Manea, under the supervision of Mariana-Atena Poiana.

Institutional Review Board Statement: Not applicable.

Informed Consent Statement: Not applicable.

Data Availability Statement: The original data presented in the study are openly available at the University of Life Sciences "King Mihai I" from Timisoara.

Acknowledgments: The authors express their gratitude for the technical support provided by the University of Life Sciences "King Mihai I" from Timisoara, where the analyses were conducted. They also thank S.C. Cavarantana, Cavaran, Caras-Severin County, Romania, for granting access to their facilities and equipment, which were used in the production of the salami formulations examined in this study.

Conflicts of Interest: The authors declare no conflicts of interest.

References

1. Manessis, G.; Kalogianni, A.I.; Lazou, T.; Moschovas, M.; Bossis, I.; Gelasakis, A.I. Plant-Derived Natural Antioxidants in Meat and Meat Products. *Antioxidants* **2020**, *9*, 1215. [CrossRef]
2. Domínguez, R.; Pateiro, M.; Gagaoua, M.; Barba, F.J.; Zhang, W.; Lorenzo, J.M. A Comprehensive Review on Lipid Oxidation in Meat and Meat Products. *Antioxidants* **2019**, *8*, 429. [CrossRef]
3. Shah, M.A.; Bosco, S.J.D.; Mir, S.A. Plant extracts as natural antioxidants in meat and meat products. *Meat Sci.* **2014**, *98*, 21–33. [CrossRef]
4. Kumar, Y.; Yadav, D.N.; Ahmad, T.; Narsaiah, K. Recent Trends in the Use of Natural Antioxidants for Meat and Meat Products. *Compr. Rev. Food Sci. Food Saf.* **2015**, *14*, 796–812. [CrossRef]
5. Sebranek, J.G.; Sewalt, V.J.H.; Robbins, K.; Houser, T.A. Comparison of a natural rosemary extract and BHA/BHT for relative antioxidant effectiveness in pork sausage. *Meat Sci.* **2005**, *69*, 289–296. [CrossRef] [PubMed]

6. Efenberger-Szmechtyk, M.; Nowak, A.; Czyzowska, A. Plant extracts rich in polyphenols: Antibacterial agents and natural preservatives for meat and meat products. *Crit. Rev. Food Sci. Nutr.* **2021**, *61*, 149–178. [CrossRef] [PubMed]
7. Ferysiuk, K.; Wójciak, K.M. Reduction of Nitrite in Meat Products through the Application of Various Plant-Based Ingredients. *Antioxidants* **2020**, *9*, 711. [CrossRef] [PubMed]
8. Karwowska, M.; Kononiuk, A. Nitrates/Nitrites in Food—Risk for Nitrosative Stress and Benefits. *Antioxidants* **2020**, *9*, 241. [CrossRef]
9. Kaloudi, A.-N.; Karageorgou, A.; Goliomytis, M.; Simitzis, P. Preliminary Exploration of Natural Polyphenols and/or Herbal Mix Replacing Sodium Nitrate to Improve the Storage Quality of Pork Sausage. *Appl. Sci.* **2025**, *15*, 789. [CrossRef]
10. Di Vita, G.; Blanc, S.; Mancuso, T.; Massaglia, S.; La Via, G.; D'Amico, M. Harmful Compounds and Willingness to Buy for Reduced-Additives Salami. An Outlook on Italian Consumers. *Int. J. Environ. Res. Public Health* **2019**, *16*, 2605. [CrossRef]
11. Teixeira, A.; Rodrigues, S. Consumer perceptions towards healthier meat products. *Curr. Opin. Food Sci.* **2021**, *38*, 147–154. [CrossRef]
12. Karre, L.; Lopez, K.; Getty, K.J.K. Natural antioxidants in meat and poultry products. *Meat Sci.* **2013**, *94*, 220–227. [CrossRef]
13. Kausar, T.; Azad, Z.R.A.A.; Anwar, S.; Shahid, S.M.A.; Kausar, M.A. Application of natural antioxidants for the formulation of functional meat products. *Neuro Pharmac J.* **2021**, *6*, 269–276. [CrossRef]
14. Tang, T.; Zhang, M.; Law, C.L.; Mujumdar, A.S. Novel strategies for controlling nitrite content in prepared dishes: Current status, potential benefits, limitations and future challenges. *Food Res. Int.* **2023**, *170*, 112984. [CrossRef]
15. Zhu, W.; Han, M.; Bu, Y.; Li, X.; Yi, S.; Xu, Y.; Li, J. Plant polyphenols regulating myoglobin oxidation and color stability in red meat and certain fish: A review. *Crit. Rev. Food Sci. Nutr.* **2024**, *64*, 2276–2288. [CrossRef] [PubMed]
16. Kalogianni, A.I.; Lazou, T.; Bossis, I.; Gelasakis, A.I. Natural Phenolic Compounds for the Control of Oxidation, Bacterial Spoilage, and Foodborne Pathogens in Meat. *Foods* **2020**, *9*, 794. [CrossRef]
17. Lee, J.A.; Kang, D.Y.; Kim, J.K.; Lim, S.M.; Choi, N.W.; Choi, M.W.; Kim, H.Y. Effects of Addition of Strawberry and Red Beet Powder on the Antioxidant Activity and Quality Characteristics of Beef Patty. *Resour. Sci. Res.* **2020**, *2*, 1–8. [CrossRef]
18. Menchetti, L.; Brecchia, G.; Branciari, R.; Barbato, O.; Fioretti, B.; Codini, M.; Bellezza, E.; Trabalza-Marinucci, M.; Miraglia, D. The effect of Goji berries (*Lycium barbarum*) dietary supplementation on rabbit meat quality. *Meat Sci.* **2020**, *161*, 108018. [CrossRef]
19. Uearreeloet, P.; Konsue, N. Effect of gac fruit powder on quality and nitrosation activity of meat product. *J. Microbiol. Biotechnol. Food Sci.* **2016**, *6*, 786. [CrossRef]
20. Stoica, M.; Antohi, V.M.; Alexe, P.; Ivan, A.S.; Stanciu, S.; Stoica, D.; Zlati, M.L.; Stuparu-Cretu, M. New Strategies for the Total/Partial Replacement of Conventional Sodium Nitrite in Meat Products: A Review. *Food Bioprocess Technol.* **2022**, *15*, 514–538. [CrossRef]
21. Huang, X.; Sun, L.; Dong, K.; Wang, G.; Luo, P.; Tang, D.; Huang, Q. Mulberry fruit powder enhanced the antioxidant capacity and gel properties of hammered minced beef: Oxidation degree, rheological, and structure. *LWT* **2022**, *154*, 112648. [CrossRef]
22. Najeeb, A.P.; Mandal, P.K.; Pal, U.K. Efficacy of fruits (red grapes, gooseberry and tomato) powder as natural preservatives in restructured chicken slices. *Int. Food Res. J.* **2014**, *21*, 2431–2436. Available online: http://www.ifrj.upm.edu.my/21%20(06)%202014/52%20IFRJ%2021%20(06)%202014%20Mandal%20533.pdf (accessed on 5 February 2025).
23. Duthie, G.; Campbell, F.; Bestwick, C.; Stephen, S.; Russell, W. Antioxidant Effectiveness of Vegetable Powders on the Lipid and Protein Oxidative Stability of Cooked Turkey Meat Patties: Implications for Health. *Nutrients* **2013**, *5*, 1241–1252. [CrossRef] [PubMed]
24. Fathy, A.M.; Abdel-Rahman, A.M.A.; Ramadan, E.S.A.; Darwish, S.M.I. Sustainable Preservation of Beef Burgers Using Blueberry. *Assiut J. Agric. Sci.* **2023**, *54*, 1–22. [CrossRef]
25. Anton, D.; Koskar, J.; Raudsepp, P.; Meremäe, K.; Kaart, T.; Püssa, T.; Roasto, M. Antimicrobial and Antioxidative Effects of Plant Powders in Raw and Cooked Minced Pork. *Foods* **2019**, *8*, 661. [CrossRef]
26. Martínez-Zamora, L.; Peñalver, R.; Ros, G.; Nieto, G. Substitution of synthetic nitrates and antioxidants by spices, fruits and vegetables in Clean Label Spanish chorizo. *Food Res. Int.* **2021**, *139*, 109835. [CrossRef]
27. Cadariu, A.I.; Cocan, I.; Negrea, M.; Alexa, E.; Obistioiu, D.; Hotea, I.; Radulov, I.; Poiana, M.-A. Exploring the Potential of Tomato Processing Byproduct as a Natural Antioxidant in Reformulated Nitrite-Free Sausages. *Sustainability* **2022**, *14*, 11802. [CrossRef]
28. Cocan, I.; Cadariu, A.-I.; Negrea, M.; Alexa, E.; Obistioiu, D.; Radulov, I.; Poiana, M.-A. Investigating the Antioxidant Potential of Bell Pepper Processing By-Products for the Development of Value-Added Sausage Formulations. *Appl. Sci.* **2022**, *12*, 12421. [CrossRef]
29. Larrauri, J.A.; Rupérez, P.; Saura-Calixto, F. Effect of Drying Temperature on the Stability of Polyphenols and Antioxidant Activity of Red Grape Pomace Peels. *J. Agric. Food Chem.* **1997**, *45*, 1390–1393. [CrossRef]

30. FAO Agricultural Services Bulletin No.119. Fruit and Vegetable Processing—Ch05 General Procedures for Fruit and Vegetable Preservation. 1995. Available online: http://www.fao.org/3/V5030E/V5030E0b.htm#5.2%20Preservation%20by%20reduction%20of%20water%20content:%20drying/dehydration%20and%20concentration (accessed on 15 March 2025).
31. Litwinek, D.; Gumul, D.; Łukasiewicz, M.; Zięba, T.; Kowalski, S. The Effect of Red Potato Pulp Preparation and Stage of Its Incorporation into Sourdough or Dough on the Quality and Health-Promoting Value of Bread. *Appl. Sci.* **2023**, *13*, 7670. [CrossRef]
32. Daukšienė, J.; Burdulienė, K.; Daukšys, G.; Paulauskaitė, L.; Burdulis, D. The Influence of Raw Material Preparation on the Yield of Bioactive Substances in Lingonberry (*Vaccinium vitis-idaea* L.) Fruits. *Chemija* **2024**, *35*, 1–8. [CrossRef]
33. Al-Farsi, M.; Al-Amri, A.; Al-Hadhrami, A.; Al-Belushi, S. Color, flavonoids, phenolics and antioxidants of Omani honey. *Heliyon* **2018**, *4*, e00874. [CrossRef]
34. Darniadi, S.; Ifie, I.; Ho, P.; Murray, B.S. Evaluation of total monomeric anthocyanin, total phenolic content and individual anthocyanins of foam-mat freeze-dried and spray-dried blueberry powder. *J. Food Meas. Charact.* **2019**, *13*, 1599–1606. [CrossRef]
35. Zhaoa, W.; Caoa, P.; Zhua, Y.; Liua, S.; Gaob, H.W.; Huang, C. Rapid detection of vitamin c content in fruits and vegetables using a digital camera and color reaction. *Quim. Nova* **2020**, *43*, 1421–1430. [CrossRef]
36. Žlabur, J.Š.; Mikulec, N.; Doždor, L.; Duralija, B.; Galić, A.; Voća, S. Preservation of Biologically Active Compounds and Nutritional Potential of Quick-Frozen Berry Fruits of the Genus *Rubus*. *Processes* **2021**, *9*, 1940. [CrossRef]
37. Babbar, N.; Oberoi, H.S.; Uppal, D.S.; Patil, R.T. Total phenolic content and antioxidant capacity of extracts obtained from six important fruit residues. *Food Res. Int.* **2011**, *44*, 391–396. [CrossRef]
38. Poiana, M.-A.; Alexa, E.; Radulov, I.; Raba, D.-N.; Cocan, I.; Negrea, M.; Misca, C.D.; Dragomir, C.; Dossa, S.; Suster, G. Strategies to Formulate Value-Added Pastry Products from Composite Flours Based on Spelt Flour and Grape Pomace Powder. *Foods* **2023**, *12*, 3239. [CrossRef]
39. Metzner Ungureanu, C.-R.; Poiana, M.-A.; Cocan, I.; Lupitu, A.I.; Alexa, E.; Moigradean, D. Strategies to Improve the Thermo-Oxidative Stability of Sunflower Oil by Exploiting the Antioxidant Potential of Blueberries Processing Byproducts. *Molecules* **2020**, *25*, 5688. [CrossRef]
40. AOAC International. *Official Methods of Analysis of AOAC International*, 20th ed.; AOAC International: Rockville, MD, USA, 2016.
41. Forouzesh, A.; Forouzesh, F.; Foroushani, S.S.; Forouzesh, A. A New Method for Calculating Fat Content and Determining Appropriate Fat Levels in Foods. *Iran. J. Public Health* **2023**, *52*, 1038–1047. [CrossRef]
42. *ISO 3960:2009*; Animal and Vegetable Fats and Oils—Determination of Peroxide Value—Iodometric Visual Endpoint Determination. International Organization for Standardization (ISO): Geneva, Switzerland, 2009.
43. Seo, J.K.; Parvin, R.; Park, J.; Yang, H.S. Utilization of Astaxanthin as a Synthetic Antioxidant Replacement for Emulsified Sausages. *Antioxidants* **2021**, *10*, 407. [CrossRef]
44. *ISO 6885:2008*; Animal and Vegetable Fats and Oils—Determination of Anisidine Value. International Organization for Standardization (ISO): Geneva, Switzerland, 2008.
45. Mariod, A.A.; Matthäus, B.; Eichner, K.; Hussein, H.I. Antioxidant activity of extracts from ***Sclerocarya birrea*** kernel oil cake. *Grasas Aceites* **2006**, *57*, 361–366. [CrossRef]
46. Jones, T. Methods and Their Applications for Measuring and Managing Lipid Oxidation: Meat, Poultry, and Seafood Products. In *Natural Antioxidants. Applications in Foods of Animal Origin*; Banerjee, R., Verma, A.K., Siddiqui, M.W., Eds.; Apple Academic Press, Inc.: Boca Raton, FL, USA, 2017; pp. 203–260.
47. Tran, T.T.T.; Ton, N.M.N.; Nguyen, T.T.; Le, V.V.M.; Sajeev, D.; Schilling, M.W.; Dinh, T.T.N. Application of natural antioxidant extract from guava leaves (*Psidium guajava* L.) in fresh pork sausage. *Meat Sci.* **2020**, *165*, 108106. [CrossRef]
48. Kim, J.S. Antioxidant Activities of Selected Berries and Their Free, Esterified, and Insoluble-Bound Phenolic Acid Contents. *Prev. Nutr. Food Sci.* **2018**, *23*, 35–45. [CrossRef]
49. Sadowska, A.; Rakowska, R.; Świderski, F.; Kulik, K.; Hallmann, E. Properties and microstructure of blackcurrant powders prepared using a new method of fluidized-bed jet milling and drying versus other drying methods. *CyTA-J. Food* **2019**, *17*, 439–446. [CrossRef]
50. Borowiec, K.; Stachniuk, A.; Szwajgier, D.; Trzpil, A. Polyphenols composition and the biological effects of six selected small dark fruits. *Food Chem.* **2022**, *391*, 133281. [CrossRef] [PubMed]
51. Papp, N.; Blázovics, A.; Fébel, H.; Salido, S.; Altarejos, J.; Fehér, E.; Kocsis, I.; Szentmihályi, K.; Abrankó, L.; Hegedűs, A.; et al. Antihyperlipidemic Effects of Sour Cherries Characterized by Different In Vitro Antioxidant Power and Polyphenolic Composition. *Plant Foods Hum. Nutr.* **2015**, *70*, 408–413. [CrossRef]
52. Sokół-Łętowska, A.; Kucharska, A.Z.; Hodun, G.; Gołba, M. Chemical Composition of 21 Cultivars of Sour Cherry (*Prunus cerasus*) Fruit Cultivated in Poland. *Molecules* **2020**, *25*, 4587. [CrossRef]
53. Milić, A.; Daničić, T.; Tepić Horecki, A.; Šumić, Z.; Bursać Kovačević, D.; Putnik, P.; Pavlić, B. Maximizing Contents of Phytochemicals Obtained from Dried Sour Cherries by Ultrasound-Assisted Extraction. *Separations* **2021**, *8*, 155. [CrossRef]

54. Yaman, M. Evaluation of genetic diversity by morphological, biochemical and molecular markers in sour cherry genotypes. *Mol. Biol. Rep.* **2022**, *49*, 5293–5301. [CrossRef]
55. Blejan, A.M.; Nour, V.; Păcularu-Burada, B.; Popescu, S.M. Wild bilberry, blackcurrant, and blackberry by-products as a source of nutritional and bioactive compounds. *Int. J. Food Prop.* **2023**, *26*, 1579–1595. [CrossRef]
56. Lee, S.G.; Vance, T.M.; Nam, T.G.; Kim, D.O.; Koo, S.I.; Chun, O.K. Evaluation of pH differential and HPLC methods expressed as cyanidin-3-glucoside equivalent for measuring the total anthocyanin contents of berries. *J. Food Meas. Charact.* **2016**, *10*, 562–568. [CrossRef]
57. Ozola, L.; Kampuse, S. The influence of drying method to the changes of bioactive compounds in lingonberry by-products. *Agron. Res.* **2018**, *16*, 1781–1795. [CrossRef]
58. Dróżdż, P.; Šėžienė, V.; Pyrzynska, K. Phytochemical Properties and Antioxidant Activities of Extracts from Wild Blueberries and Lingonberries. *Plant Foods Hum. Nutr.* **2017**, *72*, 360–364. [CrossRef] [PubMed]
59. Koponen, J.M.; Happonen, A.M.; Mattila, P.H.; Törrönen, A.R. Contents of Anthocyanins and Ellagitannins in Selected Foods Consumed in Finland. *J. Agric. Food Chem.* **2007**, *55*, 1612–1619. [CrossRef]
60. Urbonaviciene, D.; Bobinaite, R.; Viskelis, P.; Viskelis, J.; Petruskevicius, A.; Puzeryte, V.; Cesoniene, L.; Daubaras, R.; Klavins, L.; Bobinas, C. Nutritional and Physicochemical Properties of Wild Lingonberry (*Vaccinium vitis-idaea* L.)—Effects of Geographic Origin. *Molecules* **2023**, *28*, 4589. [CrossRef] [PubMed]
61. Wojdyło, A.; Nowicka, P.; Laskowski, P.; Oszmiański, J. Evaluation of sour cherry (*Prunus cerasus* L.) fruits for their polyphenol content, antioxidant properties, and nutritional components. *J. Agric. Food Chem.* **2014**, *62*, 12332–12345. [CrossRef]
62. Order No. 210 of 31 August 2006 for the Approval of the Norms Regarding the Marketing of Meat Products. Published in Official Gazette No. 10 of 8 January 2007. Available online: http://www.legex.ro/Ordin-210-2006-75963.aspx (accessed on 20 March 2024).
63. Mahapatra, G.; Biswas, S.; Banerjee, R. Improving the quality and shelf-life of chevon meatballs by incorporating fruit and fruit by-products. *Indian J. Anim. Sci.* **2021**, *90*, 1655–1662. [CrossRef]
64. Fu, Q.; Song, S.; Xia, T.; Wang, R. Effects of Cherry (*Prunus cerasus* L.) Powder Addition on the Physicochemical Properties and Oxidation Stability of Jiangsu-Type Sausage during Refrigerated Storage. *Foods* **2022**, *11*, 3590. [CrossRef]
65. Zaini, H.B.M.; Sintang, M.D.B.; Pindi, W. The roles of banana peel powders to alter technological functionality, sensory and nutritional quality of chicken sausage. *Food Sci. Nutr.* **2020**, *8*, 5497–5507. [CrossRef] [PubMed]
66. López-Vargas, J.H.; Fernández-López, J.; Pérez-Álvarez, J.Á.; Viuda-Martos, M. Quality characteristics of pork burger added with albedo-fiber powder obtained from yellow passion fruit (*Passiflora edulis* var. *flavicarpa*) Co-products. *Meat Sci.* **2014**, *97*, 270–276. [CrossRef]
67. Zanardi, E.; Ghidini, S.; Conter, M.; Ianieri, A. Mineral composition of Italian salami and effect of NaCl partial replacement on compositional, physico-chemical and sensory parameters. *Meat Sci.* **2010**, *86*, 742–744. [CrossRef]
68. Yi, G.; Haug, A.; Nyquist, N.F.; Egelandsdal, B. Hydroperoxide formation in different lean meats. *Food Chem.* **2013**, *141*, 2656–2665. [CrossRef]
69. Gray, J.I.; Gomaa, E.A.; Buckley, D.J. Oxidative quality and shelf life of meats. *Meat Sci.* **1996**, *43*, 111–123. [CrossRef] [PubMed]
70. Sun, L.; Senecal, A.; Chinachoti, P.; Faustman, C. Effect of Water Activity on Lipid Oxidation and Protein Solubility in Freeze-Dried Beef During Storage. *J. Agric. Food Chem.* **2002**, *50*, 3959–3965. [CrossRef]
71. Feiner, G. *Meat Products Handbook: Practical Science and Technology*; Woodhead Publishing Limited: Boca Raton, FL, USA, 2006; pp. 19–32.
72. CODEX-STAN 211-1999; Codex Alimentarius Commission, Codex Standard for Named Animal Fats, Food and Agricultural Organization of the United Nations and the World Health Organization. Available online: https://www.fao.org/4/y2774e/y2774e05.htm#bm5.1 (accessed on 11 February 2025).
73. Ahmad, S.R.; Gokulakrishnan, P.; Giriprasad, R.; Yatoo, M.A. Fruit-based Natural Antioxidants in Meat and Meat Products: A Review. *Crit. Rev. Food Sci. Nutr.* **2015**, *55*, 1503–1513. [CrossRef] [PubMed]
74. Hadidi, M.; Orellana-Palacios, J.C.; Aghababaei, F.; Gonzalez-Serrano, D.J.; Moreno, A.; Lorenzo, J.M. Plant by-Product Antioxidants: Control of Protein–Lipid Oxidation in Meat and Meat Products. *LWT* **2022**, *169*, 114003. [CrossRef]
75. Reddy, D.M.; Reddy, G.V.B.; Mandal, P.K. Application of Natural Antioxidants in Meat and Meat Products—A Review. *Food Nutr. J.* **2018**, *3*, 173. [CrossRef]
76. Mansour, H.M.M.; El-Sohaimy, S.A.; Zeitoun, A.M.; Abdo, E.M. Effect of Natural Antioxidants from Fruit Leaves on the Oxidative Stability of Soybean Oil during Accelerated Storage. *Antioxidants* **2022**, *11*, 1691. [CrossRef]
77. Grompone, M.A. Anisidine value as a measurement of the latent damage of fats. *Grasas Aceites* **1991**, *42*, 8–13. [CrossRef]

78. Bekdeşer, B.; Esin Çelik, S.; Bener, M.; Dondurmacıoğlu, F.; Yıldırım, E.; Nida Yavuz, E.; Apak, R. Determination of primary and secondary oxidation products in vegetable oils with gold nanoparticle based fluorometric turn-on nanosensor: A new total oxidation value. *Food Chem.* **2024**, *434*, 137426. [CrossRef]
79. Global Organization for EPA and DHA Omega-3. GOED Voluntary Monograph, Version 8.1. 6 January 2022. Available online: https://goedomega3.com/goed-monograph (accessed on 17 March 2025).
80. Jin, S.K.; Kim, G.D.; Jeong, J.Y. Evaluation of the Effect of Inhibiting Lipid Oxidation of Natural Plant Sources in a Meat Model System. *J. Food Qual.* **2021**, *2021*, 6636335. [CrossRef]
81. Hęś, M.; Korczak, J.; Gramza, A. Changes of lipid oxidation degrees and their influence on protein nutritive value of frozen meat products. *Pol. J. Food Nutr. Sci.* **2007**, *57*, 323–328. Available online: https://journal.pan.olsztyn.pl/pdf-98069-30649?filename=CHANGES%20OF%20LIPID.pdf (accessed on 10 March 2025).
82. Reitznerová, A.; Šuleková, M.; Nagy, J.; Marcinčák, S.; Semjon, B.; Čertík, M.; Klempová, T. Lipid Peroxidation Process in Meat and Meat Products: A Comparison Study of Malondialdehyde Determination between Modified 2-Thiobarbituric Acid Spectrophotometric Method and Reverse-Phase High-Performance Liquid Chromatography. *Molecules* **2017**, *22*, 1988. [CrossRef] [PubMed]
83. McKenna, D.R.; Mies, P.D.; Baird, B.E.; Pfeiffer, K.D.; Ellebracht, J.W.; Savell, J.W. Biochemical and physical factors affecting discoloration characteristics of 19 bovine muscles. *Meat Sci.* **2005**, *70*, 665–682. [CrossRef] [PubMed]
84. Campo, M.M.; Nute, G.R.; Hughes, S.I.; Enser, M.; Wood, J.D.; Richardson, R.I. Flavour perception of oxidation in beef. *Meat Sci.* **2006**, *72*, 303–311. [CrossRef] [PubMed]
85. Hughes, J.M.; McPhail, N.G.; Kearney, G.; Clarke, F.; Warner, R.D. Beef longissimus eating quality increases up to 20 weeks of storage and is unrelated to meat colour at carcass grading. *Anim. Prod. Sci.* **2015**, *55*, 174–179. [CrossRef]

foods

MDPI

Article

Fattening by Dietary Replacement with Fly Maggot Larvae (*Musca domestica*) Enhances the Edible Yield, Antioxidant Capability, Nutritional and Taste Quality of Adult Chinese Mitten Crab *Eriocheir sinensis*

Xiao Liang [1,†], Changle Qi [2,†], Jinyu Tang [1], Ting Ye [1], Bao Lou [1] and Fuyong Huang [1,*]

1 State Key Laboratory for Quality and Safety of Agro-Products, Institute of Hydrobiology, Zhejiang Academy of Agricultural Sciences, Desheng Middle Road 298, Hangzhou 310021, China; liangx@zaas.ac.cn (X.L.); tangjy@zaas.ac.cn (J.T.); yet@zaas.ac.cn (T.Y.); loubao@zaas.ac.cn (B.L.)

2 National-Local Joint Engineering Laboratory of Aquatic Animal Genetic Breeding and Nutrition, Zhejiang Provincial Key Laboratory of Aquatic Resources Conservation and Development, College of Life Science, Huzhou University, Huzhou 313000, China; qichangle1989@163.com

* Correspondence: huangfy@zaas.ac.cn

† These authors contributed equally to this work.

Abstract: Housefly maggot larvae (HML) have been identified as a potential alternative animal diet for the fattening process of the Chinese mitten crab (*Eriocheir sinensis*). However, the feasibility and potential impacts of HML supplementation require further investigation. The present study evaluated the effects of dietary HML on the growth indices, nutrient compositions, antioxidant activity, and texture profiles of edible tissues of *E. sinensis*. The results showed that dietary HML supplementation effectively improved the hepatic steatosis index of both genders and sweet amino acid content of edible tissues (except for male gonad) ($p < 0.05$). Additionally, dietary HML significantly increased the total antioxidant capacity in the gonad and female muscle ($p < 0.05$). For the textural properties, HML feeding significantly improved the adhesiveness in the male muscle, and the cohesiveness, chewiness, and resilience in the female muscle ($p < 0.05$). Furthermore, HML feeding significantly decreased the levels of the equivalent umami concentration of the male gonad, male muscle, and hepatopancreas ($p < 0.05$). Conversely, HML feeding significantly increased the sweetness value in the muscle, hepatopancreas, and female gonad ($p < 0.05$). Our findings indicate that HML could serve as a viable alternative feed for fattening to improve the edible yield and change the flavor characteristics in *E. sinensis*.

Keywords: housefly maggot larvae; *Eriocheir sinensis*; fattening; textural properties; flavor characteristics

1. Introduction

The Chinese mitten crab (*Eriocheir sinensis*) is an important aquaculture species and highly prized by consumers in China due to its high nutritional value and unique flavor [1,2]. The total production of *E. sinensis* in China ranks first globally, with an annual yield nearing 1 million tons [3]. Prior to market harvest in September, the fattening of adult *E. sinensis* is typically required, as it directly enhances the overall quality and yield of crabs, thus ensuring better market prices [4,5]. The duration of the fattening period and efficient dietary feed are crucial factors influencing both nutrition deposition and gonad development in crabs [5,6]. Previous research suggests that the optimal fattening period for pond-reared *E. sinensis* ranges from 40 to 60 days [4,6]. Common dietary options for fattening include iced trash fish (IF), soybean, cornmeal, and formulated diets (FDs) [5]. IF, once considered the "standard food" for crab fattening, has been associated with several issues, including source instability, unpredictable quality, nutritional imbalances, and water pollution [7]. As a result, numerous studies have focused on developing high-quality FDs or alternative protein sources to enhance fattening practices [5,8].

The crude protein level in most insects ranges from 40% to 63%, which have been recognized as stable sources of dietary proteins for aquaculture [8–11]. Housefly maggots (*Musca domestica*; HM) are promising and highly nutritious resources, offering significant potential as a protein source for crabs [11–13]. The crude protein content of HM ranges from 43% to 62% [14], and HM are rich in monounsaturated fatty acids, B-complex vitamins, phosphorus, and trace elements [13,15]. Previous studies have demonstrated that insect meal can effectively replace fish meal in aquaculture diets without adversely affecting the growth of various fish species, including barramundi (*Lates calcarifera*), gilthead seabream (*Sparus aurata*), and European sea bass (*Dicentrarchus labrax* L.) [16–18]. Moreover, incorporating HM larvae (HML) in fish diets has the potential to improve growth rates and feed efficiency without inducing physiological stress [11,19]. Dietary HML supplementation has been shown to improve the hardness and moisture in fish muscles, enhancing the flesh quality of Nile tilapia (*Oreochromis niloticus*) [20]. Additionally, HML-supplemented diets promote the growth of swamp eel (*Monopterus albus*) and support gonadal development in *Clarias gariepinus* (Burchell, 1822) [21,22]. In crustaceans, replacing fish meal with defatted insect meal, such as yellow mealworm (*Tenebrio molitor*), significantly increased the growth and survival rates of juvenile Pacific white shrimp (*Litopenaeus vannanmei*) [23], while dietary supplementation with black soldier fly (*Hermetia illucens*) enhanced the immunity, gut microbiota, and protein content in freshwater crayfish marron (*Cherax cainii*) [24]. Furthermore, the exoskeleton of insects contains chitin, a structural polysaccharide that may contribute to the formation of the exoskeletons in crustaceans [25]. However, studies on the effects of HML on the growth and health of crustaceans remain limited.

Although preliminary investigations have revealed the regulatory effects of HML on growth and nutrient metabolism in decapod crustaceans [12,19], their efficacy in enhancing nutritional quality remains insufficiently explored. Therefore, this study aimed to systematically evaluate the impacts of HML supplementation on the growth indices and nutrient composition of edible tissues in *E. sinensis* during the fattening period. Specifically, we assessed various growth parameters, the antioxidant status, amino acid (AA) and nucleotide composition, texture profiles, and flavor characteristics of different edible tissues. The findings will validate its application potential as a functional feed additive and establish a theoretical foundation for precision nutrient regulation in crustacean aquaculture.

2. Materials and Methods

2.1. Experimental Setup and Culture Management

This study was conducted in Changxing Town, Huzhou, Zhejiang Province (China). Healthy, active, and intact *E. sinensis* were randomly selected from a local farm in September. A total of approximately 240 female and 240 male crabs were stocked into two separate outdoor earthen ponds. Each pond was subdivided into six parallel sections (length $\times$ width $\times$ depth = 10 m $\times$ 8 m $\times$ 1.2 m) using purse seines, with a stocking density of 40 crabs per section. Each pond contained two experimental groups, with three sections designated per group. The initial body weights of the adult female and male crabs, post-puberty molt, were 150 $\pm$ 17 g and 200 $\pm$ 19 g, respectively. Based on a previous study [6], this experiment commenced on 5 September 2023, and lasted for 40 days. During the fattening period, water quality parameters, including dissolved oxygen (>5 mg/L), ammonia nitrogen (<0.3 mg/L), nitrite (<0.01 mg/L), and pH (7.5–9.0), were maintained within normal reference ranges. *Elodea canadensis* was transplanted to cover more than 60% of the area in each region, providing shelter for the crabs and assisting in the maintenance of water quality.

In this study, both the component percentage in the fattening diet combinations and administration protocols were designed in accordance with practical aquaculture production practices. The current aquaculture operations commonly employ formula diets supplemented with plant-based ingredients (soybean, corn, etc.) as standard practice [5]. Consequently, the composition of the fattening diets is shown in Figure 1, and the nutritional composition of HML is shown in Supplementary Table S1. The crabs were fed once daily, with the feeding amount comprising approximately 2–4% of their body weight, at 17:00. In the HML supplementation (HMLS) group, the crabs were provided with formula diets and HML on the first day, followed by non-HML supplementation diets (a combination of formula diets, soybean, and corn) on the second day, and this alternated daily thereafter. The control group received non-HML supplementation diets continuously. The feeding amount was adjusted based on residual feed, temperature, and water quality.

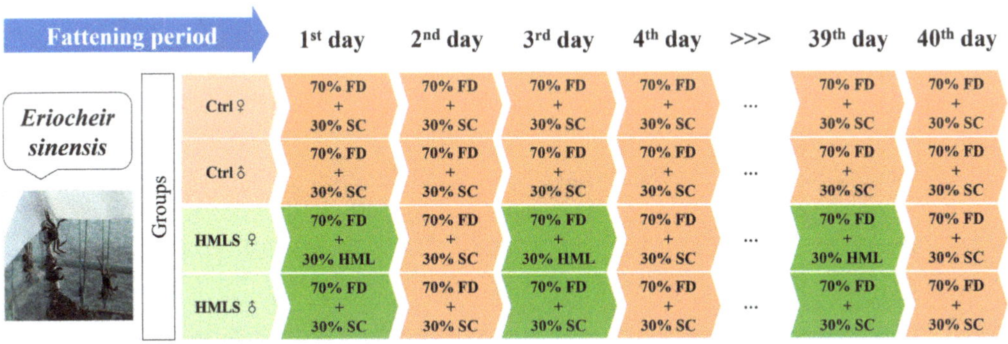

Figure 1. General experimental design of the feeding trial. Note: This feeding trial started on September 5th in 2023 and lasted for 40 days. Experimental groups consisted of control (Ctrl) group and HML supplementation (HMLS) group. FD: formula diets; SC: soybean and corn combinations.

2.2. Sample Collection and Processing

At the end of the experiment, the crabs were fasted for 24 h prior to sampling. Referring to classical studies on sample size selection [4,6], eighteen crabs per group (six individuals from each section) were randomly selected, then dissected. The meat, gonad, and hepatopancreas were weighed to calculate the meat yield (MY), hepatosomatic index (HSI), and gonadosomatic index (GSI). Nine female and nine male crabs were randomly selected for the measurement of antioxidation capacity and astaxanthin contents. For AA composition and nucleotide analysis, another set of nine female and nine male crabs were selected, gently blotted with a towel to remove surface moisture, and their edible tissues were carefully removed from the carapace cavity and body parts of the freshly uncooked crabs. All edible parts were steamed for 15 min over boiling water, after which the homogenates were stored for later analysis. All samples were stored separately at −80 °C. Claw muscle samples were obtained, naturally cooled to room temperature, and subsequently analyzed for texture.

The MY, HSI, GSI, and total edible yield (TEY) of the crabs were calculated using the following formulas:

$$\text{MY (\%)} = 100 \times \text{Meat wet weight/Body wet weight.}$$

$$\text{HSI (\%)} = 100 \times \text{Hepatopancreas wet weight/Body wet weight.}$$

$$\text{GSI (\%)} = 100 \times \text{Gonad system wet weight/Body wet weight.}$$

$$\text{TEY (\%)} = \text{MY} + \text{HSI} + \text{GSI.}$$

2.3. *Antioxidant Activity Analysis*

The activities of superoxide dismutase (SOD) and total antioxidant capacity (T-AOC), and the malonaldehyde (MDA) content in the muscle, gonad, and hepatopancreas tissues were measured using specific commercial kits (Nanjing Jiancheng Biotech Co., Ltd., Nanjing, China), following the manufacturer's protocols. The SOD was determined at 550 nm by the hydroxylamine method. The T-AOC was determined at 405 nm by the 2,2′-amino-di (2-ethylbenzothiazoline sulfonic acid-6) ammonium salt (ABTS) method. The content of MDA was determined at 532 nm by the thiobarbituric acid (TBA) method. The total protein content in tissue samples was measured at 595 nm by the Coomassie brilliant blue method to calculate the above indicators. Three replicates for each sample were analyzed.

2.4. *Determination of Astaxanthin Content*

Tissues from all individuals (n = 18) were sampled, stored in liquid nitrogen, and then transferred to −80 °C. According to the method of Su et al. [26], astaxanthin content was determined using a high-performance liquid chromatography (HPLC) system (Shimadzu, LC20AD, Kyoto, Japan). Firstly, all samples were dried in a vacuum freeze dryer (Scan Vac Cool Safe, Labo Gene, Lynge, Denmark) and then ground into powder (under light protection). About 0.2 g of tissue powder was placed in a 10 mL centrifuge tube and extracted with 4 mL of acetone (analytical grade), shaken in the dark at 25 °C (200 rpm) for about 10. Then, the supernatant was filtered through a 0.20 μm filter membrane, and 20 μL of the supernatant was injected into the system. Separation was performed using a C30 column (4.6 mm × 250 mm, 3 μm) (Thermo Fisher Scientific, Waltham, MA, USA), with a mobile phase of n-hexane–acetone (83:17, v/v) at a flow rate of 1 mL/min and a column temperature of 25 °C. Detection of astaxanthin was carried out using a UV–Vis detector at a wavelength of 478 nm.

2.5. *Texture Profile Analysis (TPA) of Muscles*

To assess the TPA of muscle, nine individuals were randomly selected from three experimental cages (three crabs per cage), and the propodus muscle of the crab claw was dissected for TPA measurement. TPA was performed using a TA-XT Plus Micro TPA (Stable Micro Systems, Surrey, UK) equipped with a flat-bottomed cylindrical probe P/50. The sample collection method followed the protocol outlined in previous studies [27,28]. The propodus muscle samples were first heated and then allowed to naturally cool to room temperature. Each sample was then trimmed into cubes (16 mm diameter, 10 mm height), and the TPA of muscle was conducted immediately. The mean TPA values obtained from the two claws of each crab were used as the final result for that individual.

The Universal TA Texture Analyzer (Tengba, Shanghai, China) was employed for the analysis, measuring various texture attributes, including hardness, springiness, chewiness, gumminess, adhesiveness, cohesiveness, and resilience. A cylindrical probe with a diameter of 36 mm and a speed of 1 mm/s was used. The distance mode was selected with a movement distance of 1.5 mm. The contact induction force was set to 5 gf, and the interval between two depressions was 5 s. Two consecutive compression cycles were performed, with the deformation set to 50% of the sample's initial height.

2.6. Free AA Determination

The free AAs in edible tissues were analyzed using the method described in a previous study [8].

2.7. Nucleotide Analysis

The nucleotides in edible tissues were analyzed according to the methodology outlined in a previous study [29]. The nucleotides responsible for eliciting the umami sensation include adenylate (AMP), guanylate (GMP), and inosinate (IMP).

2.8. Flavor Evaluation Indices

The threshold values of AAs and nucleotides were obtained from prior studies [30–32]. The taste activity value (TAV) was calculated as the ratio of the measured concentration of each AA or nucleotide to its respective threshold value. Compounds with a TAV greater than 1 were considered contributors to taste.

To assess the intensity of the umami flavor resulting from the synergistic effects of AAs and nucleotides, the equivalent umami concentration (EUC) was calculated as follows:

$$EUC = \Sigma\, a_i b_i + 1218\, (\Sigma\, a_i b_i)\, (\Sigma\, a_j b_j),$$

a_i is the concentration of umami amino acid (UAA) (Asp or Glu, g/100 g); b_i is the ratio of the umami concentration of Asp or Glu to that of monosodium glutamate (MSG) (Glu, 1; Asp, 0.077); a_j is the concentration of umami 5′-nucleotide (IMP, GMP, or AMP, g/100 g); b_j is the ratio of the umami concentration of a 5′-nucleotide to that of IMP (IMP, 1; GMP, 2.3; AMP, 0.18); and 1218 is a synergistic constant based on the concentration of g/100 g used.

2.9. Statistical Analyses

Statistical analysis was conducted using SPSS 24.0 (IBM, New York, NY, USA) using one-way ANOVA. Data are presented as the mean ± standard error (SE). To assess differences between two groups, multiple unpaired *t*-tests were performed and validated by Turkey multiple comparison tests (Turkey HSD). A p-value of < 0.05 was considered statistically significant. Radar plots and correlation analysis among the variables assessed with Spearman's correlation, as well as other graphs, were generated using OmicShare tools, a free online platform for data analysis and visualization (http://www.omicshare.com/tools, accessed on 30 April 2024).

3. Results

3.1. Growth Performance in Adult E. sinensis

After a 40-day fattening period, the HSI in both the male and female crabs of the HMLS group was significantly higher than that of the Ctrl group ($p < 0.05$). However, no significant differences were observed between the HMLS and Ctrl groups in terms of the MY, GSI, or TEY of crabs (Figure 2A,B).

3.2. Astaxanthin Content in Different Edible Tissues of E. sinensis

Astaxanthin content was measured in adult crabs from both the Ctrl group and HMLS group. In comparison to the Ctrl group, the HMLS group exhibited a significant increase in astaxanthin content in the carapace ($p < 0.01$). However, no significant differences were observed in the hepatopancreas between the Ctrl and HMLS groups ($p > 0.05$) (Figure 2C). Additionally, in female crabs, the astaxanthin content was significantly higher in the carapace ($p < 0.01$), hepatopancreas ($p < 0.05$), and gonad ($p < 0.001$) of the HMLS group compared to the Ctrl group (Figure 2D).

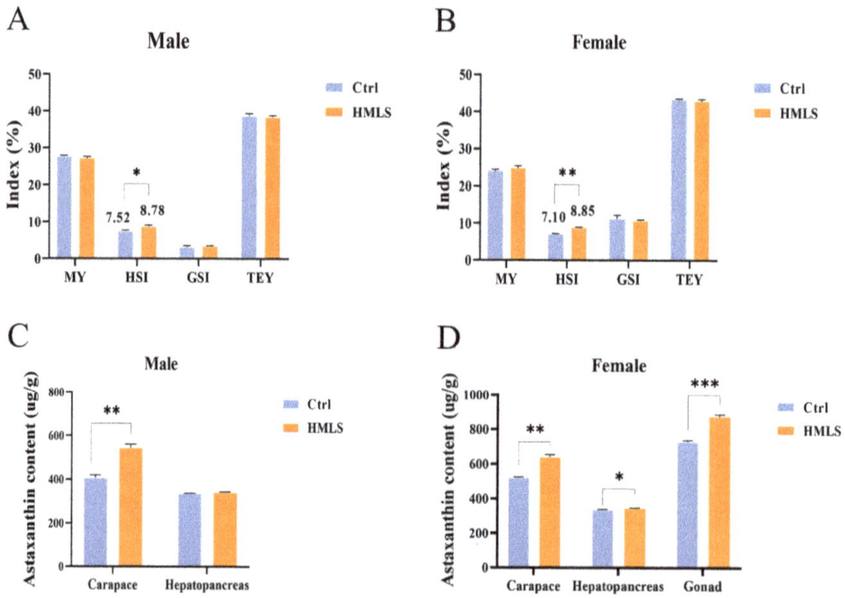

Figure 2. Effects of dietary HML on edible tissues and astaxanthin content in adult *E. sinensis*. (**A**,**B**) The results are presented as the means ± SE (n = 18). (**C**,**D**) The results are presented as the means ± SE (n = 9). An asterisk (*) indicates significance at the 0.05 level, ** at the 0.01 level, and *** at the 0.001 level.

3.3. Antioxidant Capacity in Different Edible Tissues of E. sinensis

As illustrated in Figure 3A,B, dietary supplementation with HML significantly enhanced the T-AOC in the gonad ($p < 0.05$) and female muscle ($p < 0.01$) of *E. sinensis*. No significant differences were observed in the SOD activities of male edible tissues between the Ctrl group and HMLS group (Figure 3C). However, dietary HML significantly increased the SOD activities in the muscle and hepatopancreas of female crabs ($p < 0.05$) (Figure 3D). Furthermore, dietary HML significantly decreased the MDA content in the hepatopancreas of both sexes ($p < 0.05$) (Figure 3E,F).

3.4. Textural Properties

Texture profile analysis (TPA) was conducted to assess the impact of dietary HML on the textural characteristics of various edible tissues, including hardness, adhesiveness, springiness, cohesiveness, gumminess, chewiness, and resilience (Table 1). In comparison to the Ctrl group, adhesiveness in the male muscle was significantly increased in the HMLS group ($p < 0.05$). Additionally, dietary HML significantly increased the cohesiveness ($p < 0.01$), chewiness ($p < 0.05$), and resilience ($p < 0.01$) of the female muscle.

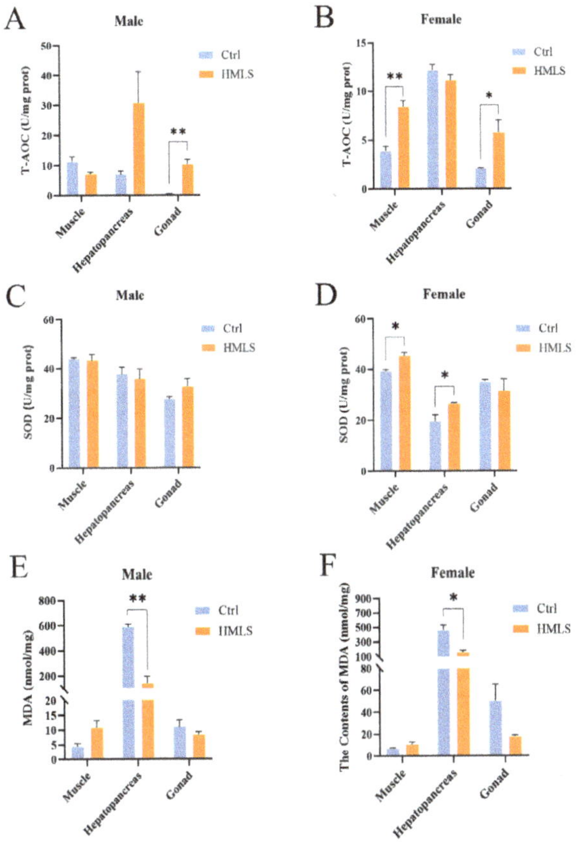

Figure 3. Effects of dietary HML on antioxidant parameters of edible tissues in adult *E. sinensis*. (**A**,**B**) T-AOC activity. (**C**,**D**) SOD activity. (**E**,**F**) MDA content. The results are presented as the means ± SE (n = 9). An asterisk (*) indicates significance at the 0.05 level and ** at the 0.01 level.

Table 1. Effects of dietary HMLS on textural properties in muscle.

Index	Female		Male	
	Ctrl	HMLS	Ctrl	HMLS
Hardness (g)	625.51 ± 90.68	805.38 ± 74.17	823.37 ± 73.21	1033.64 ± 74.39
Adhesiveness (J/m^3)	6.54 ± 0.48	8.13 ± 1.33	8.22 ± 0.71	13.71 ± 1.16 **
Springiness (mm)	0.78 ± 0.03	0.84 ± 0.07	0.76 ± 0.04	0.77 ± 0.02
Cohesiveness (N)	0.56 ± 0.02	0.63 ± 0.01 **	0.60 ± 0.01	0.60 ± 0.011
Gumminess (N)	358.75 ± 36.79	502.91 ± 58.26	495.74 ± 39.78	617.70 ± 46.98
Chewiness (mJ)	278.89 ± 26.38	414.46 ± 40.59 *	384.40 ± 31.27	481.07 ± 42.84
Resilience (N)	0.27 ± 0.01	0.31 ± 0.01 **	0.30 ± 0.01	0.30 ± 0.01

Note. The results are presented as the means ± SE (n = 9). An asterisk (*) indicates significance at the 0.05 level, ** at the 0.01 level.

3.5. Free AA Profiles in Different Edible Tissues of Steamed E. sinensis

The composition of delicious amino acids (DAAs) in edible tissues contributes to the umami taste and sweetness of crabs. To investigate the effects of dietary HML on the free AA composition in different edible tissues of steamed *E. sinensis*, a total of seventeen AAs were quantified, including six sweet amino acids (SAAs), two UAAs, and bitter amino acids (BAAs) (Tables 2–4). The total free amino acids (TAAs) content was highest in the hepatopancreas, followed by the gonad and muscle of both sexes (Figure 4A,B). Dietary HML significantly increased the SAA and TAA content in edible tissues, as well as the UAA content in the female muscle ($p < 0.05$). In male crabs, dietary HML significantly increased the SAA and TAA content in the muscle and hepatopancreas but significantly decreased the UAA content in the hepatopancreas and the SAA content in the gonad ($p < 0.05$).

Table 2. Effects of dietary HMLS on amino acids content in muscle of steamed *E. sinensis* (mg/g wet weight).

AAs	Muscle			
	Female		Male	
	Ctrl	HMLS	Ctrl	HMLS
Gly	4.63 ± 0.10	4.94 ± 0.03 **	3.39 ± 0.18	3.88 ± 0.16 *
Ala	0.02 ± 0.00	0.02 ± 0.00	0.02 ± 0.00	0.02 ± 0.00
Thr	0.25 ± 0.00	0.26 ± 0.00 *	0.19 ± 0.01	0.27 ± 0.04 *
Ser	0.02 ± 0.00	0.02 ± 0.00	0.02 ± 0.00	0.02 ± 0.00
Pro	1.07 ± 0.00	1.11 ± 0.01 **	0.91 ± 0.03	0.91 ± 0.03
Arg	0.32 ± 0.01	0.49 ± 0.23	0.31 ± 0.01	0.32 ± 0.01
SAA	6.30 ± 0.10	6.83 ± 0.27 *	4.85 ± 0.16	5.41 ± 0.22 *
Asp	0.00 ± 0.00	0.00 ± 0.00	0.01 ± 0.00	0.01 ± 0.00
Glu	0.08 ± 0.00	0.09 ± 0.00 *	0.09 ± 0.01	0.09 ± 0.01
UAA	0.08 ± 0.00	0.09 ± 0.00 *	0.09 ± 0.01	0.09 ± 0.00
Cys	0.05 ± 0.00	0.05 ± 0.00	0.04 ± 0.00	0.04 ± 0.00
Val	1.69 ± 0.01	1.74 ± 0.01 **	1.93 ± 0.05	2.29 ± 0.09 **
Met	0.15 ± 0.00	0.15 ± 0.00	0.10 ± 0.01	0.10 ± 0.01
Ile	0.12 ± 0.00	0.12 ± 0.00	0.10 ± 0.00	0.10 ± 0.00
Leu	0.19 ± 0.01	0.19 ± 0.01	0.17 ± 0.00	0.17 ± 0.00
Tyr	0.32 ± 0.01	0.35 ± 0.01 *	0.28 ± 0.01	0.27 ± 0.01
Phe	0.34 ± 0.01	0.34 ± 0.01	0.28 ± 0.03 **	0.08 ± 0.02
Lys	0.09 ± 0.00	0.09 ± 0.00	0.08 ± 0.00	0.08 ± 0.01
His	0.03 ± 0.00	0.03 ± 0.00	0.03 ± 0.00	0.03 ± 0.00
TAA	9.37 ± 0.11	9.99 ± 0.26 *	7.93 ± 0.19	8.67 ± 0.27 *

Note. The results are presented as the means ± SE (n = 9). An asterisk (*) indicates significance at the 0.05 level, ** at the 0.01 level.

Table 3. Effects of dietary HMLS on amino acids content in hepatopancreas of steamed *E. sinensis* (mg/g wet weight).

AAs	Hepatopancreas			
	Female		Male	
	Ctrl	HMLS	Ctrl	HMLS
Gly	26.79 ± 0.92	31.37 ± 0.86 **	19.70 ± 0.08	20.07 ± 0.06 **
Ala	0.02 ± 0.00	0.02 ± 0.00	0.02 ± 0.00	0.01 ± 0.00
Thr	0.46 ± 0.02	0.46 ± 0.02	0.37 ± 0.00	0.39 ± 0.00 **
Ser	0.02 ± 0.00	0.02 ± 0.00	0.02 ± 0.00	0.02 ± 0.00
Pro	0.69 ± 0.00	0.69 ± 0.00	0.64 ± 0.00	0.68 ± 0.01 **
Arg	0.15 ± 0.00	0.16 ± 0.00	0.14 ± 0.00	0.14 ± 0.00
SAA	28.13 ± 0.90	32.71 ± 0.87 **	20.89 ± 0.09	21.31 ± 0.06 **
Asp	0.01 ± 0.00	0.01 ± 0.00	0.01 ± 0.00	0.01 ± 0.00
Glu	0.19 ± 0.01	0.19 ± 0.00	0.18 ± 0.01 **	0.16 ± 0.00
UAA	0.19 ± 0.01	0.20 ± 0.00	0.19 ± 0.01 **	0.16 ± 0.00
Cys	0.04 ± 0.00	0.04 ± 0.00	0.03 ± 0.00	0.03 ± 0.00
Val	1.96 ± 0.05	2.14 ± 0.06 *	1.74 ± 0.01	1.80 ± 0.01 **
Met	0.16 ± 0.00	0.16 ± 0.00	0.16 ± 0.00	0.16 ± 0.00
Ile	0.21 ± 0.01	0.21 ± 0.01	0.18 ± 0.00	0.18 ± 0.00
Leu	0.36 ± 0.01	0.36 ± 0.01	0.33 ± 0.01	0.33 ± 0.00
Tyr	0.51 ± 0.02	0.58 ± 0.05	0.47 ± 0.01	0.50 ± 0.01 **
Phe	0.56 ± 0.02	0.56 ± 0.02	0.52 ± 0.00	0.54 ± 0.00 **
Lys	0.15 ± 0.00	0.15 ± 0.00	0.13 ± 0.00	0.13 ± 0.00
His	0.05 ± 0.00	0.05 ± 0.00	0.04 ± 0.00	0.04 ± 0.00
TAA	32.31 ± 0.92	37.15 ± 0.91 **	24.69 ± 0.08	25.19 ± 0.07 **

Note. The results are presented as the means ± SE (n = 9). An asterisk (*) indicates significance at the 0.05 level, ** at the 0.01 level.

Table 4. Effects of dietary HMLS on amino acids content in gonad of steamed *E. sinensis* (mg/g wet weight).

AAs	Gonad			
	Female		Male	
	Ctrl	HMLS	Ctrl	HMLS
Gly	11.96 ± 0.04	12.72 ± 0.08 **	13.17 ± 0.03 **	13.02 ± 0.03
Ala	0.02 ± 0.00	0.02 ± 0.00	0.02 ± 0.00	0.01 ± 0.00
Thr	0.37 ± 0.00 **	0.35 ± 0.00	0.30 ± 0.00	0.32 ± 0.00 *
Ser	0.02 ± 0.00	0.02 ± 0.00	0.02 ± 0.00	0.02 ± 0.00
Pro	0.81 ± 0.01	0.84 ± 0.01 **	0.76 ± 0.00	0.77 ± 0.00 *
Arg	0.28 ± 0.01	0.28 ± 0.00	0.23 ± 0.00	0.23 ± 0.00
SAA	13.46 ± 0.04	14.23 ± 0.08 **	14.50 ± 0.04 **	14.37 ± 0.03
Asp	0.00 ± 0.00	0.00 ± 0.00	0.01 ± 0.00	0.01 ± 0.00
Glu	0.18 ± 0.00	0.18 ± 0.00	0.13 ± 0.00	0.13 ± 0.00
UAA	0.19 ± 0.00	0.18 ± 0.00	0.14 ± 0.00	0.14 ± 0.00

Table 4. *Cont.*

AAs	Gonad			
	Female		Male	
	Ctrl	HMLS	Ctrl	HMLS
Cys	0.05 ± 0.00	0.05 ± 0.00	0.04 ± 0.00	0.04 ± 0.00
Val	1.88 ± 0.01	1.91 ± 0.01 **	1.85 ± 0.00	1.96 ± 0.04 **
Met	0.15 ± 0.00	0.15 ± 0.00	0.13 ± 0.00	0.13 ± 0.00
Ile	0.14 ± 0.00	0.14 ± 0.00	0.14 ± 0.00	0.14 ± 0.00
Leu	0.25 ± 0.00	0.25 ± 0.00	0.25 ± 0.00	0.25 ± 0.00
Tyr	0.42 ± 0.01	0.44 ± 0.00 **	0.37 ± 0.00	0.37 ± 0.00
Phe	0.45 ± 0.00	0.46 ± 0.00 **	0.39 ± 0.00	0.39 ± 0.00
Lys	0.22 ± 0.00	0.22 ± 0.00	0.11 ± 0.00	0.11 ± 0.00
His	0.05 ± 0.00	0.04 ± 0.00	0.03 ± 0.00	0.04 ± 0.00
TAA	17.26 ± 0.03	18.10 ± 0.09 **	17.96 ± 0.05	17.95 ± 0.02

Note. The results are presented as the means ± SE (n = 9). An asterisk (*) indicates significance at the 0.05 level, ** at the 0.01 level.

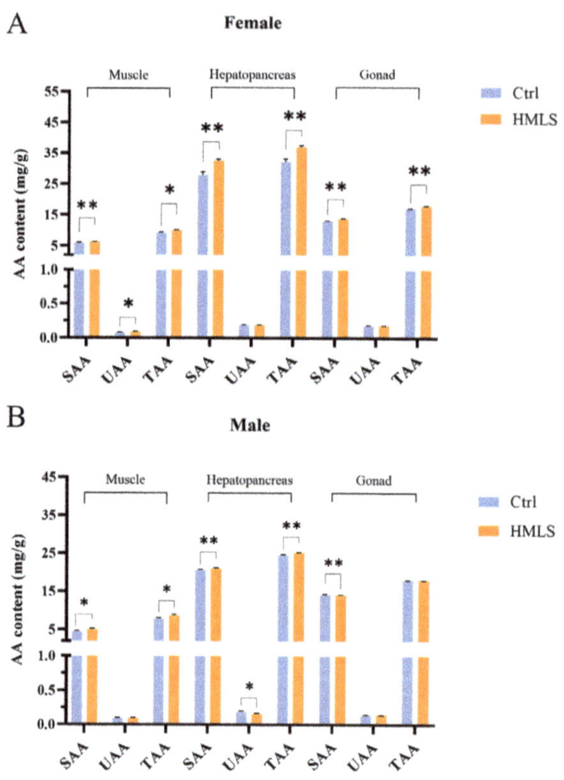

Figure 4. Effects of dietary HML on AA content of edible tissues in adult *E. sinensis*. (**A**) AA content of edible tissues in female *E. sinensis*. (**B**) AA content of edible tissues in male *E. sinensis*. The results are presented as the means ± SE (n = 9). An asterisk (*) indicates significance at the 0.05 level and ** at the 0.01 level.

As shown in Tables 2–4, glycine (Gly) was the dominant SAA in the edible tissues of crab. Dietary HML significantly increased the Gly content in the edible tissues of both sexes, except for the male gonad ($p < 0.05$). Furthermore, dietary HML significantly increased the threonine (Thr) content in the male edible tissues and female muscle, while significantly reducing the Thr content in the female gonad ($p < 0.05$). Compared with the Ctrl group, the proline (Pro) content in the gonad of both sexes, female muscle, and male hepatopancreas was significantly increased in the HMLS group ($p < 0.05$). Glutamic acid (Glu) was the dominant UAA in the edible tissues. Compared with the Ctrl group, the levels of Glu were significantly increased in the female muscle in the HMLS group but significantly decreased in the male hepatopancreas ($p < 0.05$).

The levels of BAA were also analyzed in the edible tissues of crabs. No significant differences were observed in the content of cysteine (Cys), methionine (Met), isoleucine (Ile), leucine (Leu), lysine (Lys), and histidine (His) between the Ctrl and HMLS groups ($p > 0.05$). However, the levels of valine (Val) were significantly increased in the edible tissues of crabs in the HMLS group compared to the Ctrl group. Dietary HML also significantly increased the levels of tyrosine (Tyr) in the male hepatopancreas, as well as in the female muscle and gonad. Furthermore, dietary HML significantly increased the levels of phenylalanine (Phe) in the male hepatopancreas and female gonad but significantly decreased the levels in the male muscle ($p < 0.05$).

3.6. Umami Nucleotide Content in Different Edible Tissues of Steamed *E. sinensis*

Umami nucleotides, including IMP, AMP, and GMP, were measured in various edible tissues of steamed crabs. As shown in Figure 5A, no significant differences were observed in the GMP content of the male edible tissues between the Ctrl group and HMLS group ($p > 0.05$). However, the GMP content of the female hepatopancreas in the Ctrl group was significantly higher than that in the HMLS group ($p < 0.01$) (Figure 5B). Dietary HML significantly decreased the AMP content in the muscle ($p < 0.0001$) and gonad ($p < 0.01$) of male crabs (Figure 5C), and in the muscle of female crabs ($p < 0.05$) (Figure 5D). In contrast, dietary HML significantly increased the IMP content in the muscle ($p < 0.001$) and gonad ($p < 0.05$) of male crabs (Figure 5E), as well as in the muscle of female crabs ($p < 0.01$) (Figure 5F). Furthermore, no significant differences were found in the AMP and IMP content in other edible tissues between the Ctrl group and HMLS group ($p > 0.05$) (Figure 5C–F).

3.7. Flavor Evaluation Indices in Different Edible Tissues of Steamed *E. sinensis*

The TAV > 1 was employed to assess the flavor characteristics in the edible tissues of steamed *E. sinensis*. The results indicated that Gly was the primary AA contributing to the sweetness of the edible tissues, while Val was the major AA responsible for bitterness. The highest TAV of Gly was observed in the female hepatopancreas. Additionally, nucleotide TAV analysis revealed that the TAV of AMP exceeded 1.0 only in the female gonad (Tables 5–7).

Table 5. Effects of dietary HMLS on the TAVs of amino acids and nucleotides in muscle.

AAs	Taste Threshold (mg/mL)	Muscle			
		Female		Male	
		Ctrl	HMLS	Ctrl	HMLS
Gly	1.3	1.78 ± 0.04	1.90 ± 0.01 **	1.30 ± 0.07	1.49 ± 0.06 *
Ala	0.6	0.02 ± 0.00	0.02 ± 0.00	0.01 ± 0.00	0.01 ± 0.00
Thr	2.6	0.05 ± 0.00	0.05 ± 0.00	0.04 ± 0.00	0.05 ± 0.01 *

Table 5. *Cont.*

AAs	Taste Threshold (mg/mL)	Muscle			
		Female		Male	
		Ctrl	HMLS	Ctrl	HMLS
Ser	1.5	0.01 ± 0.00	0.01 ± 0.00	0.01 ± 0.00	0.01 ± 0.00
Pro	3	0.18 ± 0.00	0.19 ± 0.00 **	0.15 ± 0.01	0.15 ± 0.00
Arg	0.5	0.32 ± 0.01	0.49 ± 0.23	0.31 ± 0.01	0.32 ± 0.01
SAA	0.5	0.00 ± 0.00	0.00 ± 0.00	0.01 ± 0.00	0.00 ± 0.00
Asp	0.2	0.20 ± 0.01	0.23 ± 0.01 *	0.22 ± 0.02	0.22 ± 0.01
Glu	0.2	0.13 ± 0.00	0.14 ± 0.00	0.10 ± 0.01	0.11 ± 0.01
UAA	0.4	2.11 ± 0.02	2.17 ± 0.01 **	2.41 ± 0.06	2.86 ± 0.11 **
Cys	0.3	0.25 ± 0.00	0.25 ± 0.00	0.16 ± 0.01	0.16 ± 0.01
Val	0.9	0.07 ± 0.00	0.07 ± 0.00	0.05 ± 0.00	0.05 ± 0.00
Met	1.4	0.07 ± 0.00	0.07 ± 0.00	0.06 ± 0.00	0.06 ± 0.00
Ile	0.7	0.23 ± 0.01	0.25 ± 0.01 *	0.20 ± 0.01	0.20 ± 0.01
Leu	0.9	0.19 ± 0.00	0.19 ± 0.01	0.15 ± 0.01 ***	0.05 ± 0.01
Tyr	0.5	0.09 ± 0.00	0.09 ± 0.00	0.08 ± 0.00	0.08 ± 0.01
Phe	0.2	0.08 ± 0.00	0.08 ± 0.00	0.07 ± 0.01	0.07 ± 0.01
Lys	0.1	0.01 ± 0.00	0.01 ± 0.00	0.01 ± 0.00	0.01 ± 0.00
His	0.2	0.01 ± 0.00	0.01 ± 0.00	0.01 ± 0.00	0.01 ± 0.00
TAA	0.5	0.14 ± 0.00	0.14 ± 0.00	0.15 ± 0.00 ***	0.10 ± 0.00

The results are presented as the means ± SE (n = 9). An asterisk (*) indicates significance at the 0.05 level, ** at the 0.01 level, and *** at the 0.001 level.

Table 6. Effects of dietary HMLS on the TAVs of amino acids and nucleotides in hepatopancreas.

AAs	Taste Threshold (mg/mL)	Hepatopancreas			
		Female		Male	
		Ctrl	HMLS	Ctrl	HMLS
Gly	1.3	10.30 ± 0.36	12.06 ± 0.33 **	7.58 ± 0.03	7.72 ± 0.02 **
Ala	0.6	0.01 ± 0.00	0.01 ± 0.00	0.01 ± 0.00	0.01 ± 0.00
Thr	2.6	0.09 ± 0.00	0.09 ± 0.00	0.07 ± 0.00	0.08 ± 0.00 **
Ser	1.5	0.01 ± 0.00	0.01 ± 0.00	0.01 ± 0.00	0.01 ± 0.00
Pro	3	0.12 ± 0.00	0.12 ± 0.00	0.10 ± 0.00	0.11 ± 0.00 **
Arg	0.5	0.15 ± 0.00	0.16 ± 0.00	0.14 ± 0.00	0.14 ± 0.00
SAA	0.5	0.01 ± 0.00	0.01 ± 0.00	0.01 ± 0.00	0.01 ± 0.00
Asp	0.2	0.47 ± 0.02	0.47 ± 0.01	0.46 ± 0.02 **	0.39 ± 0.01
Glu	0.2	0.09 ± 0.01	0.09 ± 0.01	0.09 ± 0.00	0.09 ± 0.00
UAA	0.4	2.45 ± 0.06	2.68 ± 0.08 *	2.17 ± 0.01	2.25 ± 0.01 ***
Cys	0.3	0.27 ± 0.01	0.27 ± 0.01	0.27 ± 0.01	0.27 ± 0.00
Val	0.9	0.12 ± 0.00	0.12 ± 0.00	0.10 ± 0.00	0.10 ± 0.00
Met	1.4	0.13 ± 0.00	0.13 ± 0.00	0.12 ± 0.00	0.12 ± 0.00
Ile	0.7	0.36 ± 0.01	0.41 ± 0.03	0.33 ± 0.00	0.36 ± 0.01 **

Table 6. *Cont.*

AAs	Taste Threshold (mg/mL)	Hepatopancreas			
		Female		Male	
		Ctrl	HMLS	Ctrl	HMLS
Leu	0.9	0.31 ± 0.01	0.31 ± 0.01	0.29 ± 0.00	0.30 ± 0.00 **
Tyr	0.5	0.15 ± 0.00	0.15 ± 0.00	0.13 ± 0.00	0.13 ± 0.00
Phe	0.2	0.11 ± 0.01	0.12 ± 0.01	0.10 ± 0.00	0.10 ± 0.00
Lys	0.1	0.05 ± 0.00 **	0.04 ± 0.00	0.02 ± 0.00	0.02 ± 0.00
His	0.2	0.01 ± 0.00	0.01 ± 0.00	0.00 ± 0.00	0.00 ± 0.00
TAA	0.5	0.31 ± 0.07	0.20 ± 0.02	0.04 ± 0.00	0.05 ± 0.01

The results are presented as the means ± SE (n = 9). An asterisk (*) indicates significance at the 0.05 level, ** at the 0.01 level, and *** at the 0.001 level.

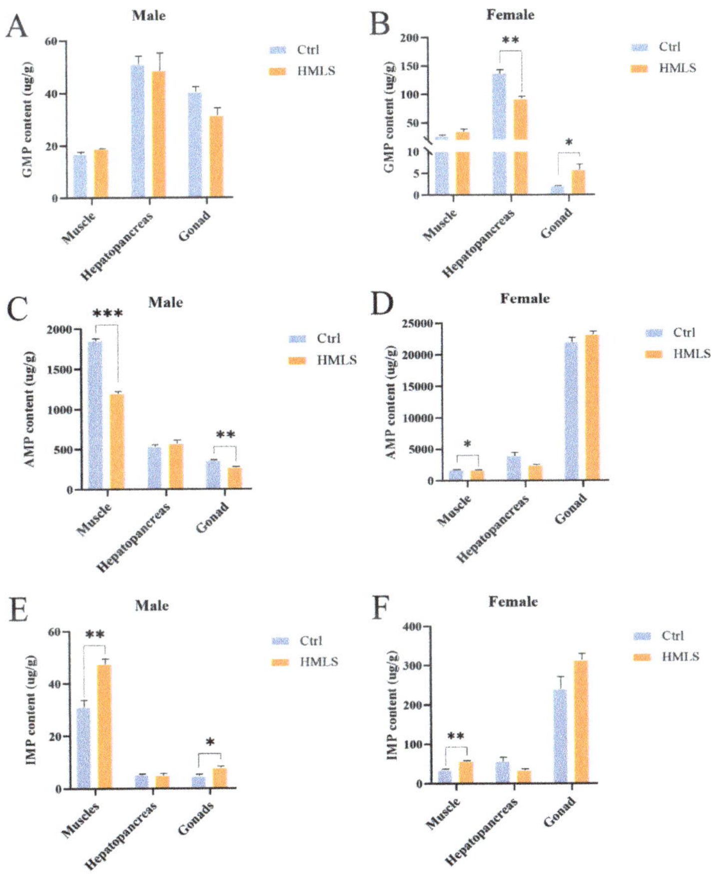

Figure 5. Effects of dietary HML on umami nucleotide content of edible tissues in adult *E. sinensis*. (**A**,**B**) GMP content. (**C**,**D**) AMP content. (**E**,**F**) IMP content. The results are presented as the means ± SE (n = 9). An asterisk (*) indicates significance at the 0.05 level, ** at the 0.01 level, and *** at the 0.001 level.

Table 7. Effects of dietary HMLS on the TAVs of amino acids and nucleotides in gonad.

AAs	Taste Threshold (mg/mL)	Gonad			
		Female		Male	
		Ctrl	HMLS	Ctrl	HMLS
Gly	1.3	4.60 ± 0.01	4.89 ± 0.03 **	5.06 ± 0.01 **	5.01 ± 0.01
Ala	0.6	0.02 ± 0.00	0.02 ± 0.00	0.01 ± 0.00	0.01 ± 0.00
Thr	2.6	0.07 ± 0.00	0.07 ± 0.00	0.06 ± 0.00	0.06 ± 0.00
Ser	1.5	0.01 ± 0.00	0.01 ± 0.00	0.01 ± 0.00	0.01 ± 0.00
Pro	3	0.13 ± 0.00	0.14 ± 0.00 **	0.12 ± 0.00	0.13 ± 0.00 *
Arg	0.5	0.28 ± 0.01	0.28 ± 0.00	0.23 ± 0.00	0.23 ± 0.00
SAA	0.5	0.00 ± 0.00	0.00 ± 0.00	0.01 ± 0.00	0.01 ± 0.00
Asp	0.2	0.45 ± 0.01	0.45 ± 0.01	0.34 ± 0.00	0.34 ± 0.00
Glu	0.2	0.13 ± 0.01	0.13 ± 0.01	0.10 ± 0.01	0.10 ± 0.00
UAA	0.4	2.35 ± 0.01	2.39 ± 0.01 **	2.31 ± 0.01	2.45 ± 0.05 **
Cys	0.3	0.26 ± 0.01	0.25 ± 0.00	0.22 ± 0.00	0.22 ± 0.00
Val	0.9	0.08 ± 0.00	0.08 ± 0.00	0.08 ± 0.00	0.08 ± 0.00
Met	1.4	0.09 ± 0.00	0.09 ± 0.00	0.09 ± 0.00	0.09 ± 0.00
Ile	0.7	0.30 ± 0.01	0.32 ± 0.00 **	0.27 ± 0.00	0.26 ± 0.00
Leu	0.9	0.25 ± 0.00	0.26 ± 0.00 **	0.22 ± 0.00	0.22 ± 0.00
Tyr	0.5	0.22 ± 0.00	0.22 ± 0.00	0.11 ± 0.00	0.11 ± 0.00
Phe	0.2	0.11 ± 0.01	0.11 ± 0.01	0.08 ± 0.00	0.09 ± 0.00
Lys	0.1	0.27 ± 0.01	0.28 ± 0.01	0.02 ± 0.00 *	0.01 ± 0.00
His	0.2	0.05 ± 0.01	0.06 ± 0.01	0.00 ± 0.00	0.00 ± 0.00
TAA	0.5	1.76 ± 0.10	1.86 ± 0.06	0.03 ± 0.00 **	0.02 ± 0.00

The results are presented as the means ± SE (n = 9). An asterisk (*) indicates significance at the 0.05 level, ** at the 0.01 level.

The SWT was utilized to quantify the sweetness imparted by SAAs in these tissues. As shown in Tables 8–10, the highest SWT was recorded in the female hepatopancreas, while the lowest SWT was found in the male muscle of both the Ctrl and HMLS groups. Following HML supplementation, SWT in the muscle and hepatopancreas of both sexes was significantly increased, whereas SWT in the male gonad was significantly decreased. These results indicated that dietary HML could significantly enhance the sweetness of the muscle and hepatopancreas.

Table 8. Effects of dietary HMLS on the SWT and EUC in muscle.

AAs	Muscle			
	Female		Male	
	Ctrl	HMLS	Ctrl	HMLS
EUC	1.80 ± 0.08 ***	0.51 ± 0.04	2.11 ± 0.12 **	0.34 ± 0.02
SWT	0.37 ± 0.01	0.39 ± 0.00 **	0.27 ± 0.01	0.31 ± 0.01 *

The results are presented as the means ± SE (n = 9). An asterisk (*) indicates significance at the 0.05 level, ** at the 0.01 level, and *** at the 0.001 level.

Table 9. Effects of dietary HMLS on the SWT and EUC in hepatopancreas.

AAs	Hepatopancreas			
	Female		Male	
	Ctrl	HMLS	Ctrl	HMLS
EUC	2.47 ± 0.38 *	1.60 ± 0.12	0.51 ± 0.03 *	0.43 ± 0.03
SWT	2.14 ± 0.07	2.51 ± 0.07 **	1.58 ± 0.01	1.61 ± 0.00 **

The results are presented as the means ± SE (n = 9). An asterisk (*) indicates significance at the 0.05 level, ** at the 0.01 level.

Table 10. Effects of dietary HMLS on the SWT and EUC in gonad.

AAs	Gonad			
	Female		Male	
	Ctrl	HMLS	Ctrl	HMLS
EUC	12.77 ± 0.86	13.56 ± 0.51	0.28 ± 0.02 *	0.23 ± 0.02
SWT	0.95 ± 0.00	1.02 ± 0.01 **	1.05 ± 0.00 **	1.04 ± 0.00

The results are presented as the means ± SE (n = 9). An asterisk (*) indicates significance at the 0.05 level, ** at the 0.01 level.

EUC is an indicator of umami taste to evaluate the synergistic effects of UAAs and umami nucleotides. The strongest umami taste was observed in the female gonad. In contrast, the EUCs of the male muscle and gonad were significantly decreased in the HMLS group compared with the Ctrl group. Furthermore, dietary HML significantly decreased Table 4.

3.8. Radar Plots and Correlation Analysis

The radar plots indicated that the EUC in the muscle of the HMLS group was significantly lower than that of the Ctrl group ($p < 0.05$), whereas the levels of IMP and MDA in the muscle of the HMLS group were notably higher ($p < 0.05$) (Figure 6A). Compared with the Ctrl group, the MDA value in the hepatopancreas of the HMLS group was the lowest ($p < 0.05$), and the T-AOC value in the male hepatopancreas of the HMLS group was the highest ($p < 0.05$). Furthermore, the values of AMP, IMP, GMP, and EUC were significantly lower in the HMLS F group compared to the Ctrl F group ($p < 0.05$) (Figure 6B). In the gonad, the MDA value of the HMLS group was significantly lower than that of the Ctrl group ($p < 0.05$), while the IMP and T-AOC values in the gonad of the HMLS group were significantly higher ($p < 0.05$) (Figure 6C).

In Pearson's correlation analysis, a correlation coefficient absolute value between 0.8 and 1.0 was considered to represent a strong correlation (Figure 6D). The SWT was positively correlated with Gly, Thr, Asp, Glu, Met, Ile, Leu, Tyr, Phe, Lys, and His, while it was negatively correlated with Ala, Pro, Arg, and Cys ($p > 0.05$). The EUC was positively correlated with Glu, Lys, His, Cys, GMP, IMP, and AMP, while it was negatively correlated with Asp ($p > 0.05$).

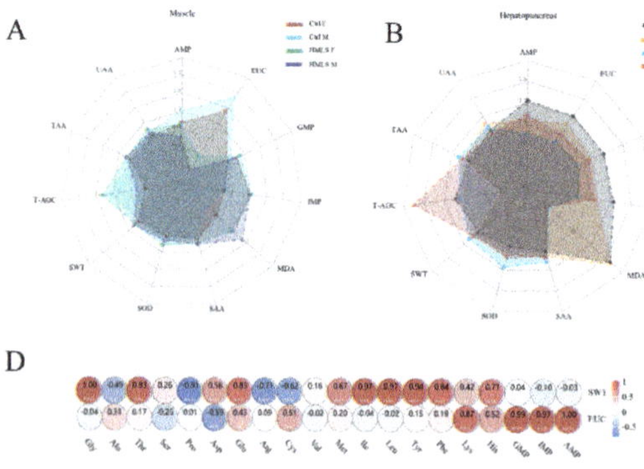

Figure 6. Radar plots and correlation analysis. (**A–C**) Radar plot of biomarker data of edible tissues (muscle, hepatopancreas, and gonad) in adult *E. sinensis*. (**D**) Correlation map of FAA/umami nucleotide content and flavor evaluation indices. The color scale indicates the correlation value, where vivid red indicates positive correlation, and vivid blue indicates negative correlation. ("*": $p < 0.05$; "**": $p < 0.01$; "***": $p < 0.001$).

4. Discussion

Currently, the demand for high-quality *E. sinensis* is increasing, and a key factor in improving quality is optimizing the fattening process [9]. Diet composition and water environmental conditions have been identified as contributing factors to the final fattening outcomes in various studies [33]. The total edible parts of *E. sinensis* include meat, the hepatopancreas, and gonads [34]. A previous study reported that black soldier fly larvae can enhance the flavor profile of edible tissues in *E. sinensis* when used to replace traditional iced trash fish in the diet [8]. This suggests that insect-based diets may offer a viable approach to improving the nutritional value of edible parts. The use of HML has gained popularity in fish aquaculture in recent years [14,21]; however, limited research has focused on the effects of dietary HML on *E. sinensis* culture. In this study, we investigated the effects of a diet supplemented with HML on growth performance, antioxidant activity, and nutritional quality in adult *E. sinensis* following 40 days of fattening.

Previous studies have reported on the yield of edible tissues in *E. sinensis* during the fattening period [2,6,35]. In this study, the TEY following dietary HML supplementation was composed of MY (male: 27.23%; female: 24.81%), HSI (male: 8.78%; female: 8.85%), and GSI (male: 3.44%; female: 10.77%), which was consistent with previous findings. Efficient nutrient feeding is crucial for achieving optimal fattening effects in *E. sinensis*. The existing research has shown no significant differences in the HSI and GSI levels of *E. sinensis* fed FDs versus IF [34]. In comparison with formulated diets, dietary supplementation with astaxanthin or *Haematococcus pluvialis* had no significant effects on the HSI and GSI [35,36]. Additionally, it was also found that dietary supplementation with 0.4% cholesterol significantly upregulated both the HSI and GSI levels but had no significant effects on the MY in *E. sinensis* [37,38]. Furthermore, supplementation with 0.33% docosahexaenoic acid (DHA) oil significantly increased the GSI in *E. sinensis* [39]. Our results suggest that HML supplementation in formulated diets effectively improved the HSI of *E. sinensis*. Similar to our results, higher HSI values were observed in *O. niloticus* fed with diets containing HM [40]. Studies have shown that HML extract reduces lipid accumulation in the hepatopancreas of rats by modulating the peroxisome proliferator-activated receptor gamma (PPARγ) [41]; however, the specific effect of HML on higher HSI values of *E. sinensis* warrants further investigation.

Astaxanthin, a carotenoid compound, offers various health benefits for crustaceans [9,42]. Currently, astaxanthin is widely used in *E. sinensis* culture to enhance immune function, antioxidant capacity, and coloration [43,44]. Our findings demonstrated that HML feeding significantly increased the astaxanthin content in the carapaces of both male and female *E. sinensis*, as well as in the female hepatopancreas and gonad. Previous studies have shown that diets incorporating insect meals (such as cricket, grasshopper, and mealworm) can elevate carotenoid levels in *Xiphophorus maculatus* (Gunther, 1866) [45]. However, there is limited research on the effects of dietary HML on the astaxanthin content in *E. sinensis*, and the underlying mechanisms warrant further investigation.

It has been demonstrated that HML exhibit significant antioxidant potential when incorporated into the diets of hybrid catfish (*Clarias gariepinus* ♀x *Heterobranchus longifilis* ♂) and (*Lithobates catesbeiana*) [14,46]. In contrast, the antioxidant capacity of prawn *Palaemon adspersus* was unaffected by the inclusion of HML in the diets (Rathke, 1837) [47]. HM have shown considerable promise in enhancing the antioxidant capacity of crustaceans by providing essential nutrients such as AAs, microelements, and chitin [11–13]. However, the effects of dietary HML on the antioxidant capacity of different edible tissues in *E. sinensis* remain underexplored. In the present study, we found that dietary HML enhanced the antioxidant capacity of the hepatopancreas and gonad in both male and female *E. sinensis*, as well as the female muscle.

Previous studies have demonstrated that a dietary insect meal has the potential to affect textural characteristics by altering the muscle fiber diameter and density [48]. Research has indicated that the functional components of black soldier fly (*Hermetia illucens*) larvae can improve the chewiness, cohesiveness, gumminess, and hardness of barramundi (*Lates calcarifer*) muscle [49]. Additionally, the inclusion of yellow mealworm (*Tenebrio molitor*) in the diet has been shown to affect muscle fiber density, which is positively correlated with muscle hardness, adhesiveness, springiness, chewiness, and gumminess [48]. Here, we observed that HML feeding enhanced the adhesiveness in the male muscle, while improving the cohesiveness, chewiness, and resilience in the female muscle. These changes can be attributed to alterations in muscle cellularity. However, the precise mechanisms by which HML-supplemented formula diets regulate muscle structure warrant further investigation.

Proteins are crucial energy sources for crustaceans, and AA composition serves as an important indicator for evaluating the nutritional value of *E. sinensis* [10]. HML are rich in AAs and are considered a more environmentally sustainable alternative to traditional protein sources such as soybean meal or fishmeal [17]. In comparison to the AA composition of soybean meal and fishmeal [14], HML in this study are particularly abundant in Leu, His, Ala, Phe, Asp, Tyr, and Cys. To date, there are limited studies exploring the effects of dietary HML on the AA composition of *E. sinensis*. However, research on other insect meals has reported that *H. illucens* is an efficient source of animal protein for crustaceans [50,51]. The substitution of IF by *H. illucens* has been shown to significantly increase the TAA content in the gonad and the DAA (SAAs and UAAs) content in the muscle and gonad of male *E. sinensis*, while significantly decreasing the TAA and DAA content in the hepatopancreas of female *E. sinensis* [8]. In contrast, our study found that dietary HML significantly increased the TAA and SAA content in the edible tissues of female *E. sinensis* and in the muscle and hepatopancreas of male *E. sinensis* but had no significant effects on the TAA content in the gonad of male *E. sinensis*. Furthermore, varying levels of dietary defatted superworm (*Zophobas atratus*) or *H. illucens* larvae as alternative protein sources did not affect the TAA content in juvenile Pacific white shrimp (*Penaeus vannamei*) [52,53]. Therefore, the effects of dietary HML on the FAA content in *E. sinensis* may be gender- and tissue-specific.

IMP serves as a taste enhancer in food, boosting sweetness while masking fishy, sour, bitter, and salty flavors [54]. A previous study demonstrated that dietary *H. illucens* significantly increases the IMP content in the muscle of both sexes and in the gonad of female *E. sinensis* [8]. Consistent with this finding, our study showed that dietary HML significantly increased the IMP content in the muscle of both sexes and in the gonads of male *E. sinensis*. However, dietary HML significantly decreased the AMP content in the muscle of both sexes and in the gonad of the male, as well as the GMP content in the hepatopancreas of the female. A previous study indicates that IMP and GMP are umami-tasting compounds commonly used as flavor enhancers, while AMP can be converted into IMP by the action of AMP deaminase [55]. Therefore, our results suggest that dietary HML in *E. sinensis* promote the conversion of AMP to IMP in the muscle of both sexes and the gonad of the male. Furthermore, proteins in the feed affect the composition of AAs and nucleotides in *E. sinensis*, which contribute to both the nutritive value and unique flavor [3,56,57]. Previous studies have highlighted that TAV serves as an indicator for evaluating the taste impacts of AAs and nucleotides on the flavor of *E. sinensis* [31,58]. The umami taste of IMP and GMP is considered stronger than that of glutamic acid (Glu) [54]. In the present study, the TAVs of UAAs, such as Glu and Asp, in the edible tissues of both the Ctrl and HMLS groups were below 1.0, with AMP primarily contributing to the umami taste in the female gonads (TAV greater than 1.0). Contrary to the results observed in *E. sinensis* fed *H. illucens* [8], IMP may be the primary contributor to the umami flavor in *E. sinensis*. Moreover, dietary HML for *E. sinensis* significantly decreased the EUC in the muscle and hepatopancreas of both sexes, as well as in the gonad of the male.

Previous studies have shown that certain SAAs, consisting of arginine (Arg), Ala, glycine (Gly), proline (Pro), and threonine (Thr), can have effects on the sweetness characteristics of edible tissues in *E. sinensis* [8]. However, limited research has focused on the impact of dietary HML on the sweetness or bitterness of edible tissues in aquatic species. Gly, for example, is known to provide a fragrant sweetness, reduce bitterness, and eliminate off-flavors in shrimp and crab [54]. In our study, the TAV of Gly was found to exceed 1.0 in all edible tissues. After HML feeding, the SWT of the muscle, hepatopancreas, and female gonad increased significantly, while the SWT in the male gonad decreased significantly. Similar changes in the SWT were observed in the muscle and male gonad of *E. sinensis* after feeding with *H. illucens* [8]. Additionally, valine (Val) is typically considered a bitter amino acid, characterized by weak bitterness and a subtle sweet taste [59]. A recent study suggested that a reduction in L-valine content in the muscle of *Megalobrama amblycephala* was significantly associated with increased sweetness [60]. However, in this study, the TAV of Val was greater than 1.0 across all edible tissues and showed a significant increase after HML feeding. Correlation analysis revealed no significant association between the Val content and either the EUC or SWT, suggesting that the increased Val content following HML feeding did not contribute to the sweetness in the edible tissues of *E. sinensis*.

5. Conclusions

In conclusion, this study demonstrated that dietary supplementation with HML significantly improved the HSI, astaxanthin content in female edible tissues, and flesh quality of *E. sinensis*. Moreover, HML feeding induced tissue- and sex-specific variations in the flavor characteristics. Specifically, dietary HML significantly reduced the EUC of the muscle, hepatopancreas, and male gonad, while significantly increasing the SWT of the muscle, hepatopancreas, and female gonad. Based on the current results, we have demonstrated that supplementation with HML is not only harmless to *E. sinensis* fattening but also significantly enhances the flavor characteristics of female crabs. Moving forward, we plan to focus on female crabs as the research subject, designing a multi-gradient HML supplementation to study its effects on edible tissues, and explore the balance between the growth of edible parts and the retention of umami.

Supplementary Materials: The following supporting information can be downloaded at: https://www.mdpi.com/article/10.3390/foods14071250/s1, Table S1: Nutritional composition of HML.

Author Contributions: Methodology, C.Q.; software, J.T.; investigation, C.Q.; resources, X.L.; data curation, T.Y.; writing—original draft, X.L.; writing—review and editing, F.H.; supervision, B.L. All authors have read and agreed to the published version of the manuscript.

Funding: This work was supported by the Major Project of Agricultural Product Standardization Production Demonstration County (ZJNY2022001-006).

Institutional Review Board Statement: The study was approved by the Committee of Laboratory Animal Experimentation at Zhejiang Academy of Agricultural Sciences (Permit Number: 2021ZA-ASLA24, approved on 26 February 2021).

Informed Consent Statement: Not applicable.

Data Availability Statement: The original contributions presented in this study are included in the article/Supplementary Materials, and further inquiries can be directed to the corresponding author.

Conflicts of Interest: The authors declare no conflicts of interest.

References

1. Zhang, L.; Zhang, R.Y.; Jiang, X.D.; Wu, X.G.; Wang, X.C. Dietary supplementation with synthetic astaxanthin and DHA interactively regulates physiological metabolism to improve the color and odor quality of ovaries in adult female *Eriocheir sinensis*. *Food Chem.* **2024**, *430*, 137020. [CrossRef] [PubMed]
2. Wang, S.H.; Wang, Y.Z.; Wu, X.G.; Zhang, X.B.; Zhao, J.S.; Yang, J.M.; Cheng, Y.X. Gonadal development and biochemical composition of Chinese mitten crabs (*Eriocheir sinensis*) from four sources. *J. Food Sci.* **2021**, *86*, 1066–1080. [CrossRef] [PubMed]
3. Ye, Y.H.; Wang, Y.L.; Liu, P.Y.; Chen, J.; Zhang, C.Z. Uncovering the nutritive profiles of adult male Chinese mitten crab (*Eriocheir sinensis*) harvested from the pond and natural water area of Qin lake based on metabolomics. *Foods* **2023**, *12*, 2178. [PubMed]
4. Wu, X.G.; Zhu, S.C.; Zhang, H.C.; Liu, M.M.; Wu, N.; Pan, J.; Luo, M.; Wang, X.C.; Cheng, Y.X. Fattening culture improves the gonadal development and nutritional quality of male Chinese mitten crab *Eriocheir sinensis*. *Aquaculture* **2020**, *518*, 734865. [CrossRef]
5. Yi, X.Y.; Gao, J.C.; Li, L.; Du, J.L.; Nie, Z.J.; Zhang, X.; Xu, G.C. Effects of fattening diets on the nutritional quality and flavor of the adult female Chinese mitten crab (*Eriocheir sinensis*). *Aquac. Rep.* **2022**, *25*, 101223. [CrossRef]
6. Long, X.W.; Guo, Q.; Wang, X.C.; Francis, D.S.; Cheng, Y.X.; Wu, X.G. Effects of fattening period on ovarian development and nutritional quality of adult female Chinese mitten crab *Eriocheir sinensis*. *Aquaculture* **2020**, *519*, 734748.
7. Feng, W.; Feng, W.; Ge, J.; Li, J.; Su, S.; Jia, R.; Yu, J.; Xu, P.; Tang, Y. Alterations of amino acid metabolism and intestinal microbiota in Chinese mitten crab (*Eriocheir sinensis*) fed on formulated diet and iced trash fish. *Comp. Biochem. Physiol. Part D Genom. Proteom.* **2021**, *40*, 100924.
8. Qiu, J.F.; Luo, C.; Ren, L.H.; Li, W.; Dai, T.M.; Wang, G.; Sun, X.N.; Moua, K.C.C.; Sima, Y.H.; Xu, S.Q. Black soldier fly larvae replace traditional iced trash fish diet to enhance the delicious flavor of Chinese mitten crab (*Eriocheir sinensis*). *Front. Mar. Sci.* **2023**, *9*, 1089421. [CrossRef]
9. Long, X.W.; Pan, J.; Wade, N.M.; Sun, Y.F.; Liu, Y.M.; Cheng, Y.X.; Wu, X.G. Effects of three feeding modes on the metabolism, antioxidant capacity, and metabolome of the adult male Chinese mitten crab *Eriocheir sinensis*. *Aquac. Int.* **2022**, *30*, 1101–1119. [CrossRef]
10. Shao, L.C.; Wang, C.; He, J.; Wu, X.G.; Cheng, Y.X. Meat quality of Chinese mitten crabs fattened with natural and formulated diets. *J. Aquat. Food Prod. Technol.* **2014**, *23*, 59–72.
11. Hua, K.; Cobcroft, J.M.; Cole, A.; Condon, K.; Jerry, D.R.; Mangott, A.; Praeger, C.; Vucko, M.J.; Zeng, C.S.; Zenger, K.; et al. The future of aquatic protein: Implications for protein sources in aquaculture diets. *One Earth* **2019**, *1*, 316–329.
12. Tran, H.Q.; Doan, H.V.; Stejskal, V. Environmental consequences of using insect meal as an ingredient in aquafeeds: A systematic view. *Rev. Aquac.* **2022**, *14*, 237–251.
13. Saleh, H.H.E. Review on using of housefly maggots (*Musca domestica*) in fish diets. *Zool. Res.* **2020**, *2*, 39–46.

14. Fawole, F.J.; Shamna, N.; Memudu, H.A.; Abdullahi, N.; Hassaan, M.S.; Gbadamosi, O.K. Housefly maggot meal complement soybean meal in a fish-free diet for hybrid catfish (*Clarias gariepinus* ♀× *Heterobranchus longifilis* ♂): Effect on growth, body composition, blood and antioxidant. *Anim. Feed Sci. Tech.* **2023**, *295*, 115543.
15. Alfiko, Y.; Xie, D.; Astuti, R.T.; Wong, J.; Wang, L. Insects as a feed ingredient for fish culture: Status and trends. *Aquac. Fish.* **2022**, *7*, 166–178.
16. Lin, Y.H.; Mui, J.J. Evaluation of dietary inclusion of housefly maggot (*Musca domestica*) meal on growth, fillet composition and physiological responses for barramundi, *Lates calcarifer*. *Aquac. Res.* **2017**, *48*, 2478–2485.
17. Mastoraki, M.; Katsika, L.; Enes, P.; Guerreiro, I.; Kotzamanis, Y.P.; Gasco, L.; Chatzifotis, S.; Antonopoulou, E. Insect meals in feeds for juvenile gilthead seabream (*Sparus aurata*): Effects on growth, blood chemistry, hepatic metabolic enzymes, body composition and nutrient utilization. *Aquaculture* **2022**, *561*, 738674.
18. Mastoraki, M.; Ferrándiz, P.M.; Vardali, S.C.; Kontodimas, D.C.; Kotzamanis, Y.P.; Gasco, L.; Chatzifotis, S.; Antonopoulou, E. A comparative study on the effect of fish meal substitution with three different insect meals on growth, body composition and metabolism of European sea bass (*Dicentrarchus labrax* L.). *Aquaculture* **2020**, *528*, 735511.
19. Maulu, S.; Langi, S.; Hasimuna, O.J.; Missinhoun, D.; Munganga, B.P.; Hampuwo, B.M.; Gabriel, N.N.; Elsabagh, M.; Van Doan, H.; Kari, Z.A.; et al. Recent advances in the utilization of insects as an ingredient in aquafeeds: A review. *Anim. Nutr.* **2022**, *11*, 334–349.
20. Wang, L.; Li, J.; Jin, J.N.; Zhu, F.; Roffeis, M.; Zhang, X.Z. A comprehensive evaluation of replacing fishmeal with housefly (*Musca domestica*) maggot meal in the diet of Nile tilapia (*Oreochromis niloticus*): Growth performance, flesh quality, innate immunity and water environment. *Aquac. Nutr.* **2017**, *23*, 983–993.
21. Ogunji, J.O.; Iheanacho, S.C.; Mgbabu, C.C.; Amaechi, N.C.; Evulobi, O.O.C. Housefly mggot meal as a potent bioresource for fish feed to facilitate early gonadal development in *Clarias gariepinus* (Burchell,1822). *Sustainability* **2021**, *13*, 921.
22. Xiang, J.H.; Qin, L.; Zhao, D.; Xiong, F.; Wang, G.T.; Zou, H.; Li, W.X.; Li, M.; Song, K.B.; Wu, S.G. Growth performance, immunity and intestinal microbiota of swamp eel (*Monopterus albus*) fed a diet supplemented with house fly larvae (*Musca domestica*). *Aquac.Nutr.* **2020**, *26*, 693–704.
23. Motte, C.; Rios, A.; Lefebvre, T.; Do, H.; Henry, M.; Jintasataporn, O. Replacing fish meal with defatted insect meal (yellow mealworm *Tenebrio molitor*) improves the growth and immunity of pacific white shrimp (*Litopenaeus vannamei*). *Animals* **2019**, *9*, 258. [CrossRef] [PubMed]
24. Foysal, M.J.; Fotedar, R.; Tay, C.Y.; Gupta, S.K. Dietary supplementation of black soldier fly (*Hermetica illucens*) meal modulates gut microbiota, innate immune response and health status of marron (*Cherax cainii*, Austin 2002) fed poultry-by-product and fishmeal-based diets. *Peer J.* **2019**, *7*, e6891.
25. Gasco, L.; Józefiak, A.; Henry, M. Beyond the protein concept: Health aspects of using edible insects on animals. *J. Insects Food Feed* **2021**, *7*, 715–741.
26. Su, F.; Xu, H.R.; Yang, N.; Liu, W.; Liu, J.G. Hydrolytic efficiency and isomerization during de-esterification of natural astaxanthin esters by saponification and enzymolysis. *Electron. J. Biotechnol.* **2018**, *34*, 37–42.
27. Shao, J.H.; Deng, Y.M.; Jia, N.; Li, R.R.; Cao, J.X.; Liu, D.Y.; Li, J.R. Low-field NMR determination of water distribution in meat batters with NaCl and polyphosphate addition. *Food Chem.* **2016**, *200*, 308–314.
28. Chang, G.L.; Jiang, S.; Song, Y.H.; Yang, L.; Zhao, H.T.; Ding, H.Y.; Xu, J.M.; Ma, A.M.; Cheng, Y.X.; Muhammad Laghari, Y. Impact of overwintering hoard on the edible tissues, muscle quality, hepatopancreas color, and proximate biochemical and amino acid compositions of male *Eriocheir sinensis*. *Aquac. Res.* **2023**, *2023*, 8143352.
29. Zhuang, K.; Wu, N.; Wang, X.; Wu, X.; Wang, S.; Long, X.; Wei, X. Effects of 3 feeding modes on the volatile and nonvolatile compounds in the edible tissues of female Chinese mitten crab (*Eriocheir sinensis*). *J. Food Sci.* **2016**, *81*, 968–981.
30. Scharbert, S.; Hofmann, T. Molecular definition of black tea taste by means of quantitative studies, taste reconstitution, and omission experiments. *J. Agric. Food Chem.* **2005**, *53*, 5377–5384.
31. Kong, Y.; Zhang, L.L.; Sun, Y.; Zhang, Y.Y.; Sun, B.G.; Chen, H.T. Determination of the free amino acid, organic acid, and nucleotide in commercial vinegars. *J. Food Sci.* **2017**, *82*, 1116–1123. [PubMed]
32. Jin, W.A.; Fan, X.R.; Jiang, C.Y.; Liu, Y.; Zhu, K.Y.; Miao, X.Q.; Jiang, P.F. Characterization of non-volatile and volatile flavor profiles of Coregonus peled meat cooked by different methods. *Food Chem. X* **2023**, *17*, 100584. [PubMed]
33. Czerniejewski, P.; Bienkiewicz, G.; Tokarczyk, G. Nutritional quality and fatty acids composition of invasive Chinese mitten crab from Odra Estuary (Baltic Basin). *Foods* **2023**, *12*, 3088. [CrossRef]
34. Shao, L.C.; Wang, C.; He, J.; Wu, X.G.; Cheng, Y.X. Hepatopancreas and gonad quality of Chinese mitten crabs fattened with natural and formulated Diets. *J. Food Qual.* **2013**, *36*, 217–227.
35. Jiang, X.D.; Wang, H.N.; Cheng, Y.X.; Wu, X.G. Growth performance, gonad development and nutritional composition of Chinese mitten crab *Eriocheir sinensis* selected for growth and different maturity time. *Aquaculture* **2020**, *523*, 735194.

36. Wu, X.; Zhao, L.; Long, X.; Liu, J.; Su, F.; Cheng, Y. Effects of dietary supplementation of *Haematococcus pluvialis* powder on gonadal development, coloration and antioxidant capacity of adult male Chinese mitten crab (*Eriocheir sinensis*). *Aquac. Res.* **2017**, *48*, 5214–5223.
37. Guo, H.; Wang, M.; Wang, X.; Xiao, K.; Huang, Y.; Hua, H.; Xiong, W.; Liu, W.; Abasubong, K.P.; Qiang, W.; et al. Effect of dietary cholesterol on ovarian development of Chinese mitten crabs (*Eriocheir sinensis*). *Front. Mar. Sci.* **2022**, *9*, 1070829.
38. Guo, H.X.; Jiang, G.Z.; Dai, Y.J.; Huang, Y.Y.; Wang, X.; Xiao, K.; Wang, M.M.; Abasubong, K.; Xu, Y.; Zhang, D.D.; et al. Effect of dietary cholesterol on growth performance, cholesterol deposition, and lipid metabolism in adult Chinese mitten crab (*Eriocheir sinensis*). *Aquac. Nutr.* **2022**, *2022*, 2012958.
39. Jiang, X.D.; Pan, K.W.; Yang, Y.H.; Shu, C.E.; Alexander, C.; Wu, X.G. Dietary DHA oil supplementation promotes ovarian development and astaxanthin deposition during the ovarian maturation of Chinese mitten crab *Eriocheir sinensis*. *Aquac. Nutr.* **2022**, *2022*, 9997317.
40. Ogunji, J.; Toor, R.U.A.S.; Schulz, C.; Kloas, W. Growth performance, nutrient utilization of nile tilapia *Oreochromis niloticus* fed housefly maggot meal (Magmeal) diets. *Turk. J. Fish. Aquat. Sc.* **2008**, *8*, 141–147.
41. Mei, H.; Xu, J.; He, Y.; Yang, X.; Liu, W.; Tian, W.; Zeng, Y.U.; Zhu, J. Protein-rich extract of Musca domestica larvae alleviated metabolic disorder in STZ-induced type 2 diabetic rat model via hepatoprotective and pancreatic β-cell protective activities. *J. Biosci.* **2018**, *43*, 969–983. [PubMed]
42. Li, Q.Q.; Zu, L.; Cheng, Y.X.; Wade, N.M.; Liu, J.G.; Wu, X.G. Carapace color affects carotenoid composition and nutritional quality of the Chinese mitten crab, *Eriochier sinensis*. *LWT-Food Sci. Technol.* **2020**, *126*, 109286.
43. Jiang, X.D.; Zu, L.; Wang, Z.Y.; Cheng, Y.X.; Yang, Y.H.; Wu, X.G. Micro-algal astaxanthin could improve the antioxidant capability, immunity and ammonia resistance of juvenile Chinese mitten crab, *Eriocheir sinensis*. *Fish Shellfish Immunol.* **2020**, *102*, 499–510.
44. Su, F.; Yu, W.J.; Liu, J.G. Comparison of effect of dietary supplementation with *Haematococcus pluvialis* powder and synthetic astaxanthin on carotenoid composition, concentration, esterification degree and astaxanthin isomers in ovaries, hepatopancreas, carapace, epithelium of adult female Chinese mitten crab (*Eriocheir sinensis*). *Aquaculture* **2020**, *523*, 735146.
45. Das, M. Effect of insect meals on the growth efficiency, survival and total carotenoid content of an ornamental fish *Xiphophorus maculatus* (Gunther, 1866). *Asian J. Fish. Aquat. Res.* **2023**, *22*, 1–9. [CrossRef]
46. Li, X.; Rahimnejad, S.; Wang, L.; Lu, K.L.; Song, K.; Zhang, C.X. Substituting fish meal with housefly (*Musca domestica*) maggot meal in diets for bullfrog Rana (*Lithobatescatesbeiana*): Effects on growth, digestive enzymes activity, antioxidant capacity and gut health. *Aquaculture* **2019**, *499*, 295–305. [CrossRef]
47. Mastoraki, M.; Vlahos, N.; Patsea, E.; Chatzifotis, S.; Mente, E.; Antonopoulou, E. The effect of insect meal as a feed ingredient on survival, growth, and metabolic and antioxidant response of juvenile prawn *Palaemon adspersus* (Rathke, 1837). *Aquac. Res.* **2020**, *51*, 3551–3562. [CrossRef]
48. Li, H.D.; Xue, R.R.; Sun, J.; Ji, H. Improving flesh quality of grass carp (*Ctenopharyngodon idellus*) by completely replacing dietary soybean meal with yellow mealworm (*Tenebrio molitor*). *Anim. Nutr.* **2023**, *12*, 375–387.
49. Chaklader, M.R.; Chung, W.H.; Howieson, J.; Fotedar, R. A mixture of full-fat and defatted *Hermetia illucens* larvae and poultry by-products as sustainable protein sources improved fillet quality traits in farmed barramundi, *Lates calcarifer*. *Foods* **2023**, *12*, 362. [CrossRef]
50. De León-Ramírez, J.J.; García-Trejo, J.F.; Sosa-Ferreyra, C.F.; Martínez-Ramos, S.A.; Bottini-Cedeño, B.S. Fly larvae (*Musca domestica*) as a protein alternative in the feeding of *Macrobrachium tenellum*. *Lat. Am. J. Aquat. Res.* **2018**, *46*, 599–603. [CrossRef]
51. Röthig, T.; Barth, A.; Tschirner, M.; Schubert, P.; Wenning, M.; Billion, A.; Wilke, T.; Vilcinskas, A. Insect feed in sustainable crustacean aquaculture. *J. Insects Food Feed* **2023**, *9*, 1115–1138.
52. Chang, T.; Lin, H.; Han, F.; Xu, C.; Li, E. Impact of defatted black soldier fly (*Hermetia illucens*) larvae meal on health, muscle texture, and intestinal microbiota in Pacific white shrimp (*Penaeus vannamei*). *Aquaculture* **2025**, *596*, 741755.
53. Lin, H.; Liang, X.; Han, F.; Luo, X.; Li, E. Growth, biochemical characteristics, flesh quality, and gut microbiota of the pacific white shrimp (*Penaeus vannamei*) fed a defatted superworm (*Zophobas atratus*) larvae meal. *Aquac. Nutr.* **2023**, *2023*, 8627246. [PubMed]
54. Tu, L.; Wu, X.; Wang, X.; Shi, W. Effects of fish oil replacement by blending vegetable oils in fattening diets on nonvolatile taste substances of swimming crab (*Portunus trituberculatus*). *J. Food Biochem.* **2020**, *44*, e13345. [CrossRef]
55. Paulino, B.N.; Sales, A.; de Oliveira Felipe, L.; Pastore, G.M.; Molina, G.; Bicas, J.L. Biotechnological production of non-volatile flavor compounds. *Curr. Opin. Food Sci.* **2021**, *41*, 26–35. [CrossRef]
56. Wang, S.; Guo, K.; Luo, L.; Zhang, R.; Xu, W.; Song, Y.; Zhao, Z. Fattening in saline and alkaline water improves the color, nutritional and taste quality of adult Chinese mitten crab *Eriocheir sinensis*. *Foods* **2022**, *11*, 2573. [CrossRef]
57. Zhang, B.Y.; Fang, W.H.; Zhu, R.; Wang, N.; Yao, Q.; Liu, H.J.; Wan, J.W.; Chen, Y.K.; Wang, Q.J.; Zhang, D.M. Comparative study on growth index and nutritional quality of female Chinese mitten crab *Eriocheir sinensis* selected at different growth periods in rice-crab culture systems. *Aquac. Nutr.* **2023**, *2023*, 4805919.

58. Chen, D.W.; Zhang, M.; Shrestha, S. Compositional characteristics and nutritional quality of Chinese mitten crab (*Eriocheir sinensis*). *Food Chem.* **2007**, *103*, 1343–1349. [CrossRef]
59. Bachmanov, A.A.; Bosak, N.P.; Glendinning, J.I.; Inoue, M.; Li, X.; Manita, S.; McCaughey, S.A.; Murata, Y.; Reed, D.R.; Tordoff, M.G.; et al. Genetics of amino acid taste and appetite. *Adv. Nutr.* **2016**, *7*, 806S–822S.
60. Wang, X.; Dong, Y.; Huang, Y.; Tian, H.; Zhao, H.; Wang, J.; Zhou, J.; Liu, W.; Cao, X.; Li, X.; et al. Docosahexaenoic acid-enriched diet improves the flesh quality of freshwater fish (*Megalobrama amblycephala*): Evaluation based on nutritional value, texture and flavor. *Food Chem.* **2024**, *460*, 140518. [CrossRef]

foods

MDPI

Article

Quality and Microbial Changes in Omega-3-Enriched Rabbit Meat Packaged with an Active Absorbent Pad in MAP

Marta Castrica [1], Michela Contò [2], Nour Elhouda Fehri [3,*], Giulio Curone [3], Claudia M. Balzaretti [3], Egon Andoni [4], Alda Quattrone [3], Daniele Vigo [3], Stella Agradi [5], Laura Menchetti [6], Olimpia Barbato [7], Dino Miraglia [7], Gabriele Brecchia [3] and Sebastiana Failla [2]

[1] Department of Comparative Biomedicine and Food Science, University of Padova, Agripolis, Viale dell'Università 16, 35020 Legnaro, Italy; marta.castrica@unipd.it
[2] Consiglio per la Ricerca in Agricoltura e l'Analisi Dell'Economia Agraria (CREA), Centro di Ricerca Zootecnia e Acquacoltura, Research Centre for Animal Production and Aquaculture, Via Salaria 31, 00015 Rome, Italy; michela.conto@crea.gov.it (M.C.); sebastiana.failla@crea.gov.it (S.F.)
[3] Department of Veterinary Medicine and Animal Sciences, University of Milan, Via dell'Università 6, 26900 Lodi, Italy; giulio.curone@unimi.it (G.C.); claudia.balzaretti@unimi.it (C.M.B.); alda.quattrone@unimi.it (A.Q.); daniele.vigo@unimi.it (D.V.); gabriele.brecchia@unimi.it (G.B.)
[4] Faculty of Veterinary Medicine, Agricultural University of Tirana, Kodër Kamëz, 1029 Tirana, Albania; eandoni@ubt.edu.al
[5] Department of Veterinary Sciences, University of Torino, Largo Paolo Braccini 2, 10095 Grugliasco, Italy; stella.agradi@unito.it
[6] School of Biosciences and Veterinary Medicine, University of Camerino, Via Circonvallazione 93/95, 62024 Matelica, Italy; laura.menchetti@unicam.it
[7] Department of Veterinary Medicine, University of Perugia, Via San Costanzo 4, 06126 Perugia, Italy; olimpia.barbato@unipg.it (O.B.); dino.miraglia@unipg.it (D.M.)
* Correspondence: nour.fehri@unimi.it

Abstract: This study evaluated the efficacy of an active absorbent pad (*a*PAD) in reducing microbial growth and enhancing the shelf life of rabbit meat stored in modified atmosphere packaging (MAP). Thigh muscles from 60 rabbits were used, divided into three dietary groups: a control group (CNT), a group supplemented with 5% extruded flaxseed (ELS5%), and a group with 3.5% extruded flaxseed and 0.2% *Padina pavonica* algae (LPP3.5%). Samples were packaged in MAP (70% O_2, 30% CO_2) with either a conventional pad (*n*PAD) or *a*PAD and analyzed at 1, 4, 7, 14, 21 days. Microbiological analysis revealed a significantly lower total viable count at 21 days in the ELS5%*a*PAD group. For coagulase-positive staphylococci, the CNT*a*PAD group showed lower microbial counts at both day 4 and day 21 ($p < 0.05$). *Enterobacteriaceae* reductions were observed at 24 h post packaging in both the CNT*a*PAD and LPP3.5%*a*PAD groups and at day 14 in ELS5%*a*PAD. Lipid oxidation (TBARS) was also lower in *a*PAD samples, particularly in LPP3.5%, which remained below 1.5 mg MDA/kg compared to >2.5 mg MDA/kg in *n*PAD ($p < 0.05$). Sensory attributes such as texture and color were better preserved with *a*PAD. These findings underscore the effectiveness of *a*PAD in MAP to control microbial growth, limit oxidation, and extend the shelf life of omega-3-enriched rabbit meat, providing a promising solution for functional meat product preservation.

Keywords: rabbit meat; active pad; microbial; lipid oxidation; linseed; *Padina pavonica*

1. Introduction

Rabbit meat production and consumption are significant in certain Mediterranean countries, where cultural and traditional factors strongly influence its demand [1–3].

Renowned for its nutritional profile, rabbit meat serves as a high-quality source of protein, characterized by low levels of cholesterol and saturated fats. Furthermore, it provides a rich source of polyunsaturated fatty acids, B vitamins, as well as minerals [4,5]. Its high digestibility and low caloric content make it especially appealing to health-conscious consumers seeking functional foods [4,6,7]. However, despite these nutritional advantages, cultural perceptions often limit its broader acceptance and consumption.

Rabbit meat also exhibits significant potential as a functional food through dietary enrichment with bioactive compounds [8–11]. Recent studies have explored the inclusion of nutraceuticals, such as flaxseed and algae-derived products, in rabbit diets. These enrichments have demonstrated substantial improvements in meat quality, particularly by enhancing the n-3 PUFA content [12–16]. The use of macroalgae in animal feed has also been shown to support gut health due to the presence of bioactive compounds with prebiotic and immunomodulatory properties [17–19]. The antioxidative capacity of macroalgae, attributed to their abundance of polyphenols and α-tocopherols, has shown promise in mitigating inflammatory and microbial challenges while preserving the sensory and nutritional quality of meat [20–22]. The lack of published data on its application in rabbit diets or its effects on meat quality and shelf life presents an opportunity for further investigation.

The shelf life of rabbit meat is critically influenced by biochemical and microbiological processes, notably lipid oxidation and microbial proliferation. During storage, fresh meat releases exudates, creating a favorable environment for microbial growth due to its high water activity and rich nutrient composition [23]. This environment significantly reduces the meat's shelf life and commercial value, contributing to substantial economic losses during the production, retail, and storage stages. Addressing the rapid oxidative degradation of PUFAs is essential, as it not only leads to rancidity but also adversely impacts the sensory and nutritional qualities of the meat [24–26]. Consequently, interventions aimed at prolonging shelf life are imperative for ensuring product quality and minimizing economic losses.

Modified atmosphere packaging (MAP) has emerged as a key technology to enhance the preservation of fresh meat. By replacing ambient air with tailored gas mixtures, MAP effectively slows microbial growth and oxidative processes [27,28]. In red meats, oxygen concentrations are crucial for maintaining a desirable appearance; however, for rabbit meat, which is a white meat, the esthetic requirement for oxygen is less critical [28]. Nonetheless, research indicates that oxygen levels of 60% or higher in MAP mixtures can yield favorable preservation outcomes for rabbit meat. In this context, Racewicz et al. [29] demonstrated that an oxygen concentration of 70% significantly reduced the *Enterobacteriaceae* population in rabbit meat samples after 21 days of storage. Moreover, several authors showed that the incorporation of carbon dioxide in the range of 20–40% further inhibits aerobic bacteria by extending microbial lag phases and increasing doubling times [29–31]. Such gas mixtures not only maintain the fresh appearance of meat but also significantly delay spoilage.

Innovative packaging technologies, such as active packaging systems, have further advanced the preservation of perishable foods [1,27,32]. These systems, defined by EU Regulation No. 450/2009, involve the integration of components that interact with the packaged food or its environment to extend shelf life and preserve quality.

Recent advancements focus on developing and enhancing absorbent pads with active and bio-based components to provide antimicrobial and antioxidant properties. Several researchers have conducted studies on this topic in recent years; Sun et al. [33] developed potassium-doped sodium alginate hydrogel pads that effectively absorbed exudates and inhibited spoilage in chilled pork, extending its shelf life by two days. Wang et al. [34] demonstrated that absorbent pads with levulinic acid and sodium dodecyl sulfate significantly reduced microbial loads and enhanced microbial diversity in ground beef. Jiang et al. [35] introduced bio-based pads from delignified wood fibers and polyvinyl alcohol that extended pork shelf life to over nine days while reducing environmental impact. Lastly, Liu et al. [36] highlighted the efficacy of pads infused with *Carum copticum* essential oil in extending the shelf life of chicken meat through antimicrobial and antioxidant activity. When combined with MAP, active absorbent pads can play a more significant role by not only absorbing liquids but also releasing antioxidant and antimicrobial agents [37]. This dual functionality enhances meat preservation while ensuring food safety, quality, and sustainability. Despite these advancements, research on packaging solutions specifically tailored for rabbit meat remains limited, likely due to its niche market and comparatively higher production costs.

Excessive drip loss is a critical parameter affecting the commercial value of rabbit meat. Beyond reducing yield, excessive exudation adversely impacts the texture, visual appeal, and overall shelf life of the product [23]. Active absorbent pads, by effectively managing moisture levels, help maintain product freshness and minimize microbial colonization [37]. Deteriorative microbial activity accelerates protein degradation and lipid rancidity, leading to a loss of nutritional properties and adverse impacts on organoleptic attributes such as discoloration, textural changes, off-flavors, and unpleasant odors [23].

The spoilage of rabbit meat is predominantly caused by psychrotrophic Gram-negative bacteria, such as *Pseudomonas* spp. [24], which thrive in aerobic conditions. These microorganisms contribute to protein degradation, lipid hydrolysis, and the production of volatile compounds, resulting in reduced consumer acceptability. Furthermore, lipid and protein oxidation are exacerbated by the reactive oxygen species (ROS) generated during storage. Oxidative protein modifications can alter primary structures and side chains, resulting in aggregation and gelation, thereby reducing the bioavailability of essential amino acids and diminishing the meat's nutritional value [38]. The dual role of oxygen supporting aerobic microbial growth while facilitating oxidative processes necessitates precise control over its levels within packaging systems to balance microbial inhibition with sensory preservation.

This study hypothesizes that the integration of an active absorbent pad within a MAP system could effectively reduce microbial growth and oxidative degradation, particularly in rabbit meat enriched with n-3 PUFAs through dietary supplementation with extruded flaxseed and *Padina pavonica* extract. While these supplementations in rabbit diet enhancements significantly improve the meat's nutritional profile, they may also increase susceptibility to oxidative damage, potentially limiting shelf life. Thus, this research aims to evaluate the efficacy of an active absorbent pad in mitigating microbial proliferation and prolonging the shelf life of PUFA-enriched rabbit meat stored under modified atmosphere conditions. By addressing these challenges, the study seeks to provide innovative strategies for preserving the quality and commercial viability of rabbit meat as well as exploring its potential as a functional food.

2. Materials and Methods

2.1. Meat Samples Preparation

The experimental trial took place at Azienda Agricola Brachino Patrizia, a commercial rabbit farm situated in Central Italy. This study was conducted as part of the PRIMA project "Omega Rabbit: food for health BenefIT", supported by funding from the European Union. The handling of animals during the trial adhered to Legislative Decree No. 146, which enforces Directive 98/58/EC.

After weaning at 35 days old, the rabbits were housed individually in standard cages (L × W × H: 75 × 38 × 25 cm) under controlled environmental conditions. They were randomly assigned to one of three experimental groups which received a specific pellet diet. The isoenergetic and isoprotein diets were briefly formulated as follows: CNT = a control diet; ELS5% = CNT diet with 5% extruded flaxseed; and LPP3.5% = CNT diet containing 3.5% extruded flaxseed and 0.2% *Padina pavonica* algae extract, as detailed in Fehri et al. [39] Fresh water was available at all times. Rabbits were slaughtered at 85 days of age, and 20 carcasses chosen randomly for each diet were used.

Both sides' hind legs (HLs) were aseptically excised from the carcasses of rabbits in each experimental group and transported under cold chain conditions to the laboratory of CREA. Upon arrival, the two HLs for each animal were deboned and sectioned into slices, which were subsequently packaged in a modified atmosphere containing 70% O_2 and 30% CO_2. This high-oxygen composition is commonly used for MAP preservation of rabbit meat [29]. The packaging consisted of polystyrene trays, with the base fitted with either a control pad (*n*PAD), made of non-woven fabric designed only to absorb liquids, or an active pad (*a*PAD). This configuration was applied across all three diet groups: CNT, ELS5%, and LPP3.5%.

2.2. Active Absorbent Pad

The *a*PAD (trade name: "Bacteria Catcher") employed in this study was designed and supplied by ANT Advanced Nonwovens Technologies s.r.l., a company that is part of the Deatex group based in Milan, Italy. The *a*PAD used in this trial is identical to the one reported by Castrica et al. [37]. Briefly, the *a*PAD, measuring 7.5 × 13.5 cm, comprises two main components: an absorbent section, constructed from non-woven fabric and anchored to the base of the packaging, and an active section impregnated with an additive mixture. This mixture consists of 30–50% polymeric cationic agents by weight, 10–20% base by weight, and 1–10% auxiliary substances by weight.

The active section of the *a*PAD is designed for direct contact with rabbit meat and exerts an attraction effect on bacterial cell walls. The *a*PAD analyzed in this investigation complies with Regulation (EC) No. 1935/2004 concerning materials and articles intended for food contact. Additionally, it does not fall under the classification of a biocidal product as defined by Regulation (EC) No. 528/2012, as its mechanism of action is strictly physical and mechanical.

2.3. Experimental Design

Then, HLs for each diet group were packaged with the *a*PAD or with the *n*PAD. Five slices obtained for each animal were packaged in the same way, and one slice was used for each experimental time.

All the packaged slices were stored at 4 ± 1 °C and analyzed at specific time points: 24 h post packaging (D1), after 4 (D4), 7 (D7), 14 (D14), and 21 days (D21) for a total of 300 samples.

Packages containing HL slices were assigned for analyses, as detailed in the experimental design, shown in Figure 1, dividing each sample into two parts: one for microbiological shelf life and the other for chemical and sensory determinations.

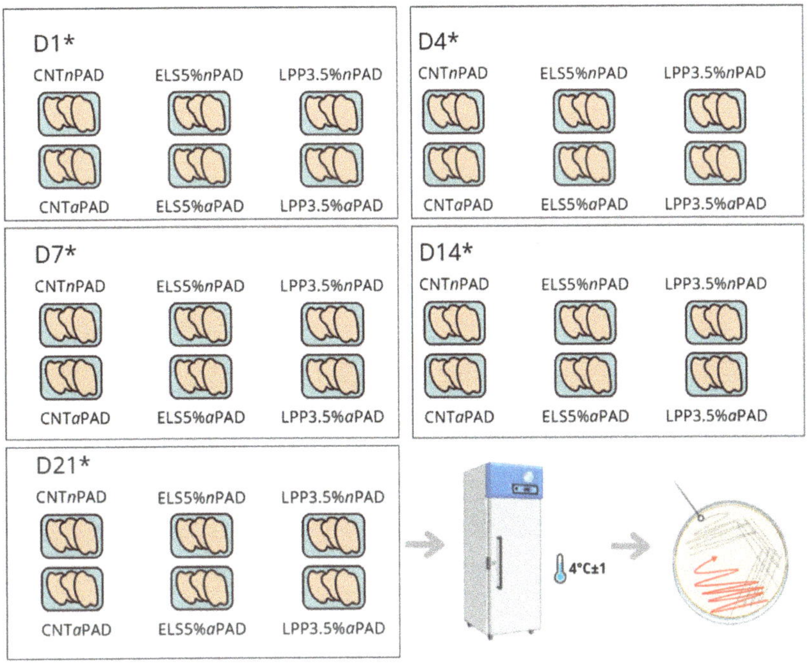

Figure 1. Microbiological and chemical shelf life experimental protocol. * For each analysis time point, 10 samples per group were analyzed.

2.4. Water-Holding Capacity (WHC) and pH

The slices were weighed before packaging and then reweighed at the established time to assess the water-holding capacity (WHC), determined by the loss of liquids during storage via the following equation: $(W0 - WT_{1,4,7,14,21})/W0 \times 100$, where W0 = weight pre-packaging and WT = weight at different storage times.

At each time point, the portion of the sample designated for chemical and sensory analysis was ground in an ice bath to ensure sample uniformity (Figure 2).

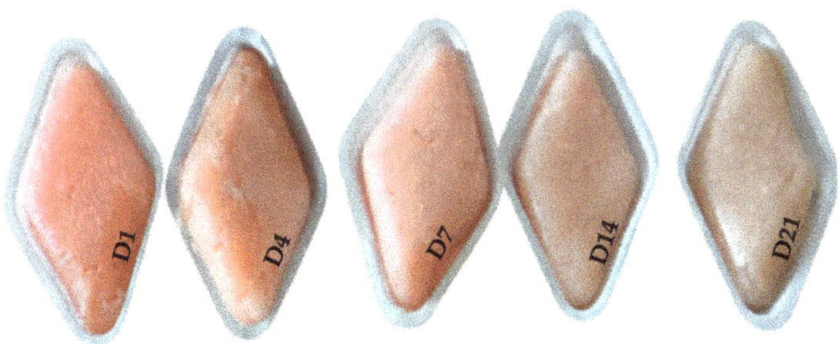

Figure 2. Minced fore leg meat in different storage times for sensorial test.

pH was measured using a pH meter with temperature compensation (XS Instrument Serie80 PC80, Giorgio Bormac s.r.l., Carpi, MO, Italy). One gram of meat was homogenized with 10 mL of NaCl 0.9%, and for each sample, three measurements were performed, and the final value was obtained as the mean.

2.5. Color

Color analyses were performed by maintaining minced meat in air for 30 min to allow for the blooming effect, and color parameters were recorded using the CIELAB system [40] to evaluate lightness (L*), redness (a*), and yellowness (b*), with D illuminant (6504 °K, daylight) using a Konica Minolta CM-3600 D (Sensing, Inc., Osaka, Japan) spectrophotometer. Chrome (C) and hue (H) were calculated using a* and b* indexes with the following equations:

$$C = \sqrt{a^2 + b^2};\ H = \mathrm{tang}^{-1}\, b^* / a^* \quad (1)$$

2.6. Microbiological Analysis

Total viable counts (TVCs) for *Enterobacteriaceae, Escherichia coli,* and coagulase-positive staphylococci were quantified using Petrifilm (3M, St. Paul, MN, USA). *Pseudomonas* spp. were cultured on Pseudomonas Agar Base (Biolife Italiana s.r.l., Milan, Italy) with CFC Pseudomonas Supplement (Biolife Italiana s.r.l., Milan, Italy) and incubated at 25 °C for 48 h. *Brocothrix thermosphacta* was grown on STAA Agar Base (Biolife Italiana s.r.l., Milan, Italy) supplemented with STAA Selective Supplement (Biolife Italiana s.r.l., Milan, Italy) and incubated at 22 °C for 48 h.

All analyses were performed in duplicate, and results were expressed as Log CFU/g. Additionally, the detection of *Salmonella* spp. and *Listeria monocytogenes* (analytical unit: 25 g) was carried out only at D1, in accordance with UNI EN ISO 6579-1:2017 [41] and AFNOR [42] BRD 07/05-09/01 standards, with results reported as either presence or absence in 25 g of the sample.

2.7. Lipid Oxidation and Carbonyl and Sulfhydryl Analysis

2.7.1. Thiobarbituric Acid Reactive Substance Assay (TBARS)

Lipid oxidation was quantified using the TBARS assay, with malondialdehyde (MDA) as the reference compound. Briefly, 2.5 g of meat was homogenized with water and 2.8% of ethanolic butylated hydroxytoluene (BHT) as an antioxidant. Subsequently, 1 mL of the homogenate was mixed with 1 mL of trichloroacetic acid (TCA) and centrifuged. The resulting supernatant was incubated at 80 °C for 30 min with 0.28% thiobarbituric acid (TBA) to form the MDA-TBA adduct. After cooling on ice, 10 μL of the solution was injected into an HPLC system (Alliance 2695, Waters Corporation, Framingham, MA, USA) equipped with a C18 reverse-phase column (Kinetex 5 μm EVO, Phenomenex, Torrance, CA, USA). Detection was carried out via fluorescence at λex = 515 nm and λem = 543 nm. The MDA-TBA sample peak was identified by comparison with an MDA standard peak. The TBARS concentration was expressed as mg of MDA/kg of meat, following the method described by Cifuni et al. [43].

2.7.2. Sulfhydryl and Carbonyl Content

Protein oxidation was investigated in terms of changes in the sulfhydryl group and carbonyl (the CO group) content in meat. The procedures were performed as reported by Valerio et al. [44].

For the SH group analysis, 1 g of meat was homogenized with 16 mL of 20 mM potassium phosphate (pH 6.0), filtered through a gauze to remove collagen, and centrifuged. The pellet was washed twice with 16 mL of 50 mM potassium chloride, and after a 1:50 dilution, with 20 mM potassium phosphate (pH 6.0), two aliquots were used. One aliquot was mixed with 8M urea in 100 mM buffer phosphate and 10 mM 2,2-ditio bis 5-nitropiridina (DTNP), and the other was mixed only with 8M urea in 100 mM buffer phosphate, which represented the blank. Samples were incubated for 1 h in the dark, and the absorbance was measured at 386 nm with a PerkinElmer spectrophotometer Lambda 25 (PerkinElmer, Shelton, CT, USA). The sulfhydryls were expressed in nmol of SH/mg of protein.

The carbonyls were quantified from the filtrate sample obtained from 1 g of meat for sulfhydryl analysis before the dilution phase. Two aliquots of 1 mL were used. After centrifugation, one pellet was used as the blank, mixed with 1 mL of 2 N HCl, and the other pellet was treated with 1 mL of 0.2% dinitro-phenylhydrazine (DNPH) in HCl 2 N (*w*/*v*). Samples were incubated for 1 h at room temperature, and after centrifugation, the pellets were collected. The pellets were washed and centrifuged three times with ethanol:ethyl acetate (1:1 *v*/*v*) to remove the DTNP traces and solubilized lipid. After the washes, the pellets were solubilized in 6M guanidine HCl and then incubated for 1 h at 90 °C. The blank samples were used to calculate the protein concentration in the sample by measuring the absorbance at 280 nm with the PerkinElmer spectrophotometer Lambda 25 (PerkinElmer, Shelton, CT, USA), using BSA as the standard. The samples treated with DNPH were read at 370 nm. The carbonyls were expressed in nmol of DNPH/mg of protein.

2.8. Sensory Tests

A sensory test was conducted by 10 semi-trained panelists. In each panel session, panelists evaluated the color, odor, and overall acceptability using a hedonic scale from 0 to 10 (0 = dislike for color/odor/overall acceptability; 10 = excellent for color/odor/overall judgment) for each experimental diet (CNT, ELS 5%, and LPP 3.5%), PAD type, and storage time (D1–D21). The samples were presented as minced raw meat in white sample holders (Figure 2), and they were evaluated in an isolated room under artificial lighting. During each panel session, one sample from each storage time belonging to the same diet but with two different PADs was presented simultaneously. Each day, three panel sessions were conducted, one for each experimental diet. The sensory test was carried out over ten consecutive days, with three sessions held each day.

2.9. Statistical Analysis

The statistical analysis in this study was performed using the PROC MIXED procedure in SAS/STAT Software Version 9.4 (SAS Institute Inc., Cary, NC, USA). The model included the feed group, PAD type, and storage time as fixed effects, while the animal was included as a random effect. For the sensory analysis, the panelist and session analysis were considered as a random effect to account for variability among individual assessors. All possible two-way and three-way interactions between the fixed effects were tested and included in the model if they were statistically significant. Post hoc comparisons of the means were carried out using Tukey's test, with a significance level set at $p < 0.05$. To ensure the reliability and validity of the sensory test, a comprehensive analysis of the distribution of sensory scores was conducted. Both distribution analysis and box plot analysis were employed for each sensory attribute to detect potential outliers and assess the homogeneity of scores among panelists. Descriptive statistics, including mean, median, standard deviation, and range, were calculated to evaluate the central tendency and the dispersion of sensory scores. In addition, histograms were generated for each sensory attribute to visualize the distribution and assess the normality of the data. The results of these analyses are reported in Figures S1 and S2 and Tables S1 and S2, which provide a graphical representation of the score distributions for each sensory parameter.

3. Results and Discussion

3.1. pH and WHC

The pH and WHC of rabbit meat during storage are critical indicators of its quality and shelf life. Their interaction provides valuable insights into the physiological and biochemical changes occurring in the meat and helps us understand the impact of storage conditions and dietary supplementation [26].

The average pH across the three feed groups was 5.97 ± 0.02 as mean ± standard error (SE), with no significant differences detected among them. Relatively high pH values around 6.0 did not appear to pose issues for the shelf life of rabbit meat, as reported by Pereira et al. [24]. However, when considering the overall effect of the PADs on pH, the *n*PAD group demonstrated a significantly lower value compared to the *a*PAD group. Notably, this difference was specifically observed in the CNT group, as highlighted in Table 1, which presents the interaction effects among the feed groups (CNT, ELS 5%, and LPP 3.5%) and the two PAD conditions on physical parameters. This difference suggests that *a*PAD effectively minimizes pH decline, potentially reducing microbial activity responsible for producing acidic metabolites [37]. For pH, the CNT*n*PAD group exhibited a significantly lower value compared to the CNT*a*PAD group (5.95 ± 0.02 vs. 6.00 ± 0.02, respectively; $p = 0.028$).

Table 1. Physical parameters of rabbit meat affected by different diets and PADs in different stored times.

	CNT		ELS5%		LPP3.5%				
	***n*PAD**	***a*PAD**	***n*PAD**	***a*PAD**	***n*PAD**	***a*PAD**	**p Diets**	**p Pad**	**RMSE**
pH	5.93 ^{a}	6.07 ^{b}	5.98	5.97	5.96	5.97	0.423	0.028	0.14
WHC%	3.01	2.88	2.79	2.72	3.08 ^{b}	2.84 ^{a}	<0.001	<0.001	0.31
L*	60.99 ^{a}	60.56 ^{a}	61.38 ^{ab}	61.85 ^{ab}	61.73 ^{b}	62.16 ^{b}	0.038	0.670	2.07
a*	1.68 ^{b}	1.71 ^{b}	1.54 ^{ab}	1.58 ^{ab}	1.29 ^{a}	0.90 ^{a}	0.041	0.495	0.54
b*	5.22	4.67	5.31	5.44	5.04	4.84	0.121	0.339	1.19
C	5.59	5.12	5.63	5.80	5.31	5.01	0.094	0.380	1.54
H	72.20	69.92	73.88	73.84	75.68	79.50	0.371	0.637	11.52

CNT = control diet; ELS5% = CNT diet with 5% extruded flaxseed; LPP3.5% = CNT diet containing 3.5% extruded flaxseed and 0.2% *Padina pavonica* algae extract; *n*PAD= control pad; *a*PAD = active pad; WHC = water-holding capacity; L* = lightness; a* = redness index; b* = yellowness index; C = chrome; H = hue; RMSE = root square error; ^{a,b} = different letters in the same row indicate a significant difference for $p < 0.05$; data are expressed as the mean.

Regarding WHC, as observed for pH, the PAD effect was evident ($p < 0.001$); the *a*PAD group generally exhibited lower values compared to the *n*PAD group (2.81 ± 0.03% vs. 2.96 ± 0.03%). The water loss was positively influenced by the use of the *a*PAD, likely due to its ability to minimize liquid loss [37]. Among feed groups, ELS5% showed significantly lower values compared to the other two ($p < 0.001$). With respect to the interaction between feed and pad types, significant differences were observed exclusively within the LPP3.5% group, where the *n*PAD group demonstrated higher water loss compared to the *a*PAD group (3.08 ± 0.05% vs. 2.84 ± 0.05%; $p < 0.001$).

pH and WHC were significantly influenced by the storage time (Table 2). The pH values exhibited a significant increase at the end of the storage period, rising from 5.96 on D1 to 6.05 on D21 ($p = 0.002$). This significant increase at 21 days is consistent with proteolytic activity typically exacerbated during extended storage periods, resulting in the release of alkaline compounds such as ammonia and amines [38]. The WHC constantly increased at each time point ($p < 0.001$), particularly from D1 to D4 (+1.33%). Protein denaturation over time during aging is closely linked to an increase in liquid loss [25,45,46], further emphasizing the importance of managing these parameters to preserve the quality and shelf life of rabbit meat.

Table 2. Effect of different storage times and PADs on the physical characteristics of rabbit meat.

	D1	D4	D7	D14	D21	*p* Value	RMSE
pH	5.96 ^{a}	5.93 ^{a}	5.98 ^{a}	5.97 ^{a}	6.05 ^{b}	0.002	0.14
WHC%	1.38 ^{a}	2.71 ^{b}	2.89 ^{c}	3.47 ^{d}	3.98 ^{e}	<0.001	0.31
L*	59.48 ^{a}	61.09 ^{b}	61.99 ^{bc}	62.93 ^{c}	61.73 ^{bc}	<0.001	2.07
a*	1.82	1.06	1.31	1.49	1.42	0.122	1.04
b*	4.21 ^{a}	4.68 ^{ab}	5.17 ^{b}	5.77 ^{c}	5.52 ^{bc}	<0.001	1.19
C	4.83 ^{a}	4.91 ^{a}	5.41 ^{ab}	6.12 ^{b}	5.78 ^{ab}	<0.001	1.54
H	71.08	73.87	77.59	73.92	74.44	0.171	11.52

D1, 4, 7, 14, and 21 = times of storage expressed in days; WHC = water-holding capacity; L* = lightness; a* = redness index; b* = yellowness index; C = chrome; H = hue. RMSE = root square error; $^{a–e}$ = different letters in the same row indicate a significant difference for $p < 0.05$; data are expressed as the mean.

3.2. Color

Meat color is one of the primary sensory parameters influencing consumer choice [1,2]. Primarily governed by myoglobin, meat color is subject to changes driven by the auto-oxidation of this heme protein as well as interactions with lipid and protein oxidation processes [29,38,46].

The color, in general, was not influenced by dietary group, showing no significant differences for most parameters, except for L* ($p = 0.038$) and a* ($p = 0.041$). Specifically, the CNT group, compared to the LPP3.5% group, exhibited lower L* (60.77 ± 0.33 vs. 61.94 ± 0.33) and higher a* (1.69 ± 0.14 vs. 1.09 ± 0.14, respectively), while the ELS5% group displayed intermediate values. Regarding the color coordinates, no significant differences were observed overall in relation to the presence or absence of an active PAD. Notably, the effect of *a*PAD was not evident within the different dietary groups. During storage, fluctuations were observed in most color parameters, except for a* and H, which remained significantly unaffected by storage times. The other color parameters (L*, b*, and C) increased up to D14, likely due to enzymatic proteolytic processes that temporarily enhance the meat's appearance by improving light reflection [46,47]. The subsequent significant decline on D21 ($p < 0.001$), indicating the onset of degradative oxidative processes. The gradual loss of color, transitioning to gray-brown, is associated with the degradation of structural and cytoplasmic proteins, including myoglobin, which is responsible for the characteristic color of fresh meat [47–49]. In rabbit meat, lipid and protein oxidation has been reported to contribute to discoloration, demonstrating a close relationship between oxidative processes and chromatic changes [9,29,50].

The trends observed in pH, WHC, and color highlight their interdependence. Higher pH and WHC values, as observed with the active absorbent pad, are associated with improved color, likely due to reduced oxidative and proteolytic degradation [37,38,46,51]. This oxidative protection is also evident, albeit to a lesser extent, in animals fed a diet supplemented with *Padina pavonica*. The presence of polyphenols and active compounds in this diet likely acted as scavengers, enhancing the oxidative state of the animal and, consequently, the meat quality [19,22].

3.3. Microbiological Profile

After 21 days of shelf life, the storage time had a significant effect on all evaluated microorganisms, with an overall increase in bacterial counts observed over time ($p < 0.005$).

Moreover, significant differences at 21 days of storage were observed for coagulase-positive staphylococci, where the control group with the active PAD exhibited lower microbial loads compared to the CNT group with the *n*PAD ($p < 0.001$). Notably, the slowdown, likely due to the effect of the *a*PAD, was already evident by day 4 (Figure 3). A similar trend was observed for the total viable count, where the ELS5% *a*PAD group exhibited lower counts compared to the ELS5%*n*PAD group at day 21. This finding aligns with the observations reported by Cullere et al. [1], where the group fed a diet supplemented with flax exhibited a reduction in the total bacterial count.

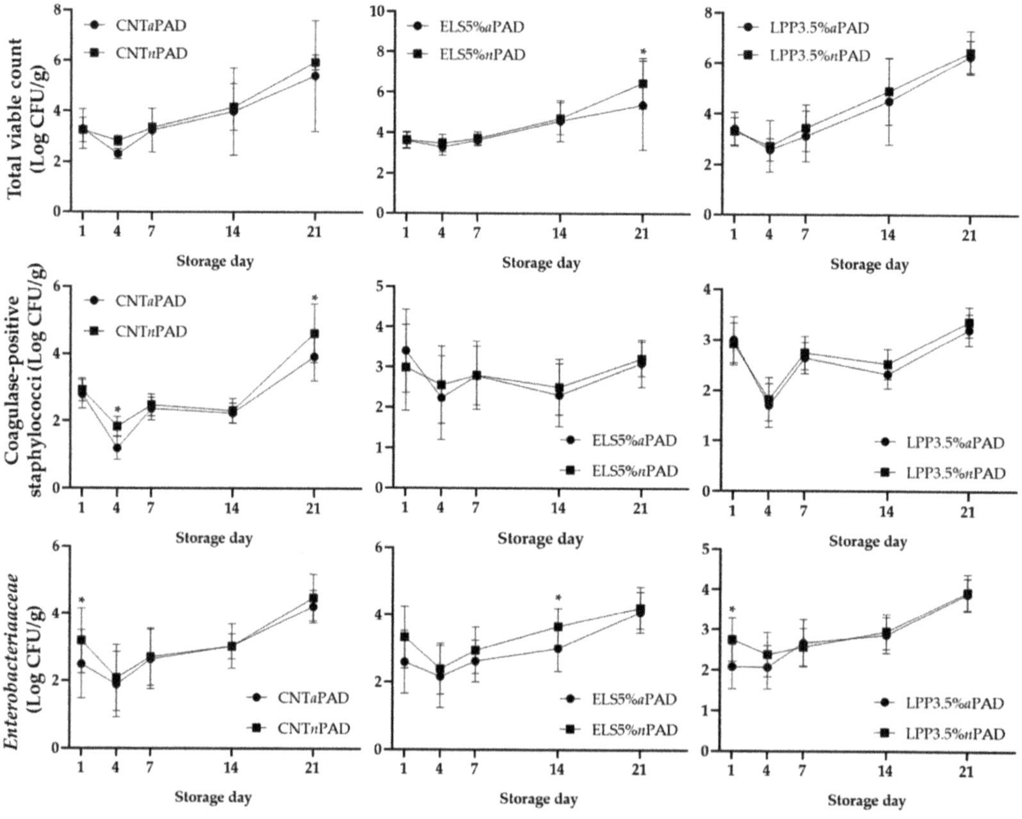

Figure 3. Trend of microbial development across storage days. The bar graphs show mean ± standard deviation. *: $p < 0.05$.

For *Enterobacteriaceae*, lower microbial loads were already evident at D14 in the ELS5% *a*PAD group compared to the ELS5% *n*PAD group ($p < 0.001$). Furthermore, for this microorganism, differences became apparent as early as 24 h of storage in both the CNT and LPP3.5% groups, with packages containing *n*PAD exhibiting higher values compared to those with *a*PAD. These differences, observed even in the early days of shelf life, could be attributed to the hygienic conditions in the slaughterhouse and during processing [52]. The variations identified from day 14 onward can be linked to the selective effects of different factors, including temperature, pH, and packaging atmosphere, on the predominant bacterial populations and spoilage dynamics during storage [1,53]. In the present study, the average pH values recorded across all experimental groups remained at or below 6.

Such low pH levels are recognized for their bacteriostatic properties, which help maintain microbial equilibrium and contribute to the preservation of meat quality [54]. Finally, no significant effect was observed against *Brochothrix thermosphacta*, *Pseudomonas* spp., and *E. coli* throughout the storage days (Figure 4). The findings of this study regarding the antimicrobial pad align with those of Komodromos et al. [55] and Fernandez et al. [56] who observed no statistically significant impact of the antimicrobial pad on the counts of *Pseudomonas* spp. and *B. thermosphacta*. A similar conclusion can also be drawn for *E. coli* in the present investigation. Moreover, when evaluating the results on the antimicrobial efficacy of the pad, it is essential to consider the substantial differences between the pads tested in this study and those reported in the literature [33–36,55,56]. These differences, primarily influenced by variations in composition, may result in significantly distinct actions depending on the bacterial species targeted or the specific parameter under investigation.

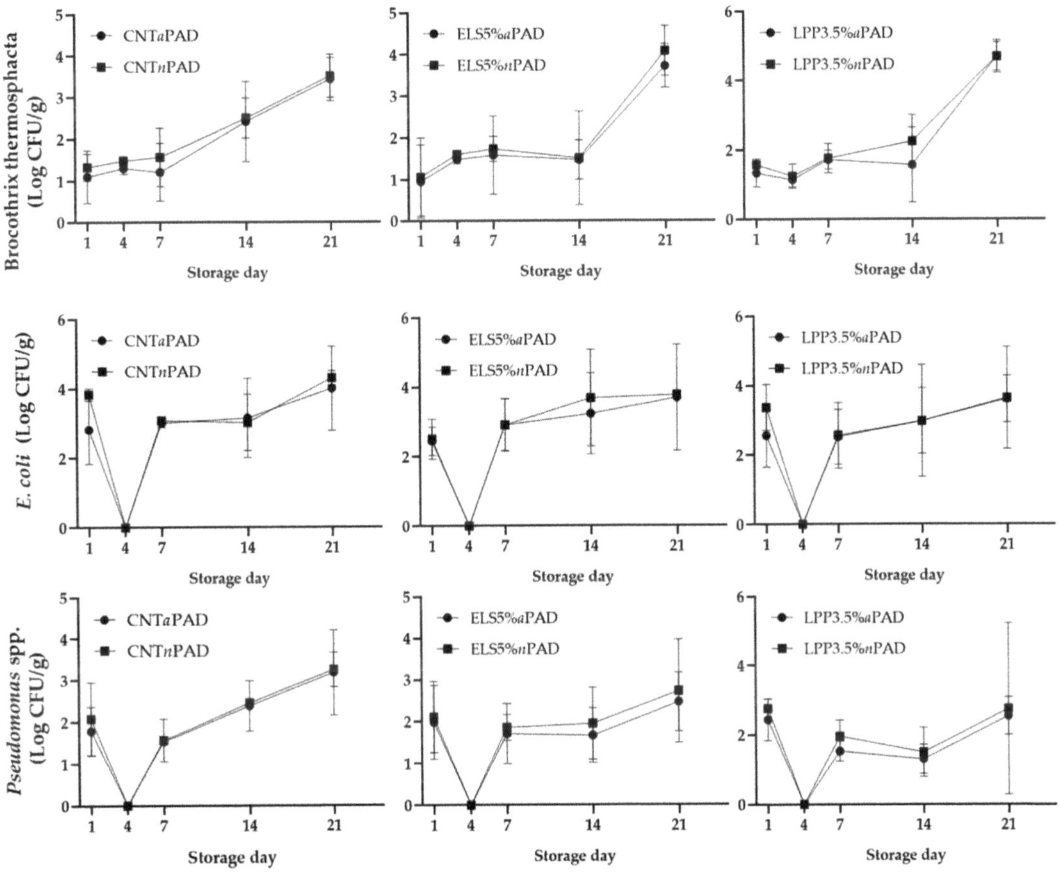

Figure 4. Trend of microbial development across storage days. The bar graphs show mean ± standard deviation. $p < 0.05$.

All samples were negative for *Salmonella* spp. and *L. monocytogenes*. The present study investigated the interplay between diet and packaging method, with particular attention to the effectiveness of MAP in comparison to air-permeable overwrapping for preserving meat quality over longer storage durations. According to Bobbitt [57], the shelf life of rabbit carcasses stored in aerobic conditions ranges between 3 and 6 days, whereas MAP storage can extend this period by up to threefold. This extension in shelf life can be attributed to the role of the gaseous environment, alongside factors such as meat pH and storage temperature, in shaping the microbial communities. For instance, it is widely recognized that carbon dioxide sensitivity in pseudomonads and *Enterobacteriaceae* [58] influences microbial growth [1]. In contrast, MAP tends to favor the proliferation of facultative anaerobes such as *Brochothrix thermosphacta* as the main spoilage microorganisms [30,59]. In the current study, sensory evaluations indicated that rabbit meat retained acceptable quality levels until day 14. This observation suggests a synergistic preservative effect arising from the combined influence of diet and packaging. Rodríguez-Calleja et al. [60] noted that rabbit meat stored under vacuum conditions reached unacceptable sensory scores by day 28. Similarly, findings by Berruga et al. [30] indicate that under various MAP gas compositions, lactic acid bacteria typically exhibit a lag phase of approximately 10 days.

3.4. Lipid Oxidation and Carbonyl and Sulfhydryl Grouphiol

TBARS levels are widely used to assess lipid peroxidation, reflecting the formation of secondary oxidative products like malondialdehyde (MDA). In rabbit meat, the enrichment of PUFA n-3, as achieved through dietary supplementation with linseed or algae, increases its susceptibility to lipid oxidation due to the higher number of double bonds in PUFA [61,62]. In terms of oxidation parameters, the TBARS value (Figure 5) was significantly higher ($p < 0.001$) in the ELS5% group (1.07 ± 0.04 mg MDA/kg) compared to the CNT and LPP3.5% groups (0.86 ± 0.04 mg MDA/kg; on average for the two groups). The presence of active PAD significantly reduced TBARS value ($p = 0.005$). With the exception of CNT, the other two groups displayed lower TBARS levels in *a*PAD compared to *n*PAD (0.99 vs. 1.15 mg MDA/kg for ELS5% and 0.78 vs. 0.93 mg MDA/kg for LPP3.5% with error standard 0.05).

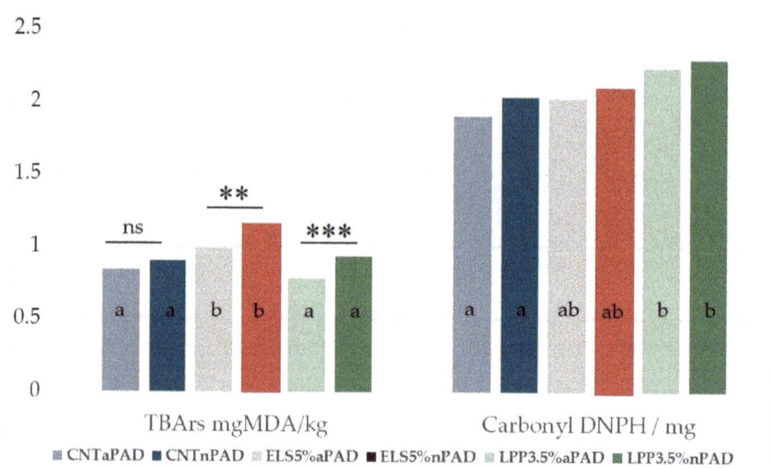

Figure 5. TBARS and the carbonyl group of rabbit meat affected by different diets and pads on average for different stored times. Significant differences between *a*PAD and *n*PAD inside the same feed group are indicated as ns: not significant, ** $p < 0.01$, and *** 0.001; different letters mean significant differences among ($p < 0.05$) the means of the diets.

The carbonyl group (C=O) is a polar group because oxygen is more electronegative than carbon, and it is formed when reactive oxygen species (ROS) attack amino acid side chains, leading to oxidative modifications in proteins. Protein oxidation, as evidenced by elevated carbonyl levels, contributes to structural changes that affect the texture and WHC of rabbit meat. For carbonyls, the CNT group exhibited significantly lower values than the LPP3.5% group ($p < 0.001$), while no significant differences were observed for the PAD type. The polyphenols from algal supplementation were expected to counteract the formation of oxidation products [22,37]. However, the elevated levels of long-chain polyunsaturated fatty acids (PUFAs) in rabbit meat, from animals fed with extruded linseed and *Padina pavonica*, as highlighted by Agati et al. [63], exhibited a heightened susceptibility to oxidation, thereby diminishing the protective efficacy of the algal polyphenols.

The sulfhydryl group showed no significant differences attributable to either the diet or the PAD type. The sulfhydryl group (-SH) is a sensitive marker of oxidative damage in proteins, primarily cysteine residues, which are critical for maintaining protein structure and enzymatic function. The loss of the sulfhydryl group is indicative of disulfide bond formation, leading to protein cross-linking and aggregation [38].

Over time, both TBARS and carbonyl levels increased significantly at each analysis storage point. TBARS levels rose from 0.06 ± 0.31 at D1 to 2.15 ± 0.31 mg MDA/kg at D21 (Table 3), while carbonyl levels increased from 0.67 ± 0.44 at D1 to 3.62 ± 0.44 nmol of DNPH/mg of protein at D21. In contrast, the sulfhydryl group, because it acts as a scavenger for oxidative processes, exhibited a significant decrease over time, declining from 69.11 ± 5.65 nmol SH/mg protein at D1 to 32.27 ± 5.65 nmol SH/mg protein at D21 ($p < 0.001$).

Table 3. Effect of different storage times on lipid oxidation and the sulfhydryl and carbonyl groups.

	D1	D4	D7	D14	D21	*p* Value	RMSE
TBARS (mg MDA/kg)	0.06 [a]	0.49 [b]	0.71 [c]	1.27 [d]	2.15 [e]	<0.001	0.314
Sulfhydryl (nmol SH/mg protein)	69.11 [e]	59.44 [d]	52.74 [c]	43.37 [b]	32.27 [a]	<0.001	5.656
Carbonyl (nmol DNPH/mg of protein)	0.67 [a]	1.27 [b]	1.99 [c]	2.93 [d]	3.62 [e]	<0.001	0.447

D1, 4, 7, 14, and 21 = times of storage expressed in days ([a–e] = different letters in the same row indicate a significant difference for $p < 0.05$); data are expressed as the mean.

The interplay between TBARS and the carbonyl and sulfhydryl groups underscores the oxidative dynamics in rabbit meat during storage. While lipid oxidation primarily affects flavor and aroma, protein oxidation alters texture, WHC, and appearance. The active pad's role in mitigating both lipid and protein oxidation, along with dietary strategies such as supplementation with *Padina pavonica*, demonstrates an integrated approach to enhancing the oxidative stability and overall quality of rabbit meat, particularly for products enriched with PUFA n-3 [1,29].

3.5. Sensory Traits

The panelists' evaluations revealed significant differences for color and overall acceptability attributes exclusively in the ELS5% group, where the ELS5%*a*PAD samples received higher scores compared to the ELS5%*n*PAD samples (color, $p = 0.006$ and overall acceptability, $p = 0.009$; Table 4).

Table 4. Sensory results of rabbit meat affected by different diets and PADs in different stored times.

	CNT		ELS5%		LPP3.5%				
	*n*PAD	*a*PAD	*n*PAD	*a*PAD	*n*PAD	*a*PAD	p Diets	p Pad	RMSE
Color	6.26	6.40	5.93 [b]	6.35 [a]	6.00	6.31	0.245	0.006	0.75
Odor	6.19	6.03	5.86	6.11	5.92	6.05	0.632	0.538	0.86
Overall acceptability	6.06	6.18	5.93 [b]	6.22 [a]	6.04	6.25	0.757	0.009	0.56

CNT = control diet; ELS5% = CNT diet with 5% extruded flaxseed; and LPP3.5% = CNT diet containing 3.5% extruded flaxseed and 0.2% *Padina pavonica* algae extract; *n*PAD= control pad, *a*PAD = active pad; RMSE = root square error, [a,b] = different letters in the same row mean a significant difference for $p < 0.05$; data are expressed as the mean.

Across all groups, irrespective of the presence or absence of an active PAD, samples received higher scores during the initial days of shelf life. However, as the shelf life progressed, the scores (Table 5) for all evaluated attributes declined significantly over time ($p < 0.001$), reflecting the cumulative impact of oxidative and microbial spoilage on sensory attributes. Sensory evaluations of color and odor further reinforced the importance of an active PAD in maintaining meat quality. The observed sensory deterioration aligns with microbial spoilage, a critical factor contributing to discoloration and the development of abnormal colors in meat [64].

Table 5. Effect of different storage times on the sensorial characteristics of rabbit meat.

	D1	D4	D7	D14	D21	*p* Value	RMSE
Color	7.31 [e]	6.63 [d]	6.41 [c]	5.67 [b]	5.04 [a]	<0.001	0.75
Odor	7.31 [e]	6.78 [d]	6.24 [c]	5.10 [b]	4.69 [a]	<0.001	0.86
Overall acceptability	7.18 [e]	6.68 [d]	6.24 [c]	5.47 [b]	5.01 [a]	<0.001	0.56

D1, 4, 7, 14, and 21 = times of storage expressed in days; [a–e] = different letters in the same row mean a significant difference for $p < 0.05$; data are expressed as the mean.

3.6. Interaction of Principals Parameters

Some of the most significant interactions between the feed groups, PAD, and storage time are illustrated in Figure 6. Considering the WHC, significant differences were observed early at D1 of storage between ELS5%*a*PAD and LPP3.5%*n*PAD (1.20 ± 0.10% vs. 1.63 ± 0.10%), with the latter showing the highest value among all groups. These differences between the two groups persisted throughout the 21 days of storage, except on D4, when all groups exhibited a sudden increase in liquid loss. Subsequently, regardless of the PAD type, the ELS5% group consistently presented the lowest liquid loss, significantly different from both LPP3.5%*n*PAD and CNT*n*PAD. The other groups exhibited intermediate values, with differences remaining constant over time, even as liquid loss increased.

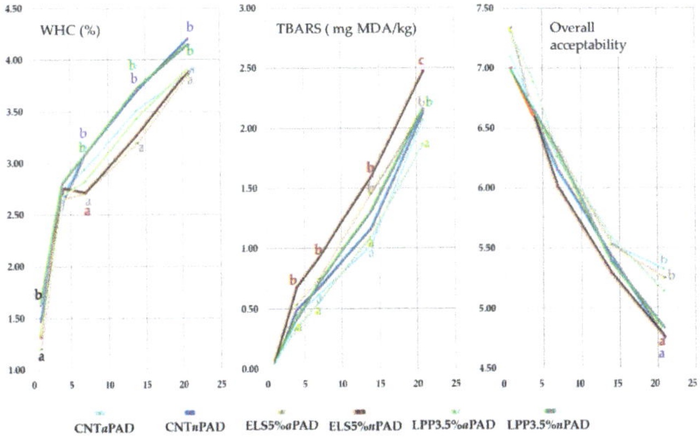

Figure 6. Trend of water-holding capacity (WHC). TBARS and overall acceptability across the storage times. Different letters at the same time mean significant differences for $p < 0.05$; the color of the letter corresponds to the color of the group; the group without letters is in the intermediate position.

Considering TBARS, all groups showed similar values at D1 (0.06 mg MDA/kg of meat). However, by D4, the ELS5% *n*PAD group was significantly different from the LPP3.5%*a*PAD group (0.68 ± 0.10 mg MDA/kg vs. 0.34 ± 0.10 mg MDA/kg). At D7, while maintaining the same differences observed at D4, the CNT*a*PAD and LPP3.5%*a*PAD groups showed the lowest TBARS values (0.58 ± 0.10 mg MDA/kg). By D14, the ELS5%*a*PAD group also exhibited high TBARS values, similar to samples with the normal PAD, significantly differing from other active PAD samples. At D21, the LPP3.5%*a*PAD group presented the lowest TBARS values compared to the others at the same time, while the ELS5%*n*PAD group showed the highest (1.87 vs. 2.48 mg MDA/kg, respectively), followed by ELS5% *a*PAD and LPP3.5%*n*PAD (2.16 mg MDA/kg on average). This suggests that the synergistic effect of algae polyphenols and the active PAD remains effective in controlling lipid oxidation over extended storage periods, highlighting its potential for enhancing the oxidative stability of PUFA-enriched rabbit meat [22]. In contrast, the ELS5%*n*PAD group showed the highest TBARS values, emphasizing the heightened susceptibility of linseed-enriched diets to lipid oxidation in the absence of adequate antioxidant interventions. The remaining groups with the active PAD showed no significant differences among themselves, displaying the lowest TBARS values overall. This indicates that the protective effect of the active PAD diminishes over prolonged storage, particularly in groups with higher PUFA content, such as those fed with extruded linseed [61].

Regarding overall acceptability, which summarizes the sensory evaluation of the color and odor of raw meat, differences among the groups were only evident at D21. Samples stored without the active PAD exhibited lower acceptability, particularly for CNT (4.75) and ELS5% (4.84). Conversely, samples stored with the active PAD received significantly higher evaluations, particularly for CNT and ELS5% (5.33 and 5.18). Nevertheless, overall sensory acceptability declined significantly over time for all six groups analyzed. In addition to oxidation reactions, microbial spoilage is another significant factor underlying the sensory deterioration of meat [64], causing discoloration and the development of abnormal colors.

4. Conclusions

The present study underscores the efficacy of an active absorbent pad in enhancing the shelf life of rabbit meat, particularly in the context of omega-3 polyunsaturated fatty acid (PUFA) enrichment achieved through dietary supplementation with extruded linseed and *Padina pavonica*. The inclusion of *a*PAD within modified atmosphere packaging effectively mitigated microbial growth and oxidative degradation, as evidenced by reduced TVCs, lipid peroxidation (TBARS), and protein oxidation indicators such as carbonyl and sulfhydryl content.

The oxidative stability of rabbit meat was significantly improved with the use of *a*PAD, particularly in the LPP3.5% dietary group, which exhibited the lowest TBARS values and enhanced sensory scores after extended storage. This suggests that the polyphenols and bioactive compounds in *Padina pavonica*, combined with the physical and chemical properties of *a*PAD, play a synergistic role in preserving meat quality. However, the elevated levels of long-chain PUFAs in enriched meat heightened its susceptibility to oxidative damage, underscoring the challenges associated with balancing nutritional enhancement and oxidative stability.

WHC and pH trends further emphasized the role of *a*PAD in maintaining the physicochemical properties of rabbit meat, with higher WHC and stable pH values correlating with reduced microbial and oxidative activity. These parameters also showed a clear interdependence with meat color stability, where active packaging systems contributed to the preservation of visual appeal, a metric crucial for consumer acceptance. Despite these advances, this study highlights the need for further exploration of optimal combinations of dietary strategies and packaging technologies to fully realize the benefits of omega-3 enrichment while minimizing the risks of oxidation. The findings offer valuable insights into innovative preservation methods, demonstrating the potential of *a*PAD as a practical and effective solution for improving the shelf life and quality of functional meat products.

5. Patents

Principi, A. and Merlotti, S. (2022). "International Patent application PCT WO 2022/029597A1—Bacteria—trapping item".

The complete disclosure could be forwarded to readers but only under an NDA agreement.

Supplementary Materials: The following supporting information can be downloaded at the following website: https://www.mdpi.com/article/10.3390/foods14030404/s1, Figure S1. Normal distribution for each sensory attribute (color, odor, and overall acceptability); Figure S2. Box plot analysis for each sensory attribute to detect potential outliers of scores among panelists. Table S1. Descriptive statistics of sensory test. Table S2. Trend of physical, chemical, and sensory parameters during times for each treatment of storage.

Author Contributions: Conceptualization, M.C. (Marta Castrica), S.F., N.E.F., M.C. (Michela Contò) and G.B.; methodology, M.C. (Marta Castrica), M.C. (Michela Contò), S.F., N.E.F. and G.B.; validation, G.B., D.V., G.C., C.M.B., S.F. and O.B.; formal analysis, M.C. (Marta Castrica), M.C. (Michela Contò), S.F., S.A. and E.A.; investigation, M.C. (Marta Castrica), S.F., N.E.F. and M.C. (Michela Contò); data curation, M.C. (Marta Castrica), S.F., M.C. (Michela Contò) and S.A.; writing—original draft preparation, M.C. (Marta Castrica), S.F., N.E.F., S.F. and M.C. (Michela Contò); writing—review and editing, C.M.B., M.C. (Marta Castrica), E.A., S.F., A.Q., S.A., N.E.F., L.M., G.C., A.Q., O.B., S.F. and G.B.; visualization, C.M.B., D.V., E.A., A.Q., S.A., S.F., L.M., D.M., O.B., N.E.F. and M.C. (Michela Contò); supervision, M.C. (Marta Castrica), D.V., S.F., N.E.F., D.M., G.B., O.B. and M.C. (Michela Contò); project administration, G.B., S.F. and C.M.B. All authors have read and agreed to the published version of the manuscript.

Funding: This research was funded by the PRIMA project "Omega rabbit: food for health benefit" (ORABBIT) grant number G15F21000810001.

Institutional Review Board Statement: This research was approved by the Ethics Committee of the Department of Veterinary Medicine of the University of Milano (Approval Code: OPBA_18_2021, Approval date: 9 April 2021).

Informed Consent Statement: Informed consent was obtained from all subjects involved in the study.

Data Availability Statement: The original contributions presented in the study are included in the article/Supplementary Material; further inquiries can be directed to the corresponding author.

Conflicts of Interest: The authors declare no conflicts of interest.

References

1. Cullere, M.; Dalle Zotte, A. Rabbit Meat Production and Consumption: State of Knowledge and Future Perspectives. *Meat Sci.* **2018**, *143*, 137–146. [CrossRef]
2. Petracci, M.; Soglia, F.; Leroy, F. Rabbit Meat in Need of a Hat-Trick: From Tradition to Innovation (and Back). *Meat Sci.* **2018**, *146*, 93–100. [CrossRef] [PubMed]
3. Siddiqui, S.A.; Gerini, F.; Ikram, A.; Saeed, F.; Feng, X.; Chen, Y. Rabbit Meat—Production, Consumption and Consumers' Attitudes and Behavior. *Sustainability* **2023**, *15*, 2008. [CrossRef]
4. Dalle Zotte, A.; Szendro, Z. The Role of Rabbit Meat as Functional Food. *Meat Sci.* **2011**, *88*, 319–331. [CrossRef] [PubMed]
5. Petrescu, D.C.; Petrescu-Mag, R.M. Consumer Behaviour Related to Rabbit Meat as Functional Food. *World Rabbit Sci.* **2018**, *26*, 321–333. [CrossRef]
6. Kumar, P.; Sharma, N.; Narnoliya, L.K.; Verma, A.K.; Umaraw, P.; Mehta, N.; Ismail-Fitry, M.R.; Kaka, U.; Yong-Meng, G.; Lee, S.J.; et al. Improving Quality and Consumer Acceptance of Rabbit Meat: Prospects and Challenges. *Meat Sci.* **2025**, *219*, 109660. [CrossRef] [PubMed]
7. Hernández, P.; Dalle Zotte, A. Influence of Diet on Rabbit Meat Quality. In *Nutrition of the Rabbit*, 2nd ed.; CABI: Wallingford, UK, 2010; pp. 163–178. [CrossRef]
8. Dalle Zotte, A.; Celia, C.; Szendro, Z. Herbs and Spices Inclusion as Feedstuff or Additive in Growing Rabbit Diets and as Additive in Rabbit Meat: A Review. *Livest. Sci.* **2016**, *189*, 82–90. [CrossRef]
9. Ribeiro, J.S.; Santos, M.J.M.C.; Silva, L.K.R.; Pereira, L.C.L.; Santos, I.A.; da Silva Lannes, S.C.; da Silva, M.V. Natural Antioxidants Used in Meat Products: A Brief Review. *Meat Sci.* **2019**, *148*, 181–188. [CrossRef] [PubMed]
10. Delis-Hechavarria, E.A.; Guevara-Gonzalez, R.G.; Ocampo-Velazquez, R.V.; Gomez-Soto, J.G.; Vargas-Hernandez, M.; Parola-Contreras, I.; Torres-Pacheco, I. Functional Food for Rabbits. Current Approaches and Trends to Increase Functionality. *Food Rev. Int.* **2023**, *39*, 2057–2074. [CrossRef]
11. Mattioli, S.; Cardinali, R.; Balzano, M.; Pacetti, D.; Castellini, C.; Dal Bosco, A.; Frega, N.G. Influence of Dietary Supplementation with Prebiotic, Oregano Extract, and Vitamin E on Fatty Acid Profile and Oxidative Status of Rabbit Meat. *J. Food Qual.* **2017**, *2017*, 3015120. [CrossRef]
12. Kouba, M.; Benatmane, F.; Blochet, J.E.; Mourot, J. Effect of a Linseed Diet on Lipid Oxidation, Fatty Acid Composition of Muscle, Perirenal Fat, and Raw and Cooked Rabbit Meat. *Meat Sci.* **2008**, *80*, 829–834. [CrossRef] [PubMed]
13. Petracci, M.; Bianchi, M.; Cavani, C. Development of Rabbit Meat Products Fortified with N-3 Polyunsaturated Fatty Acids. *Nutrients* **2009**, *1*, 111–118. [CrossRef]
14. Matics, Z.; Cullere, M.; Szín, M.; Gerencsér, Z.; Szabó, A.; Fébel, H.; Odermatt, M.; Radnai, I.; Dalle Zotte, A.; Szendrő, Z. Effect of a Dietary Supplementation with Linseed Oil and Selenium to Growing Rabbits on Their Productive Performances, Carcass Traits and Fresh and Cooked Meat Quality. *J. Anim. Physiol. Anim. Nutr.* **2017**, *101*, 685–693. [CrossRef] [PubMed]
15. Tariq, M.; Khan, M.; Sameen, A.; Nisa, M. Effect of flaxseed enrichment on quality attributes of rabbit meat and meat product. *J. Anim. Plant Sci.* **2016**, *26*, 1850–1858.
16. Mattioli, S.; Castellini, C.; Mancini, S.; Roscini, V.; Cartoni Mancinelli, A.; Cotozzolo, E.; Pauselli, M.; Dal Bosco, A. Effect of Trub and/or Linseed Dietary Supplementation on in Vivo Oxidative Status and Some Quality Traits of Rabbit Meat. *Meat Sci.* **2020**, *163*, 108061. [CrossRef]
17. Makkar, H.P.S.; Tran, G.; Heuzé, V.; Giger-Reverdin, S.; Lessire, M.; Lebas, F.; Ankers, P. Seaweeds for Livestock Diets: A Review. *Anim. Feed. Sci. Technol.* **2016**, *212*, 1–17. [CrossRef]
18. Costa, M.; Cardoso, C.; Afonso, C.; Bandarra, N.M.; Prates, J.A.M. Current Knowledge and Future Perspectives of the Use of Seaweeds for Livestock Production and Meat Quality: A Systematic Review. *J. Anim. Physiol. Anim. Nutr.* **2021**, *105*, 1075–1102. [CrossRef] [PubMed]

19. Michalak, I.; Tiwari, R.; Dhawan, M.; Alagawany, M.; Farag, M.R.; Sharun, K.; Bin Emran, T.; Dhama, K. Antioxidant Effects of Seaweeds and Their Active Compounds on Animal Health and Production—A Review. *Vet. Q.* **2022**, *42*, 48. [CrossRef]
20. Rossi, R.; Vizzarri, F.; Chiapparini, S.; Ratti, S.; Casamassima, D.; Palazzo, M.; Corino, C. Effects of Dietary Levels of Brown Seaweeds and Plant Polyphenols on Growth and Meat Quality Parameters in Growing Rabbit. *Meat Sci.* **2020**, *161*, 107987. [CrossRef] [PubMed]
21. Al-Soufi, S.; García, J.; Muíños, A.; López-Alonso, M. Marine Macroalgae in Rabbit Nutrition—A Valuable Feed in Sustainable Farming. *Animals* **2022**, *12*, 2346. [CrossRef] [PubMed]
22. Maj, D.; Migdał, Ł.; Zapletal, P. Effects of Dietary Supplementation with Algae, Sunflower Oil, or Soybean Oil, and Age on Fat Content, Fatty Acid Profile and the Expression of Related Genes in Rabbits. *Anim. Sci. Pap. Rep.* **2023**, *41*, 243–260. [CrossRef]
23. Pellissery, A.J.; Vinayamohan, P.G.; Amalaradjou, M.A.R.; Venkitanarayanan, K. Spoilage Bacteria and Meat Quality. In *Meat Quality Analysis*; Academic Press: Cambridge, MA, USA, 2020; pp. 307–334. [CrossRef]
24. Pereira, M.; Malfeito-Ferreira, M. A Simple Method to Evaluate the Shelf Life of Refrigerated Rabbit Meat. *Food Control* **2015**, *49*, 70–74. [CrossRef]
25. Cullere, M.; Dalle Zotte, A.; Tasoniero, G.; Giaccone, V.; Szendrő, Z.; Szín, M.; Odermatt, M.; Gerencsér, Z.; Dal Bosco, A.; Matics, Z. Effect of Diet and Packaging System on the Microbial Status, PH, Color and Sensory Traits of Rabbit Meat Evaluated during Chilled Storage. *Meat Sci.* **2018**, *141*, 36–43. [CrossRef]
26. Redondo-Solano, M.; Guzmán-Saborío, P.; Ramírez-Chavarría, F.; Chaves-Ulate, C.; Araya-Quesada, Y.; Araya-Morice, A. Effect of the Type of Packaging on the Shelf Life of Ground Rabbit Meat. *Food Sci. Technol. Int.* **2022**, *28*, 190–199. [CrossRef]
27. Brody, A.L.; Strupinsky, E.P.; Kline, L.R. *Active Packaging for Food Applications*; CRC Press: Boca Raton, FL, USA, 2001. [CrossRef]
28. Kandeepan, G.; Tahseen, A. Modified Atmosphere Packaging (MAP) of Meat and Meat Products: A Review. *J. Packag. Technol. Res.* **2022**, *6*, 137–148. [CrossRef]
29. Racewicz, P.; Biesiada, H.; Majewski, M.; Ludwiczak, A.; Stanisz, M.; Składanowska-Baryza, J. Impact of Packaging System on the Microbial Quality of Chilled Rabbit Meat over 21 Days of Storage. *Anim. Sci. J.* **2023**, *94*, 13852. [CrossRef]
30. Berruga, I.; Vergara, H.; Beí, M.; Linares, E. Control of Microbial Growth and Rancidity in Rabbit Carcasses by Modified Atmosphere Packaging. *J. Sci. Food Agric.* **2005**, *85*, 1987–1991. [CrossRef]
31. Holman, B.W.B.; Kerry, J.P.; Hopkins, D.L. Meat Packaging Solutions to Current Industry Challenges: A Review. *Meat Sci.* **2018**, *144*, 159–168. [CrossRef] [PubMed]
32. Liu, Y.; Sameen, D.E.; Ahmed, S.; Dai, J.; Qin, W. Antimicrobial Peptides and Their Application in Food Packaging. *Trends Food Sci. Technol.* **2021**, *112*, 471–483. [CrossRef]
33. Sun, P.; Zhu, Y.; Yin, L.; Kong, B.; Xia, X.; Liu, Q.; Wang, H. Development of Active Hydrogel Absorbent Pads Based on Sodium Alginate/Potassium-Doped Carbon Dots for Chilled Pork Preservation. *Food Hydrocoll.* **2024**, *155*, 110229. [CrossRef]
34. Wang, X.; Yan, X.; Xu, Y.; Liu, J.; Chen, D. Changes in the Quality and Microbial Compositions of Ground Beef Packaged on Food Absorbent Pads Incorporated with Levulinic Acid and Sodium Dodecyl Sulfate. *Int. J. Food Microbiol.* **2022**, *376*, 109771. [CrossRef]
35. Jiang, C.; Yu, G.; Wang, P.; Yao, Z.; Mu, L.; Li, Q.; Shi, Y. Preparation of Delignified Wood Fibers Based Absorbent Pad for Tray Package and Its Application in Preservation of Cold and Fresh Pork. *Ind. Crops Prod.* **2024**, *219*, 119097. [CrossRef]
36. Liu, G.F.; Song, H.Y.; Zhang, Q.; Wang, J.Q.; Wang, L.J.; Zhang, Z.J. Cellulose-Based Absorbent Pad Loaded with Carum Copticum Essential Oil for Shelf Life Extension of Refrigerated Chicken Meat. *Packag. Technol. Sci.* **2022**, *35*, 425–433. [CrossRef]
37. Castrica, M.; Miraglia, D.; Menchetti, L.; Branciari, R.; Ranucci, D.; Balzaretti, C.M. Antibacterial Effect of an Active Absorbent Pad on Fresh Beef Meat during the Shelf-Life: Preliminary Results. *Appl. Sci.* **2020**, *10*, 7904. [CrossRef]
38. Domínguez, R.; Pateiro, M.; Munekata, P.E.S.; Zhang, W.; Garcia-Oliveira, P.; Carpena, M.; Prieto, M.A.; Bohrer, B.; Lorenzo, J.M. Protein Oxidation in Muscle Foods: A Comprehensive Review. *Antioxidants* **2021**, *11*, 60. [CrossRef] [PubMed]
39. Fehri, N.E.; Contò, M.; Castrica, M.; Quattrone, A.; Renzi, G.; Di Giovanni, S.; Agradi, S.; Vigo, D.; Brecchia, G.; Menchetti, L.; et al. Effects of Diets Containing Extruded Linseed and Padina Pavonica Algae on Meat Rabbit: Carcass Performance and Meat Quality. *Foods* **2025**, *14*, 274. [CrossRef] [PubMed]
40. Commission Internationale de L'éclairage (CIE). *Colorimetry: Official Recommendations of the Commission Internationale de l'Éclairage (CIE)*, 2nd ed.; Commission Internationale de L'éclairage (CIE): Vienna, Austria, 1986.
41. *ISO 6579-1:2017*; Microbiology of the Food Chain-Horizontal Method for the Detection, Enumeration and Serotyping of Salmonella-Part 1: Detection of Salmonella spp. International Organization for Standardization (ISO): Geneva, Switzerland, 2017.
42. *AFNOR BRD 07/05-09/01*; Enumeration of Listeria monocytogenes and Listeria spp. AFNOR (Association Française de Normalisation): Paris, France, 2001.
43. Cifuni, G.F.; Contò, M.; Failla, S. Potential Use of Visible Reflectance Spectra to Predict Lipid Oxidation of Rabbit Meat. *J. Food Eng.* **2016**, *169*, 85–90. [CrossRef]

44. Valerio, F.; Skandamis, P.N.; Failla, S.; Contò, M.; Di Biase, M.; Bavaro, A.R.; Pirovano, M.P.; Lavermicocca, P. Microbiological and Physicochemical Parameters for Predicting Quality of Fat and Low-Fat Raw Ground Beef during Refrigerated Aerobic Storage. *J. Food Sci.* **2020**, *85*, 465–476. [CrossRef]
45. Dal Bosco, A.; Mattioli, S.; Cullere, M.; Szendrő, Z.; Gerencsér, Z.; Matics, Z.; Castellini, C.; Szin, M.; Dalle Zotte, A. Effect of Diet and Packaging System on the Oxidative Status and Polyunsaturated Fatty Acid Content of Rabbit Meat during Retail Display. *Meat Sci.* **2018**, *143*, 46–51. [CrossRef]
46. Wang, Z.; Tu, J.; Zhou, H.; Lu, A.; Xu, B. A Comprehensive Insight into the Effects of Microbial Spoilage, Myoglobin Autoxidation, Lipid Oxidation, and Protein Oxidation on the Discoloration of Rabbit Meat during Retail Display. *Meat Sci.* **2021**, *172*, 108359. [CrossRef]
47. Guo, X.; Wang, Y.; Lu, S.; Wang, J.; Fu, H.; Gu, B.; Lyu, B.; Wang, Q. Changes in Proteolysis, Protein Oxidation, Flavor, Color and Texture of Dry-Cured Mutton Ham during Storage. *LWT* **2021**, *149*, 111860. [CrossRef]
48. Adeyemi, K.D.; Shittu, R.M.; Sabow, A.B.; Abubakar, A.A.; Karim, R.; Karsani, S.A.; Sazili, A.Q. Comparison of Myofibrillar Protein Degradation, Antioxidant Profile, Fatty Acids, Metmyoglobin Reducing Activity, Physicochemical Properties and Sensory Attributes of Gluteus Medius and Infraspinatus Muscles in Goats. *J. Anim. Sci. Technol.* **2016**, *58*, 23. [CrossRef] [PubMed]
49. Wang, Z.; He, Z.; Gan, X.; Li, H. Interrelationship among Ferrous Myoglobin, Lipid and Protein Oxidations in Rabbit Meat during Refrigerated and Superchilled Storage. *Meat Sci.* **2018**, *146*, 131–139. [CrossRef]
50. Traore, S.; Aubry, L.; Gatellier, P.; Przybylski, W.; Jaworska, D.; Kajak-Siemaszko, K.; Santé-Lhoutellier, V. Higher Drip Loss Is Associated with Protein Oxidation. *Meat Sci.* **2012**, *90*, 917–924. [CrossRef] [PubMed]
51. Carrillo-Lopez, L.M.; Robledo, D.; Martínez, V.; Huerta-Jimenez, M.; Titulaer, M.; Alarcon-Rojo, A.D.; Chavez-Martinez, A.; Luna-Rodriguez, L.; Garcia-Flores, L.R. Post-Mortem Ultrasound and Freezing of Rabbit Meat: Effects on the Physicochemical Quality and Weight Loss. *Ultrason. Sonochem.* **2021**, *79*, 105766. [CrossRef]
52. Koutsoumanis, K.P.; Sofos, J.N. Comparative Acid Stress Response of Listeria Monocytogenes, Escherichia Coli O157:H7 and Salmonella Typhimurium after Habituation at Different PH Conditions. *Lett. Appl. Microbiol.* **2004**, *38*, 321–326. [CrossRef] [PubMed]
53. Doulgeraki, A.I.; Ercolini, D.; Villani, F.; Nychas, G.J.E. Spoilage Microbiota Associated to the Storage of Raw Meat in Different Conditions. *Int. J. Food Microbiol.* **2012**, *157*, 130–141. [CrossRef] [PubMed]
54. Dalle Zotte, A. Perception of Rabbit Meat Quality and Major Factors Influencing the Rabbit Carcass and Meat Quality. *Livest. Prod. Sci.* **2002**, *75*, 11–32. [CrossRef]
55. Komodromos, D.; Sergelidis, D.; Amvrosiadis, I.; Kontominas, M.G. Combined Effect of an Active AgIon® Absorbent Pad and a Chitosan Coating on the Preservation of Fresh Beef. *Foods* **2024**, *13*, 1387. [CrossRef] [PubMed]
56. Fernández, A.; Picouet, P.; Lloret, E. Reduction of the Spoilage-Related Microflora in Absorbent Pads by Silver Nanotechnology during Modified Atmosphere Packaging of Beef Meat. *J. Food Prot.* **2010**, *73*, 2263–2269. [CrossRef] [PubMed]
57. Bobbitt, J. Shelf Life and Microbiological Safety of Selected New and Emerging Meats Destined for Export Markets. *RIRDC* **2002**, *2*, 038.
58. Jeremiah, L.E. Packaging Alternatives to Deliver Fresh Meats Using Short- or Long-Term Distribution. *Food Res. Int.* **2001**, *34*, 749–772. [CrossRef]
59. Nakyinsige, K.; Sazili, A.Q.; Aghwan, Z.A.; Zulkifli, I.; Goh, Y.M.; Abu Bakar, F.; Sarah, S.A. Development of Microbial Spoilage and Lipid and Protein Oxidation in Rabbit Meat. *Meat Sci.* **2015**, *108*, 125–131. [CrossRef] [PubMed]
60. Rodríguez-Calleja, J.M.; Santos, J.A.; Otero, A.; García-López, M.L. Effect of Vacuum and Modified Atmosphere Packaging on the Shelf Life of Rabbit Meat del Envasado al Vacío y en Atmósfera Modificada en la Vida Útil de la Carne de Conejo. *CyTA-J. Food* **2010**, *8*, 109–116. [CrossRef]
61. Mattioli, S.; Collodel, G.; Signorini, C.; Cotozzolo, E.; Noto, D.; Cerretani, D.; Micheli, L.; Fiaschi, A.I.; Brecchia, G.; Menchetti, L.; et al. Tissue Antioxidant Status and Lipid Peroxidation Are Related to Dietary Intake of N-3 Polyunsaturated Acids: A Rabbit Model. *Antioxidants* **2021**, *10*, 681. [CrossRef]
62. Nieto, G.; Ros, G.; Nieto, G.; Ros, G. Modification of Fatty Acid Composition in Meat Through Diet: Effect on Lipid Peroxidation and Relationship to Nutritional Quality—A Review. *Lipid Peroxidation* **2012**, *12*, 239–258. [CrossRef]
63. Agati, G.; Brunetti, C.; Di Ferdinando, M.; Ferrini, F.; Pollastri, S.; Tattini, M. Functional Roles of Flavonoids in Photoprotection: New Evidence, Lessons from the Past. *Plant Physiol. Biochem.* **2013**, *72*, 35–45. [CrossRef]
64. Sabow, A.B.; Sazili, A.Q.; Aghwan, Z.A.; Zulkifli, I.; Goh, Y.M.; Ab Kadir, M.Z.A.; Nakyinsige, K.; Kaka, U.; Adeyemi, K.D. Changes of Microbial Spoilage, Lipid-Protein Oxidation and Physicochemical Properties during Post Mortem Refrigerated Storage of Goat Meat. *Anim. Sci. J.* **2016**, *87*, 816–826. [CrossRef] [PubMed]

foods

MDPI

Article

Effects of Diets Containing Extruded Linseed and *Padina pavonica* Algae on Meat Rabbit: Carcass Performance and Meat Quality

Nour Elhouda Fehri [1], Michela Contò [2], Marta Castrica [3,*], Alda Quattrone [1], Gianluca Renzi [2], Sabrina Di Giovanni [2], Stella Agradi [4], Daniele Vigo [1], Gabriele Brecchia [1], Laura Menchetti [5], Claudia Maria Balzaretti [1], Doriana Beqiraj [6], Egon Andoni [6], Giulio Curone [1,†] and Sebastiana Failla [2,†]

[1] Department of Veterinary Medicine and Animal Sciences, University of Milan, Via dell'Università 6, 26900 Lodi, Italy; nour.fehri@unimi.it (N.E.F.); alda.quattrone@unimi.it (A.Q.); daniele.vigo@unimi.it (D.V.); gabriele.brecchia@unimi.it (G.B.); claudia.balzaretti@unimi.it (C.M.B.); giulio.curone@unimi.it (G.C.)
[2] Consiglio per la Ricerca in Agricoltura e l'Analisi Dell'Economia Agraria (CREA), Centro di Ricerca Zootecnia e Acquacoltura, Research Centre for Animal Production and Aquaculture, Via Salaria 31, 00015 Rome, Italy; michela.conto@crea.gov.it (M.C.); gianluca.renzi@crea.gov.it (G.R.); sabrina.digiovanni@crea.gov.it (S.D.G.); sebastiana.failla@crea.gov.it (S.F.)
[3] Department of Comparative Biomedicine and Food Science, University of Padova, Agripolis, Viale dell'Univesità 16, 35020 Legnaro, Italy
[4] Department of Veterinary Sciences, University of Torino, Largo Paolo Braccini 2, 10095 Grugliasco, Italy; stella.agradi@unito.it
[5] School of Biosciences and Veterinary Medicine, University of Camerino, Via Circonvallazione 93/95, 62024 Matelica, Italy; laura.menchetti@unicam.it
[6] Faculty of Veterinary Medicine, Agricultural University of Tirana, Kodër Kamëz, 1029 Tirana, Albania; dkalamishi@yahoo.com (D.B.); eandoni@ubt.edu.al (E.A.)
* Correspondence: marta.castrica@unipd.it
† These authors contributed equally to this work.

Abstract: This study investigated the effects of dietary supplementation with extruded linseed (ELS) and *Padina pavonica* algae extract (PP) on rabbit carcass and meat quality. Ninety-six rabbit carcasses from two production cycles were analyzed. In the first cycle (C1), rabbits were fed a control diet (1CNT), the same diet supplemented with 5% ELS (1ELS5%), and supplemented with 3.5% ELS and 0.2% PP (1LPP3.5%). In the second cycle (C2), the diets varied in composition and supplementation levels: a different control diet (2CNT), the same diet with 5% ELS (2ELS5%), and with 5% ELS and 0.2% PP (2LPP5%). Meat analyses were performed on *Longissimus thoracis* et *lumborum* (LTL) muscle for physical properties and on thigh meat (THM) for proximate composition, vitamin E, coenzyme-Q10, cholesterol, fatty acid profile, and mineral content. No significant differences in LTL physical quality were observed in C1, although LTL was brighter in C2 ($p < 0.001$). THM in C2 had higher fat content ($p < 0.001$). Dietary supplementation with ELS and PP extract significantly increased polyunsaturated fatty acids (n-3 PUFAs) and improved the n-6/n-3 ratio ($p < 0.001$) in rabbit meat, demonstrating their positive impact on meat quality.

Keywords: extruded linseed; *Padina pavonica*; meat; rabbit; PUFA; vitamin E

1. Introduction

Currently, the rabbit industry is currently facing a severe economic crisis, driven by structural weaknesses, a steady decline in consumer demand, and increasing criticism from Western consumers regarding animal welfare and ethical concerns [1]. Additionally, rabbit meat is priced higher than other meats, such as poultry or pork [2]. Despite these challenges, rabbit meat is a valuable source of bioactive compounds with potential nutraceutical benefits [3], offering exceptional nutritional advantages. It is rich in polyunsaturated fatty acids (PUFAs) [4], high-quality proteins, and essential amino acids [5], while providing moderate energy values, having low fat and cholesterol, and being an excellent source of B vitamins, particularly B12. Rabbit meat is also low in sodium [6] but high in phosphorus [3,7]. However, limited consumer awareness and inefficiencies in marketing and communication often overshadow these benefits. To enhance both consumer acceptance and economic viability, the rabbit meat industry must focus on innovative marketing strategies and better dissemination of its health benefits [6].

Moreover, to address and mitigate the health risks associated with unbalanced diets, research has increasingly focused on developing new functional foods by incorporating high-quality nutraceuticals into the animals' diets [8]. This approach is particularly effective in monogastric species, such as poultry, rabbits, and pigs, where the absence of the rumen prevents microbial alterations of key nutrients, like vitamins and PUFAs. As a result, these nutrients can be directly incorporated into muscle tissue without significant modification [9–12]. Beyond enriching animal products, these dietary supplements also improve animal health by reducing oxidative stress and limiting the formation of free radicals [13]. These reactive compounds, if transferred to animal-derived foods, can negatively impact human health. This interconnected relationship highlights the "One Health" concept, which emphasizes the intrinsic link between the proper management of animal health, nutrition, and living conditions (key components of animal welfare) and human well-being [13].

Recent research has examined the effects of dietary PUFA supplementation on the growth performance and meat quality of rabbits [14]. Specifically, studies [15,16] have shown that dietary intake significantly influences the fatty acid composition of both fat depots and intramuscular fat. Furthermore, increasing emphasis has been placed on dietary strategies to enhance the n-3 PUFA content in rabbit feed, aiming not only to improve animal performance but also to meet the growing demand for these essential fatty acids in human diets. These efforts align with recommendations from the Food and Agriculture Organization (FAO) and the World Health Organization (WHO), which emphasize the importance of increasing n-3 intake while reducing n-6 consumption to achieve a balanced n-6/n-3 fatty acid ratio [17]. An imbalance in this ratio, combined with insufficient essential fatty acid intake in modern diets, has been associated with the rising prevalence of cardiovascular diseases in developed countries [17].

Rabbit diets have been enriched with various sources of n-3 polyunsaturated fatty acids (PUFAs) to enhance the nutritional value of their meat as a functional food. Among the strategies employed, linseed has been widely studied as a primary source of alpha-linolenic acid (ALA) [18]. Supplementing rabbit diets with 8% linseed significantly improved meat quality by increasing n-3 PUFA levels, and reducing the n-6/n-3 ratio. Other studies found that 3% linseed supplementation similarly enhanced the nutritional quality of rabbit meat without compromising product characteristics [19]. However, higher supplementation levels (6% or 9%) were associated with increased meat oxidation and cooking losses, although sensory acceptability remained unaffected after storage [20].

On the other hand, fewer studies have investigated the use of fish oil [21,22] and marine algae supplementation as sources of n-3 PUFAs in rabbit diets [23–27], with results varying significantly depending on the specific type of algae incorporated. Despite the long-established use of marine algae as a feed additive in livestock production [28], research specifically focusing on the application of *Padina pavonica* in animal diets remains limited [29].

In general, marine algae possess specific enzymatic pathways [30,31] that enable them to synthesize long-chain polyunsaturated fatty acids (PUFAs), including arachidonic acid (20:4 n-6), eicosapentaenoic acid (20:5 n-3), and docosahexaenoic acid (22:6 n-3). Among the advantages of incorporating seaweed into animal feed is its ability to exert significant biological effects, even at low inclusion levels, due to the presence of numerous bioactive antioxidant and antimicrobial compounds that support critical cellular functions [32]. Specifically, *Padina pavonica*, a brown alga (*Phaeophyta*), is known for its richness in bioactive compounds that enhance the absorption of phosphorus and calcium. Additional benefits are attributed to its variety of phenolic compounds and flavonoids, such as myricetin, quercetin, and resveratrol [33], along with fat-soluble vitamins like vitamins E and D, and water-soluble vitamins such as niacin, thiamine, pantothenic acid, and folic acid. Furthermore, *Padina pavonica* is abundant in phytosterols, including fucosterol, ergosterol, and β-sitosterol, which are vital for cellular functions and contribute to cholesterol reduction [32]. Its antifungal properties are also notable, as they can inhibit the growth of foodborne pathogens and spoilage microorganisms, including mycotoxin-producing species such as *Aspergillus*, *Fusarium*, and *Penicillium* [33]. This inhibition offers dual benefits: preventing food spoilage and reducing exposure to harmful mycotoxins, such as aflatoxins and ochratoxin A, which are among the most concerning toxins.

Supplementing rabbit diets with marine algae could provide numerous benefits for rabbits, acting as a detoxifying agent and promoting the accumulation of nutraceutical compounds in the meat, including antioxidants, fat-soluble vitamins, and essential minerals. Additionally, it helps reduce cholesterol accumulation and supports the production of high levels of very long-chain polyunsaturated fatty acids (VLC-PUFAs) of both the n-3 and n-6 series [30,32–34].

The present study aims to evaluate the effects of dietary supplementation with extruded linseed (ELS) and *Padina pavonica* algae (PP) extract on rabbit meat quality and carcass traits. Given their unique properties, these supplements are expected to influence carcass performance and nutritional quality of rabbit meat.

2. Materials and Methods

2.1. Animals and Diets

This study was conducted as part of the PRIMA project entitled "Omega Rabbit: food for health BenefIT", funded by the European Union. The experimental trial received approval from the Italian Ministry and the Ethical Committee of the Department of Veterinary Medicine of the University of Milano (OPBA_18_2021). The experimental trial took place in a commercial rabbit farm located in Central Italy. The rabbits were reared in compliance with Legislative Decree No. 146, implementing Directive 98/58/EC, regarding the minimum standards for the protection of animals kept for farming purposes. Efforts were made to reduce animal distress and use only the minimum number of animals required to ensure reliable and consistent results. Additionally, the health and welfare of the rabbits were monitored daily by the farm's responsible veterinarian.

The experiment was conducted over two separate rearing cycles, each carried out in a different year. For each cycle, the trial encompassed four weekly sessions, with 36 male New Zealand White rabbits included in each session, resulting in a total of 144 rabbits per cycle. After weaning (at 35 days of age), the animals had an initial weight of 816 g. The rabbits were housed in conventional individual cages (L × W × H: 75 × 38 × 25 cm) and were kept in a controlled environment, where temperature was maintained between +18 °C and +23 °C, relative humidity was kept between 60% and 75%, and the lighting cycle consisted of 16 h of light followed by 8 h of darkness. The animals were randomly divided in three experimental groups (n = 48 animals for each group per cycle), each receiving a distinct pelleted diet. The experimental diets were formulated based on current nutritional recommendations [35] for fattening rabbits, with the ingredient composition detailed in Table 1.

In the first cycle (C1), the three groups received three different diets formulated as follows: 1CNT = control diet; 1ELS5% = 1CNT supplemented with 5% ELS, and 1LPP3.5% = 1CNT supplemented with 3.5% ELS and 0.2% PP algae extract.

In the second cycle (C2), the CNT diet was modified in the fatty acid profile by increasing monounsaturated fatty acid (MUFA) levels and reducing n-6 PUFA content to improve fatty acid characteristics. In addition, the group that received PP was supplemented with 5% of ELS. Despite these modifications, the proximate composition and fiber content were consistent across all diets. Therefore, the groups of the second cycle received the diets formulated as follows: 2CNT = control diet; 2ELS5% = 2CNT supplemented with 5% ELS; 2LPP5% = 2CNT supplemented with 5% ELS and 0.2% PP extract. All the ingredients were adjusted to achieve isoenergetic and isoproteic formulations. The chemical composition and fatty acid profiles of the six diets are detailed in Tables 2 and 3. Rabbits were slaughtered at 85 days of age. During the study (35 to 85 days of age), the rabbits were provided with a daily ration gradually increasing from 100 g/day to 160 g/day. Fresh water was always available.

Table 1. Formulation of the experimental diets of the different groups in the two cycles.

Ingredients (%)	First Cycle			Second Cycle		
	1CNT	1ELS5%	1LPP3.5%	2CNT	2ELS5%	2LPP5%
Extruded linseed	-	5	3.5	-	5	5
"*Padina pavonica*" Algae extract	-	-	0.2	-	-	0.2
Wheat Bran	23.16	23.09	23.08	24	23	23
Beet pulp	11.5	9.33	11	13.5	12.5	11.48
Wheat straw	11	11	11	11	11	11
Alfalfa	10	12.5	10.17	10	10	10
Sunflower husks	9.95	6	10.78	-	-	-
Barley	9.5	9	9.5	5.6	5.8	8.25
Sunflower seed meal	8.83	14	9.17	18.5	17.41	17.97
Soybean hulls	0.17	-	-	3.82	4.3	2.22
Toasted soybean seed	5	-	1.50	1.7	-	-
Molasses cane	3	3	3	2.5	2.5	2.5
Wheat	2.5	2.5	2.5	3	3	3
Grape seed meal	2.17	1.83	1.87	2.7	2.8	2.68
Soybean oil	0.55	-	-	0.99	-	-
Palm oil	0.33	0.33	0.33	0.33	0.33	0.33
Carboxymethylcellulose	0.2	0.2	0.2	0.2	0.2	0.2
Liquid acidifier [1]	0.15	0.15	0.15	0.15	0.15	0.15
Calcium carbonate	0.8	0.8	0.8	0.8	0.8	0.8
Sodium chloride	0.4	0.4	0.4	0.42	0.42	0.42
Magnesium oxide	0.15	0.15	0.15	0.15	0.15	0.15
Oligo-vitamin supplement [2]	0.25	0.25	0.25	0.25	0.25	0.25
Methionine hydroxyanalog	0.15	0.14	0.15	0.15	0.13	0.12
Lysine	0.14	0.21	0.19	0.14	0.16	0.17
L Threonine	0.07	0.08	0.08	0.07	0.07	0.07
Vitamin E 50%	0.03	0.03	0.03	0.03	0.03	0.03

[1] Liquid acidifier composition = Formic acid 75%. [2] Oligo-vitamin supplement: Vitamin A 6,000,000 IU, D3 600,000 IU, E 20,000 IU, K3 1200 mg, B1 800 mg, B2 1600 mg, B6 800 mg, B12 6.0 mg, biotin 60.0 mg, niacinamide 16,000 mg, folic acid 400 mg, calcium pantothenate 6666 mg.; 1CNT = control diet of the first cycle; 1ELS5% = 1CNT+ 5% ELS; 1LPP3.5% = 1CNT+ 3.5% ELS and 0.2% *PP* extract. 2CNT = control diet of second cycle; 2ELS5% = 2CNT+ 5% ELS; 2LPP5% = 2ELS5% with 5% ELS and 0.2% *PP* extract.

Table 2. Chemical composition of the experimental diets of the different groups in the two cycles.

Components g/100 g	First Cycle			Second Cycle		
	1CNT	1ELS5%	1LPP3.5%	2CNT	2ELS5%	2LPP5%
Dry matter	88.61	88.89	88.85	89.96	89.05	88.89
Crude proteins	15.03	15.28	15.13	15.02	15.25	15.11
Crude fat	3.56	3.75	3.57	3.74	3.81	3.88
Ash	7.95	8.04	8.10	7.88	7.95	8.09
Crude fiber	18.64	18.63	18.71	18.70	18.45	18.34
Nitrogen-free extract	43.43	43.19	432.34	43.62	43.59	43.47
Calcium	0.79	0.80	0.81	0.80	0.79	0.81
Phosphate	0.51	0.52	0.53	0.50	0.51	0.53
Starch	12.07	12.07	11.87	12.10	12.05	12.03
Digestible energy (kcal)	2189.1	2190.4	2188.7	2196.4	2198.7	2197.9
Met + Cys [1]	0.6	0.6	0.6	0.6	0.6	0.6
Lys [2]	0.7	0.7	0.7	0.7	0.7	0.7
Trp [3]	0.19	0.19	0.19	0.19	0.19	0.19
Thr [4]	0.58	0.58	0.58	0.58	0.58	0.58
Vitamin E (ppm)	200	200	200	200	200	200

[1] Met + Cys = Methionine + cysteine; [2] Lys = lysine; [3] Trp = tryptophan; [4] Thr = threonine., 1CNT = control diet of first cycle; 1ELS5% = 1CNT+ 5% ELS; 1LPP3.5% = 1CNT+ 3.5% ELS and 0.2% *PP* extract. 2CNT = control diet of second cycle; 2ELS5% = 2CNT+ 5% ELS; 2LPP5% = 2CNT + 5% ELS and 0.2% *PP* extract.

Table 3. Percentage of fatty acids on total FAME of the experimental diets of the different groups in the two cycles.

Ingredients (%)	First Cycle			Second Cycle		
	1CNT	1ELS5%	1LPP3.5%	2CNT	2ELS5%	2LPP5%
14:0	0.33	0.29	0.32	0.40	0.31	0.28
16:0	16.75	15.74	16.11	16.38	16.08	15.76
16:1 n-7	0.19	0.23	0.21	0.12	0.09	0.07
17:0	0.12	0.11	0.09	0.11	0.10	0.11
18:0	6.80	6.94	7.18	6.86	6.98	7.01
18:1 n-9	19.41	18.63	18.60	26.37	23.88	23.82
18:1 n-7	1.25	1.01	1.20	1.90	1.85	1.79
18:2 n-6, LA [1]	47.25	33.32	33.54	41.12	25.89	26.60
20:0	0.32	0.29	0.26	0.28	0.28	0.22
18:3 n-6, γ-ALA [2]	0.23	0.30	0.25	0.20	0.21	0.22
18:3 n-3, α-ALA	5.35	21.71	20.42	3.90	22.90	22.44
22:1 n-11	0.17	0.17	0.17	0.19	0.22	0.23
20:4 n-6, AA [3]	0.06	0.05	0.19	0.12	0.12	0.16
20:5 n-3, EPA [4]	-	-	0.09	-	-	0.09
22:5 n-3 DPA [5]	-	-	0.06	-	-	0.05
SFA [6]	25.26	24.21	24.85	25.00	24.46	24.15
MUFA [7]	21.79	20.36	20.54	28.97	26.35	26.25
PUFA n-6 [8]	47.60	33.72	34.04	42.14	26.29	27.02
PUFA n3 [9]	5.35	21.71	20.58	3.90	22.90	22.58

[1] LA = linoleic acid; [2] ALA = linolenic acid; [3] AA= arachidonic acid, [4] EPA = eicosapentaenoic acid; [5] DPA = docosapentaenoic acid; [6] SFA = saturated fatty acids ∑14:0, 16:0, 17:0, 18:0, 20:0; [7] MUFA = monounsaturated fatty acids ∑ 16:1, 18:1n-9, 18:1 n-7, 22:1 n-11; [8] PUFA n-6 = polyunsaturated fatty acid omega 6 ∑ 18:2 n-6, 18:3 n-6, 20:4 n-6; [9] PUFA n-3 = polyunsaturated fatty acid omega 3 ∑ 18:3 n-3, 20:5 n-3, 22:5 n-3; 1CNT = control diet of first cycle; 1ELS5% = 1CNT+ 5% ELS; 1LPP3.5% = 1CNT+ 3.5% ELS and 0.2% *PP* extract. 2CNT = control diet of second cycle; 2ELS5% = 2CNT+ 5% ELS; 2LPP5% = 2CNT + 5% ELS and 0.2% *PP* extract.

2.2. Carcass Dissection and Meat Sampling

Twelve rabbits from each group were individually weighed before slaughter during each weekly session. The handling of animals during the trial adhered to Legislative Decree No. 146, which enforces Directive 98/58/EC. The slaughtering process was conducted in a certified slaughterhouse, located on the same farm, so there was no transport involved for the animals. The animals were stunned by electronarcosis. Carcasses were immediately chilled at 4 °C, and four carcasses per group were randomly selected for analysis, resulting in a total of 64 carcasses per cycle (16 per experimental group). Each week, the 12 selected carcasses (without head) were transported in refrigerated boxes to CREA-ZA laboratory. Twenty-four hours after slaughter, the carcasses were weighed to determine refrigerated carcass weight and were dissected according to the recommendations of the Worlds Rabbit Science Association (WRSA) [36]. The weights of the carcass excluding the liver, lungs, thymus, esophagus, heart, and kidneys were recorded as the reference carcass weight (RCW). Additionally, the weights of the liver (LvW), kidneys (KiW), abdominal fat (AFW), and scapular fat (SFW), as well as the percentage of the carcass regions defined as fore leg (FLP), thoracic part (TP), and hind part (HPP), were also recorded. For each carcass, the two *Longissimus thoracis et lumborum* (LTL) muscles were collected for physical analyses, including pH, color, drip and cooking loss, and shear force on cooked meat. The right hind leg was carefully dissected to separate and weigh the bone, muscle, and the combined intermuscular fat and connective tissues. This process allowed for the estimation of the percentage distribution of tissues, classified as hind leg bone percentage (HLBP), hind leg meat percentage (HLMP), and hind leg fat and other tissue percentage (HLFP). Additionally, the meat-to-bone ratio (M/B) of the hind leg was calculated using the formula: M/B = HLMP/HLBP. A sample from the thigh was obtained during dissection from the first portion of muscles near the distal part of the femur, as this area contains several muscles composed of both white and red fibers. The hind leg muscles, after removing the visible fat and tendons (THM), were finely ground under controlled conditions at 4 °C to prevent overheating.

Fresh thigh samples were analyzed for proximate composition, including moisture, protein, fat, and ash content (expressed as g/100 g of meat). The remaining samples were vacuum-packed and frozen at −80 °C for subsequent chemical analyses, including vitamin E, coenzyme Q10, cholesterol content, fatty acids profile, and mineral concentrations (macroelements: phosphorus, potassium, sodium, magnesium; microelements: iron, copper, zinc).

2.3. Physical Analysis

2.3.1. pH

Immediately after dissection, two samples of 1 g of minced LTL muscle were mixed with 10 mL of NaCl solution (0.1 mol/L) at pH 7. The mixture was homogenized using a homogenizer (T 25 digital ULTRA-TURRAX®—IKA) for 30 s at 6000 rpm. The pH of each sample was measured using a digital pH meter with temperature compensation (XS Instrument Serie 80 PC80, Giorgio Bormac S.r.l., Carpi, Italy), calibrated at pH 7 and pH 4. The final value was obtained as the mean of the two measurements.

2.3.2. Color

Color was recorded on the LTL muscle using the CIELAB system [37] to estimate lightness (L*), redness (a*), and yellowness (b*), with D illuminant (6504°K, daylight) using a Konica Minolta CM-3600 D (Sensing, Inc., Osaka, Japan) color spectrophotometer. The muscle was cut along its longitudinal line and maintained to air for 30 min to allow a blooming effect. Color data were obtained as the means of four measurements per sample.

2.3.3. Water Holding Capacity

Water holding capacity (WHC) in the LTL muscle was determined using two samples of approximately 1 g each. The samples were finely minced, placed between two paper discs, and subjected to an external force of 1 kg for 5 min. Following compression, the samples were reweighed, and the weight difference was expressed as a percentage of WHC. This analysis was performed in duplicate [38].

Cooking loss was evaluated according to the method described by Honikel and Hamm [38]. Two weighed portions of about 5 cm, one from each LTL muscle (right and left) at the same anatomical position, were sealed in plastic bags under light vacuum conditions and cooked in a water bath until reaching 75 °C at the core of the sample for approximately 20 min and then cooked in ice water. From this determination, we use a portion of both LTL muscles measuring approximately 5 cm in length.

2.3.4. Shear Force

Shear force was determined using a Warner–Bratzler Shear (WBS) apparatus on a dynamometer (Instron 5543 Single Column Table Top Tensile Tester System, Instron, Norwood, MA, USA) on cooked meat samples measuring 1 × 1 cm in thickness and 2 cm in length, cut along the fiber direction of muscles. The speed of the crosshead was set at 100 mm/min, as reported in Contò et al. [39]. Cooked meat samples were obtained as described above for the cooking losses on LTL muscles. The final shear force value was the mean of four determinations and expressed in Newton (N).

2.4. Chemical Analysis

2.4.1. Proximate Composition

The proximate composition of THM and pellet was analyzed according to AOAC [40] methods for dry matter (DM, method 934.01), ether extract (EE, method 920.39), ash (method 942.05) and crude protein (CP, method 984.13). Immediately after dissection, 8 g of samples was used for dry matter determination at 102 °C for 24 h. On the same sample, the ash % was determined at 540 °C for 8 h. Ether extract was identified using a traditional Soxhlet extraction method with diethyl ether (Tecator Soxtec System HT 1043 extraction unit Gemini, Apeldoorn, Sweden). Protein content (N × 6.25) was measured using the Kjeldahl method with Tecator Digestion System and Kjeltec Auto 1030 Analyzer (FOSS, Hillerød, Denmark).

2.4.2. Vitamin E

Vitamin E (tocopherol) was extracted from minced THM after saponification phase with 60% KOH at 70 °C per 1 h, as reported in Gąsior et al. [41]. The vitamin was recovered with hexane/ethyl acetate (9:1 vol:vol), dried and suspended in methanol. The extract was injected into an HPLC Alliance 2695, Waters (Waters Corporation, Framingham, MA, USA) with a Kinetex 5 µm EVO (Phenomenex, Torrance, CA, USA) C18 column with isocratic mobile phase, methanol/Waters (95:5 vol:vol) and read in fluorescence at λEX = 295 nm, λEM = 350 nm. The vitamin E peak was identified by comparison with the standard tocopherol peak and expressed as µg/g of meat [42].

2.4.3. Coenzyme Q10

The analysis of coenzyme Q10 was performed on minced THM following the method described by Mattila et al. [43]. Briefly, 1 g of meat was homogenized in a sodium chloride solution, and the sample was washed twice with an organic solvent mixture (ethanol/n-hexane 1:1 *v*/*v*) to separate the coenzyme Q10. The extract was then filtered, evaporated, and re-suspended in n-propanol. A sample of 20 µL was recovered and injected into an HPLC Alliance 2695, Waters (Waters Corporation, Framingham, MA, USA), with a Gemini NX 5 µm (Phenomenex, Torrance, CA, USA) C18 column. The isocratic phase consisted (methanol/ethanol/2-propanol/ammonium acetate buffer at concentrations of 53:21:21:1, respectively), and detection was performed at 275 nm. The peak representing the coenzyme Q10 was identified compared to the coenzyme Q10 standard peak and expressed as mg of Q10/100 g of meat.

2.4.4. Cholesterol

Cholesterol content was performed following the procedures of Maraschiello et al. [44]. Briefly, 0.1 g of minced THMt was saponified with 0.5 N KOH methanolic solution, with

BHT, at 80 °C for 1 h. After cooling, a saturated sodium chloride solution was added, and the unsaponifiable fraction, containing cholesterol, was extracted twice with a mixture of ethylic ether/hexane (1:1 vol:vol). Subsequently, the extracted fraction was then dried under nitrogen flow and re-suspended in acetonitrile/isopropanol (1:1 vol:vol) and injected in HPLC Alliance 2695, Waters (Waters Corporation, Framingham, MA, USA) with Synergi 4µ (Phenomenex, Torrance, CA, USA) C18 column with isocratic phase acetonitrile/isopropanol (55:45 vol:vol) and read at 210 nm. The cholesterol was identified by comparing the sample peak with standard cholesterol peak and expressed in mg/100 g of meat.

2.4.5. Fatty Acids Profile

The fat was extracted from approximately 6 g of minced THM or 2 g of pellet using a chloroform/methanol mixture (2:1 vol:vol). After suspension in hexane, the fat was methylated following the IUPAC procedure [45] using a 2 M of methanolic KOH solution, obtaining fatty acid methyl esters (FAMEs) solubilized in hexane. The FAMEs were quantified using a gas-chromatography (GC) system (GC 6890N, Agilent, Inc., Santa Clara, CA, USA) equipped with a flame ionization detector (FID) and a CP-Sil88 fused silica capillary column (100 m 0.25 mm (internal diameter) with 0.2 µm film thickness; Agilent Technologies). Fatty acid extraction, methylation, and gas chromatograph conditions were reported in detail by Failla et al. [46]. An internal standard C19:0 was added to the samples before the fatty acid extraction.

Fatty acid methyl esters were identified by comparing the retention time peaks of each compound with standard peaks from the Supelco mix 37 FAME and docosapentaenoic acid (Sigma-Aldrich Merck, Darmstadt, Germany). The different classes of fatty acids, including saturated fatty acids (SFAs), MUFAs, PUFAs, PUFA n-6, PUFA n-3 and n-6/n-3 ratio, were calculated and expressed as a % of total FAMEs.

2.4.6. Mineral Content

The mineral composition was estimated using Near-Infrared Spectroscopy (NIR). Samples of THM were lyophilized, and the NIR spectra were acquired using the NIRFLEX 500 (Büchi, Flawil, Switzerland). The data were processed using a previously developed Partial Least Squares (PLS) model, which was calibrated on 120 rabbit meat samples. Before application, the model was revalidated using 18 meat samples (three from each diet group), for which mineral content was analyzed following the methods specified below.

Minerals such as potassium, sodium, iron, copper and zinc were analyzed using PerkinElmer atomic absorption spectroscopy (AAS, PerkinElmer, Shelton, CT, USA) and a microwave Ethos 900 (Milestone, Shelton, CT, USA) for the mineralizing phase. Samples of 1 g of meat were mineralized with 64% nitric acid and hydrogen peroxide using the Ethos 900 microwave, and the solution was then read in AAS using a multi-channel lamp. A characteristic wavelength for each mineral was used to measure absorbance, and the results were expressed as mg/100 g or mg/kg of meat for microelements. For phosphorus, a photometric procedure was used [47]. After mineralization phase, the sample was reacted with a molybdate reagent (1%) and hydrazine sulfate (0.2%), and the phosphorus content was read and quantified in a PerkinElmer spectrophotometer Lambda 25 (PerkinElmer, Shelton, CT, USA) at 730 nm, with results expressed as mg/100 g of meat.

The dataset obtained from these analyses was used as a validation set. The results showed a coefficient of determination (R^2) greater than 0.83 for all minerals, with an RPD (Ratio of Performance to Deviation) of approximately 3. The RPD was calculated as SD/SEP × 100, where SD is the standard deviation of the mineral data and SEP is the standard error of prediction.

2.5. Additional Chemical Analysis on Pelleted Feed

A sample of about 1 kg was finely ground using a mill (Cemotec 1090 sample mill, FOSS, Hilleroed, DK, EU), to obtain a uniform sample both in composition and particle size. The following analytical determinations were performed on feedstuff used as proximate composition, including dry matter (DM), ash, crude protein (CP), and crude fat (EE), using the same methods and instruments as those described for meat analysis [48]. Calcium, phosphorus, vitamin E, and fatty acids were also analyzed following the procedures described for meat samples [41].

2.5.1. Crude Fiber

Crude fiber (CF) was determined as reported in AOAC [40] methods 962.09 ISO 5498:2000, using a fiber analyzer (FIWE Fiber Extractor, Velp Scientifica, New York, NY, USA). A minced feed sample of 1 g ± 0.1 was weighed and placed in a pre-weighed crucible and then boiled in 0.12 M sulfuric acid for 30 min. Afterward, the sample was washed with distilled water to neutralize the pH and remove residues. The sample was then digested in a 0.22 M alkaline f potassium hydroxide solution. After alkaline digestion, the sample was washed with distilled water to neutralize the pH, and then defatted with acetone. The crucible was dried in an oven at 100 °C, and the weight was recorded. The sample was incinerated in an oven at 550 °C for 2 h and weighed again. CF is expressed in g/100 g of sample.

2.5.2. Starch

Starch was quantified by the enzymatic colorimetric method as reported in AOAC method 996.11 [40]. The procedure was performed using an assay kit from Megazyme (Megazyme International, Wicklow, IE, EU). Briefly, 100 mg of feed was mixed with an 80% ethanol solution (vol:vol), followed by the addition of 3 mL of α-amylase and incubation for 11 min. Afterward, 0.1 mL of amyloglucosidase was added, and the sample was incubated again. The sample was then centrifuged, and a portion of the supernatant was incubated for 20 min at 50 °C. The starch content was read at 510 nm, and expressed as g/100 g of sample.

2.5.3. Amino Acids

Total amino acids were determined using AccQ•Tag Fluor reagent kit (Waters, Milford, MA, USA) [Water manual] on an HPLC system (Waters Alliance 2695, Waters Corporation, Framingham, MA, USA) with an AccQ•Tag column (3.9 × 150 mm, Waters Corporation, Framingham, MA, USA). A sample of about 1 g was hydrolysed with HCl 6 N for 14 h at 110 °C, except for tryptophan, for which NaOH 4 M was used. After hydrolysis, the samples were appropriately diluted and derivatized with the AccQ•Tag reagent, then detected by fluorescence at λEX = 205 nm and λEM = 395 nm. The gradient mobile phase consisted of Waters buffer/acetonitrile/water (vol:vol:vol). Peak identification was performed by comparing the sample peaks with the standard peaks from the AccQ•Tag kit, and results were expressed as g/100 g of meat.

2.5.4. Nitrogen-Free Extract and Digestible Energy

The nitrogen-free extract (NFE) content was calculated according to the following formula:

$$NFE(g/100) = DM - (CF + EE + CF + ash) \quad (1)$$

Based on the identified chemical composition, the digestible energy (kcal/kg) of the feed was calculated using the equation provided by the EU guidelines for rabbit feed [48].

$$\text{Digestible energy (kcal/kg)} = -1801 + 7.10\ \text{CP} + 12.01 \times \text{EE} + 5.59 \times \text{NFE} \quad (2)$$

2.6. Statistical Analysis

To evaluate the differences among the six groups of rabbits, a mixed model was employed, incorporating two fixed factors and their interaction, with all physical and chemical composition data used as dependent variables. In this model, feed group and cycle were treated as fixed factors, while the weekly session was considered a random effect. The analysis was conducted using the MIXED procedure, on the SAS statistical package (SAS Institute Inc., Cary, NC, USA). Mean comparisons were performed using the Student's *t*-test, with the significance threshold set at $p < 0.05$. This approach ensures a robust and statistically sound evaluation of the physical and chemical composition differences across feed groups while accounting for potential confounding factors and individual variability.

3. Results and Discussion

3.1. Carcass Characteristics

The data presented in Table 4 provide a comparative analysis of live weight and carcass characteristics of rabbits fed different diets across two cycles. During the growth stage, the rabbits had an average daily weight gain of 32.4 ± 3.2 g for C1 and 35.6 ± 2.8 g for C2 across the three experimental groups. Daily pellet consumption averaged 133.5 g, with no significant differences observed between cycles or groups.

Rabbit live weight (LW) in C1 ranged from 2398 g (1ELS5%) to 2477 g (1CNT), while in C2, it varied from 2467 g (2ELS5%) to 2541 g (2CNT). However, these differences were not statistically significant ($p > 0.05$) either among groups or between the two cycles. These results are consistent with those of Pla et al. [35], who did not observe detrimental effects of sunflower oil and linseed on rabbit LW. Similarly, the chilled carcass weight (CCW) in C1 ranged from 1345.3 g (1ELS5%) to 1390.1 g (1CNT), and in C2 from 1399.9 g (2ELS5%) to 1441.0 g (2CNT), with no significant differences observed among the groups and cycles. However, slight variability was observed between cycles for CCW, in which C2 reported a trend ($p = 0.086$) higher than C1 (1425.8 g vs. 1368.4 in average for the groups, respectively, for C2 vs. C1).

The LvW, SFaW, and PFaW did not show significant differences among groups in either cycle. The average values for all groups were 79.6 ± 11.7 g for liver weight, 15.3 ± 1.5 g for scapular fat, and 7.9 ± 2.6 g for perirenal fat. Regarding carcass composition, significant differences were observed between cycles for fore leg (FLP) and hind part (HPP). However, in C2, the groups receiving ELS supplementation (2ELS5% and 2LPP5%) tended to show a higher percentage of HPP ($p = 0.063$).

The tissue composition of the thighs showed significant differences in HLFP between cycles, with all three groups in C2 exhibiting higher values ($p = 0.042$). For HLBP and HLMP, significant differences were observed among groups only in C2 ($p = 0.002$ and $p = 0.015$, respectively). The 2CNT group exhibited lower HLBP and higher HLMP compared to the other groups. This trend also impacted the M/B ratio, where significant differences were observed in C2, with the 2CNT group showing a significantly higher M/B ratio ($p = 0.005$) compared to the other two groups. This observed tendency toward greater muscle trophism in the control group in C2, reflected in the enhanced development of the HPP region, was due to the higher dietary intake of oleic and linoleic acids, which appear to have a confirmed ability to promote muscle growth as reviewed by Abreu et al. [49]

Table 4. Live weight and carcass characteristics of rabbits fed on different diets.

	First Cycle (C1)				Second Cycle (C2)						
	1CNT	1ELS5%	1LPP3.5%	*p* Value [1]	2CNT	2ELS5%	2LPP5%	*p* Value [2]	RMSE	*p* Value Cycles	*p* Value Groups
LW (g)	2477	2398	2631	ns	2541	2467	2511	ns	233.1	0.137	0.428
CCW (g)	1390.1	1345.3	1369.8	ns	1441.0	1399.9	1436.4	ns	131.5	0.086	0.428
LvW (g)	80.74	76.27	79.26	ns	79.64	81.28	80.69	ns	11.94	0.467	0.877
SFaW (g)	7.68	7.55	8.82	ns	7.69	7.76	7.98	ns	2.67	0.709	0.454
PFaW (g)	9.61	9.31	9.40	ns	10.15	10.07	9.96	ns	2.93	0.304	0.957
FLP (%)	25.22	25.77	25.53	ns	25.04	24.55	24.48	ns	1.25	0.033	0.575
TP (%)	37.49	37.11	37.41	ns	37.23	36.92	37.27	ns	1.54	0.209	0.232
HPP (%)	37.29	37.12	37.06	ns	37.74 ^{b}	38.53 ^{a}	38.25 ^{a}	0.063	1.21	0.008	0.071
HLBP (%)	18.98	19.39	19.45	ns	18.55 ^{b}	19.56 ^{a}	19.31 ^{a}	0.002	0.681	0.603	<0.001
HLFP (%)	1.55	1.46	1.42	ns	1.61	1.51	1.58	ns	0.135	0.042	0.288
HLMP (%)	79.48	79.15	79.14	ns	79.84 ^{a}	79.03 ^{b}	79.11 ^{b}	0.015	0.713	0.321	0.018
M/B	4.19	4.08	4.07	ns	4.30 ^{a}	4.04 ^{b}	4.10 ^{ab}	0.005	0.184	0.378	0.005

[1] *p* = value between groups of first cycle; [2] = *p* value between groups of second cycle; RMSE = root mean square error; a, b = different letters in the same row mean significant difference for $p < 0.05$; ns = non-significant. 1CNT = control diet of C1; 1ELS5% = 1CNT+ 5% ELS; 1LPP3.5% = 1CNT+ 3.5% ELS and 0.2% *PP* extract. 2CNT = control diet of C2; 2ELS5% = 2CNT+ 5% ELS; 2LPP5% = 2CNT with 5% ELS and 0.2% *PP* extract. LW = live weight; CCW = chilled carcass weight; LvW = liver weight; SFaW = scapular fat weight; PFaW= dissectible perirenal and abdominal fat weight; FLP = fore legs; TP = thoracic cage; HPP = hind part; HLBP = hind part separate bone; HLMP = hind part separate meat; HLFP= hind part separate fat and other tissues; M/B = meat-to-bone ratio of the hind leg (HLMP/HLBP).

In contrast, Dal Bosco et al. [50] reported that whole linseed supplementation had no effect on rabbit carcass traits and that the diet did not influence the meat-to-bone ratio (4.06 ± 0.06) or the percentage of perirenal (3.33 ± 0.16%) and scapular fat (0.76 ± 0.03%).

In this study, the supplemented diets had a limited impact on overall rabbit growth but contributed to the enhancement of specific high-value carcass traits, such as muscle development, particularly in the C2. This effect is likely attributable to the lower linoleic acid content and the higher oleic fatty acid levels in the C2 diets. These findings highlight the importance of long-term dietary strategies and targeted nutritional interventions in enhancing rabbit meat production. Moreover, these results align with previous studies on the inclusion of n-3 PUFA-rich vegetable sources in rabbit diets, which have consistently shown no significant impact on major carcass traits [35,50].

Further investigation is needed to better understand the relationship between diet and meat quality, particularly concerning the bioavailability of nutrients from experimental feed ingredients [1].

3.2. Physical Characteristics of LTL of Rabbit Feed on Different Diets

Table 5 shows the physical characteristics of the LTL muscle from rabbits fed with different diets across two rearing cycles. In both cycles, the different diets did not significantly affect the physical composition of this muscle. These data agreed with some authors that reported no linseed effect on rabbit meat physical analyses [20]. However, when comparing the two cycles, the groups of C2 on average showed higher values for pH, L* (lightness), and WHC (p= 0.002, 0.001, and 0.002, respectively). In contrast, lower values were observed for cooking loss and shear force in cooked meat ($p = 0.014$ and $p = 0.018$, respectively).

The pH of meat plays a crucial role in influencing various quality attributes, such as WHC, color, and shear force [51]. In this study, pH values remained consistent across groups within each cycle, indicating that the diets containing ELS and PP extract did not affect the muscle's post-mortem biochemical properties. These pH values were similar to those reported by Mattioli et al. [52] on rabbits fed diets supplemented with linseed. In our study, the significant difference in pH between cycles suggests that pre- or post-slaughter factors may have influenced this parameter [51].

Table 5. Physical characteristics of *Longissimus thoracis et lumborum* of rabbits fed different diets.

	First Cycle (C1)				Second Cycle (C2)						
	1CNT	1ELS5%	1LPP3.5%	*p* Value [1]	2CNT	2ELS5%	2LPP5%	*p* Value [2]	RMSE	*p* Value Cycles	*p* Value Groups
pH	5.73	5.75	5.72	Ns	5.83	5.84	5.77	ns	0.123	0.002	0.208
L*	48.32	46.91	47.20	Ns	50.94	51.05	52.39	ns	3.08	<0.001	0.139
a*	−2.45	−2.60	−2.49	Ns	−1.88	−2.17	−2.44	ns	0.728	0.081	0.225
b*	3.28	3.04	3.81	Ns	3.60	3.90	3.21	ns	1.344	0.485	0.979
WHC (%)	13.02	13.13	13.69	Ns	13.98	14.16	14.45	ns	1.34	0.002	0.218
Cooking loss (%)	28.06	27.02	27.38	Ns	26.53	26.34	26.62	ns	1.93	0.014	0.448
WBS (N) raw	12.47	11.78	14.47	Ns	12.93	13.30	13.13	ns	2.75	0.701	0.133
WBS (N) Cooked	21.69	26.56	24.47	Ns	21.99	22.09	20.59	ns	5.47	0.018	0.183

[1] p = value between groups of the first cycle; [2] = p value between groups of second cycle; RMSE = root mean square error; 1CNT = control diet of C1; 1ELS5% = 1CNT+ 5% ELS; 1LPP3.5% = 1CNT+ 3.5% ELS and 0.2% *PP* extract. 2CNT = control diet of C2; 2ELS5% = 2CNT+ 5% ELS; 2LPP5% = 2CNT with 5% ELS and 0.2% *PP* extract. WHC = water holding capacity, L* = lightness; a* = a index, redness; b* = b index, yellowness; WBS = shear force with Warner–Bratzler dispositive.

The WHC was lower in C1 than in C2 (13.28% vs. 14.20%, respectively, as the average of the three groups). This difference was likely due to the higher pH of C2 meat, which may have affected protein structure and degradation, consequently influencing WHC [4]. As a result, cooking loss was lower in C2 than in C1 (27.49% vs. 26.59%, on average for the groups in each cycle), likely due to the higher drip loss in C2.

In general, meat pH and lightness are negatively correlated. However, in the LTL of the C2, despite having a significantly higher pH, greater lightness was observed. The reason for this may lie in the changes induced by the diet. Specifically, the higher percentage of MUFAs and the lower levels of n-6 PUFAs in the diets of the second cycle increased the fat content, which is known to result in a paler meat color [53]. Color is usually an indicator of freshness and safety, influencing consumer purchase intention [54,55]. The higher presence of fat in the animals and a higher pH that promote greater proteolysis likely affected the WBS values of cooked meat (p = 0.014), making the LTM more tender in the second cycle compared to the first.

Previous studies on rabbit meat quality following dietary supplementation with linseed [56] and brown algae [57] found no significant effects of these ingredients on meat color [18,53,58].

3.3. Chemical Characteristics

The chemical composition of thigh meat is reported in Table 6. In both C1 and C2, no significant differences were observed in proximate composition among groups.

Between the two cycles, the percentage of ash and fat were significantly higher in C2 compared to C1 (p = 0.019 and 0.001, respectively), whereas moisture content was higher in C1 than C2 (75.57 vs. 74.98% in means for the three groups for each cycle). This trend was due to a negative correlation between moisture and fat percentage content [59]. These differences between the two cycles are consistent with the greater amount of HLFP observed in the thigh in C2, probably due to the higher percentage of MUFAs in the diet.

Significant differences were reported between diets for vitamin E and Q10 (Table 6), while for cholesterol, only a trend toward significance was observed (p = 0.109). Among the different diets within the same cycle, the highest vitamin E values were observed in the carcasses belonging to the rabbits fed with *PP* extract (p < 0.001 in both cycles). As reported by Caf et al. [33], marine algae such as *PP* are naturally rich in phenolic compounds, antioxidants, and fat-soluble vitamins, including vitamin E. Prates [60] found that algae significantly contribute to vitamin fortification, particularly vitamin E, which acts as an antioxidant to improve meat stability and shelf life. In this regard, Ribeiro et al. [61] reported that supplementing broiler diets with 2% microalgae increased vitamin E content in chicken meat, enhancing its nutritional value and health benefits for consumers.

Table 6. Chemical composition of thigh meat of rabbits fed different diets.

	First Cycle (C1)				Second Cycle (C2)						
	1CNT	1ELS5%	1LPP3.5%	*p* Value [1]	2CNT	2ELS5%	2LPP5%	*p* Value [2]	RMSE	*p* Value Cycles	*p* Value Groups
Moisture (%)	75.56	75.63	75.51	ns	75.04	75.08	74.83	ns	0.785	0.005	0.653
ash (%)	1.26	1.21	1.23	ns	1.30	1.28	1.28	ns	0.106	0.019	0.474
Fat (%)	2.50	2.59	2.40	ns	2.80	2.92	2.87	ns	0.443	<0.001	0.504
Protein (%)	20.68	20.57	20.85	ns	20.86	20.72	21.01	ns	0.718	0.255	0.284
Vitamin E (mg/100 d)	2.38 b	2.56 b	3.09 a	<0.001	2.26 c	2.61 b	3.31 a	<0.001	0.426	0.547	<0.001
Q10 (mg/100 g)	19.24 b	20.33 a	20.09 ab	0.046	19.61 b	20.34 a	20.54 a	0.031	1.251	0.289	0.006
Cholesterol (mg/100 g)	50.87	50.36	50.24	ns	50.65 ab	50.98 a	48.83 b	0.054	2.76	0.551	0.109

[1] *p* value between groups of first cycle; [2] *p* value between groups of second cycle; RMSE = root mean square error; a, b = different letters in the same raw means significant difference for $p < 0.05$; ns = non-significant value. 1CNT = control diet of C1; 1ELS5% = 1CNT+ 5% ELS; 1LPP3.5% = 1CNT+ 3.5% ELS and 0.2% *PP* extract. 2CNT = control diet of C2; 2ELS5% = 2CNT+ 5% ELS; 2LPP5% = 2CNT with 5% ELS and 0.2% *PP* extract.

In C1, coenzyme Q10 levels were higher in 1ELS5% compared to 1CNT (20.33 mg/100 g vs. 19.24 mg/100 g, $p = 0.043$), but not significantly different from 1LPP3.5%. In C2, both 2ELS5% and 2LPP5% showed the highest coenzyme Q10 levels compared to the 2CNT diet (20.44 mg/100 g as average of diets with ELS vs. 19.61 mg/100 g, $p = 0.031$). Previous studies have reported high concentrations of coenzyme Q10 in vegetables and algae. However, the inclusion of ELS and PP in the diets could explain the lower levels observed in the CNT diets [62].

In C1, cholesterol levels did not significantly differ among groups. However, in C2, the 2LPP5% diet tended to reduce cholesterol levels in the meat. The observed differences in cholesterol content in the 2LPP5% group may be due to the PP supplementation, although this effect may have been limited due to the low inclusion level of PP in the diet [32].

Similarly, Vlaicu et al. [63], in their study on the nutritional composition and bioactive compounds of basil, thyme, and sage, demonstrated that plant additives significantly reduced cholesterol concentration in broiler thigh meat. Martin et al. [64], in their study on the effect of dietary inclusion of Spirulina on meat quality traits in piglets, reported a reduction in cholesterol concentration.

These results reinforce the notion that incorporating bioactive-rich plant additives, such as PP and other functional compounds, into animal diets can improve meat nutritional quality by lowering cholesterol levels, thus meeting consumer expectations for healthier meat products.

3.4. Fatty Acid Composition

A fatty acid profile on THM revealed significant effects of diet and cycle on the principal fatty acid classes (Table 7). SFAs were the predominant components, although rabbit meat contains high amounts of MUFAs and PUFAs, compared to other livestock meats [65].

SFA levels were significantly lower in both cycles for rabbits fed the supplemented diets compared to their respective control groups. The lowest SFA levels were observed in the groups fed 2LPP5% ($p < 0.001$) followed by 2ELS5%, 1ELS5% and 1LPP3.5%.

Table 7. Principal fatty acid classes expressed as percentage of total FAMEs on thigh meat of rabbits fed different diets.

	First Cycle (C1)				Second Cycle (C2)						
	1CNT	1ELS5%	1LPP3.5%	*p* Value [1]	2CNT	2ELS5%	2LPP5%	*p* Value [2]	RMSE	*p* Value Cycles	*p* Value Groups
SFA	36.30 [a]	34.35 [b]	35.09 [ab]	<0.001	35.83 [a]	34.52 [b]	33.34 [c]	<0.001	1.223	0.007	<0.001
MUFA	23.77	24.88	24.07	ns	29.88 [a]	28.11 [b]	27.08 [c]	<0.001	1.576	<0.001	0.006
PUFA n-6	35.63 [a]	29.09 [b]	29.83 [b]	<0.001	30.18 [a]	24.58 [c]	26.38 [b]	<0.001	1.601	<0.001	<0.001
PUFA n-3	4.06 [b]	11.46 [a]	10.78 [a]	<0.001	3.93 [b]	12.57 [a]	12.86 [a]	<0.001	0.954	<0.001	<0.001
PUFA	39.88	40.72	40.79	ns	34.20 [c]	37.26 [b]	39.44 [a]	<0.001	1.545	<0.001	<0.001
n-6/n-3	9.05 [a]	2.57 [b]	2.80 [b]	<0.001	8.10 [a]	1.98 [b]	2.06 [b]	<0.001	1.243	0.003	<0.001

[1] *p* = value between groups of the first cycle; [2] = *p* value between groups of second cycle; RMSE = root mean square error; a, b, c = different letter in the same row means significant difference for $p < 0.05$; ns = non-significant value. SFA = saturated fatty acid (10:00, 12:00, 13:00, 14:00, 15:0iso, 14:01, 15:0anteiso, 15:00, 16:0iso, 16:00, 17:0iso, 17:0anteiso, 17:00, 18:0iso, 18:00, 20:00, 21:00, 22:00 23:00, 24:00); MUFA = monounsaturated fatty acid (14:1, 15:1, 16:1n-9, 16:1n-7, 17:1 n-7, 18:1trans-9, 18:1trans-11, 18:1n-9, 18:1n-7, 20:1n-9, 20:1n-7, 22:1n-11, 24:1n-15); PUFA n-6 = polyunsaturated fatty acids omega-6 (18:2n6, 18:2n6 trans isomer, conjugate linoleic acids (CLAs), 18:3 n-6, 20:2 n-6, 20:3 n-6, 20:4 n-6, 22:2 n-6, 22:4 n-6); PUFA n-3 = polyunsaturated fatty acids omega 3 (18:3 n-3, 20:3 n-3, 20:5 n-3, 22:5 n-3, 22:6 n-3); 1CNT = control diet of C1; 1ELS5% = 1CNT+ 5% ELS; 1LPP3.5% = 1CNT+ 3.5% ELS and 0.2% *PP* extract. 2CNT = control diet of C2; 2ELS5% = 2CNT+ 5% ELS; 2LPP5% = 2ELS5% with 5% ELS and 0.2% *PP* extract.

In the C1, the THM muscle exhibited the highest SFA content among the three groups compared to C2 ($p = 0.007$) likely due to differences in the fatty acid composition of the diets between the two cycles. This change naturally led to a greater accumulation of MUFAs in rabbit meat of C2 compared to C1 (28.36% vs. 24.24%, respectively, on average of the three groups within cycles; $p < 0.001$). Significant differences in MUFA content among groups within each cycle were observed only in C2. The highest MUFA value was recorded for the 2CNT group, followed by progressively lower values for the 2ELS5% and 2LPP5% diets, ranging from 29.88% to 27.08%.

Contrasting percentages of n-6 PUFA were observed between the two cycles, with C1 showing a higher average across diets compared to C2 (32.52% vs. 27.04%, respectively, as average of the three groups within cycles). This difference was attributed to variations in the n-6 fatty acid composition of the diets used in the two cycles. Additionally, the n-6 PUFA percentage in THM was significantly influenced by groups within each cycle ($p < 0.001$), with the highest levels observed in the CNT groups of both cycles, followed by the PP-supplemented groups. Furthermore, differences in n-3 PUFA content were observed between cycles, although the increase was relatively modest (approximately +1 percentage point).

Similar trends have been reported in broiler meat, as shown by Ivanova et al. [66], where the inclusion of flaxseed oil in the diet decreased SFA levels and increased PUFA levels, particularly due to the higher content of linoleic and linolenic acids in the lipids. In their study, a diet supplemented with 3% flaxseed oil had the greatest influence on the lipid composition of broiler meat, aligning with the findings of our study.

The n-3 PUFA content in rabbit meat significantly increased ($p < 0.001$) in rabbits fed diets containing ELS and PP extract compared to CNT diets. These findings highlight the potential of linseed and algae supplementation to improve the lipid profile of rabbit meat, making it more beneficial for human health. The results are consistent with previous research showing that dietary n-3 PUFA supplementation can improve the fatty acid profile of rabbit meat [18,28,32,58,67]. For example, linseed supplementation in rabbits has been shown to significantly increase n-3 PUFA content while reducing SFA and MUFA levels in both the *longissimus dorsi* muscle and perirenal fat [68].

The total PUFA content showed lower differences in the C1 cycle, as the opposing trends observed for n-6 and n-3 PUFAs balanced each other out. In the C2 cycle, the large differences in n-3 PUFA content among groups influenced the total PUFA level, reflecting the trend of n-3 PUFA. The lowest total PUFA value was recorded in 2CNT, followed by 2ELS5%, with the highest value observed in 2LPP5%. Similarly, Marino et al. [69] observed that including linseed in the diet increased the content of n-3 PUFA and MUFA, while reducing the proportions of SFA and n-6 PUFA in beef meat. Likewise, Atti et al. [70] observed a significant increase in n-3 PUFA content in lamb meat with the incorporation of extruded linseed in the diet. These findings further support the efficacy of linseed supplementation in improving the lipid profile of meat across different species.

In this study, the n-6/n-3 ratio was significantly different between cycles and groups ($p < 0.001$ for both). This ratio is a crucial health indicator, and its reduction is considered beneficial for human health [50,71]. Notably, in rabbits fed diets supplemented with ELS, this ratio decreased from 9.05 in the 1CNT group to 2.68 as average of the other two groups. Although the 1ELS5% and 1LPP3.5% diets differed in ELS supplementation (5 vs. 3.5%), they did not show significant differences for this ratio. Similar reductions were observed in the second cycle (from 8.10 in the 2CNT to 2.02 on average for the other two groups).

These findings highlight the potential of linseed-enriched diets to improve the fatty acid profile of rabbit meat, particularly by increasing the relative proportion of n-3 PUFA, while lowering the n-6/n-3 ratio. A few several studies have investigated the impact of fish oil-enriched diets on the fatty acid composition of rabbit meat, consistently reporting a reduction in the n-6/n-3 ratio, with some values reaching as low as 1.61 [21,72,73]. These findings highlight the potential of dietary strategies aimed at reducing the n-6/n-3 ratio to improve the nutritional quality of rabbit meat for human consumption. In contrast, a study found a reduction in the n-6/n-3 ratio in lambs, which decreased from 5.23 in the control group to 0.71 and 1.1 in the group fed extruded linseed [70].

3.5. Macro- and Micromineral Elements

The analysis of macro- and micromineral content (Table 8) in rabbit THM revealed no significant differences in mineral content among groups, except for phosphorus, zinc, and copper between the two cycles. These results were anticipated, as the control diet was formulated to meet the recommended concentrations of macrominerals.

Table 8. Macro- and micromineral elements on rabbit thigh meat in two cycles and different diets.

	First Cycle (C1)				Second Cycle (C2)						
	1CNT	1ELS5%	1LPP3.5%	*p* Value [1]	2CNT	2ELS5%	2LPP5%	*p* Value [2]	RMSE	*p* Value Cycles	*p* Value Groups
P (mg/100 g)	238.83	240.65	240.72	ns	238.92 ^{b}	242.98 ^{ab}	246.92 ^{a}	0.014	7.555	0.116	0.035
K (mg/100 g)	377.79	380.37	379.97	ns	380.08	383.65	382.12	ns	8.36	0.135	0.328
NA (mg/100 g)	56.69	57.40	57.23	ns	56.91	56.35	56.05	ns	2.196	0.138	0.909
Mg (mg/100 g)	24.94	25.20	25.15	ns	25.03	25.51	25.39	ns	1.07	0.336	0.311
Mn (mg/kg)	0.34	0.33	0.33	ns	0.35	0.33	0.33	ns	0.029	0.094	0.233
Fe (mg/kg)	4.32	4.17	4.29	ns	4.36	4.23	4.42	ns	0.291	0.193	0.08
Zn (mg/kg)	11.27 ^{b}	11.90 ^{ab}	12.26 ^{a}	0.022	11.42 ^{b}	11.87 ^{ab}	12.52 ^{a}	0.012	1.015	0.538	<0.001
Cu (mg/kg)	0.83	0.91	0.89	ns	0.97	0.90	0.94	ns	0.096	0.004	0.843

[1] p = value between groups of first cycle; [2] P = value between groups of second cycle. RMSE = root mean square error; ^{ab} = different letter in the same row means significant difference for $p < 0.05$; ns = non-significant value. P mg/100 g = Phosphorus; K mg/100 g = Potassium; Na mg/100 g = Sodium; Mg mg/100 g = Magnesium; Mn mg/kg = Manganese; Fe mg/kg = Iron; Zn mg/kg = Zinc; Cu mg/kg = Copper; 1CNT = control diet of C1; 1ELS5% = 1CNT+ 5% ELS; 1LPP3.5% = 1CNT+ 3.5% ELS and 0.2% *PP* extract. 2CNT = control diet of C2; 2ELS5% = 2CNT+ 5% ELS; 2LPP5% = 2ELS5% with 5% ELS and 0.2% *PP* extract.

Phosphorus levels were significantly higher in the 2LPP5% diet compared to the control diet (246.92 mg/100 g vs. 238.92 mg/100 g) in C2. Zinc content also increased significantly with dietary supplementation in both cycles, with the highest levels observed in the 1LPP3.5% and 2LPP5% groups compared to their respective controls (12.26 mg/kg and 12.52 mg/kg vs. 11.27 mg/kg and 11.42 mg/kg, respectively). Marine macroalgae are well known for their high mineral content, particularly zinc [74]. Moreover, brown algae may have higher bioavailability of phosphorus, likely due to lower levels of anti-nutritional compounds, such as phytic acid [74]. Improved mineral bioavailability enhances the absorption and utilization of these nutrients by rabbits, potentially reducing the need for additional mineral supplementation in their diets. This approach could help maintain or even improve the nutritional quality of the meat [24].

Elevated levels of phosphorus and zinc not only enhance the nutritional value of rabbit meat but also align with the growing demand for functional foods that provide additional health benefits beyond basic nutrition. Furthermore, rabbit meat is uniquely characterized by its low sodium content and high levels of potassium and phosphorus compared to other meats such as pork, beef, and chicken [3,75]. These attributes make rabbit meat an excellent choice for diets aimed at preventing hypertension, further solidifying its role as a functional and health-promoting food [3].

4. Conclusions

This study demonstrates the significant effects of dietary supplementation with extruded linseed (ELS) and Padina pavonica (PP) algae extract on carcass performance and meat quality of rabbits. These dietary interventions notably improved the fatty acid profile of rabbit meat by reducing saturated fatty acids and the n-6/n-3 ratio, while increasing n-3 PUFA levels. Additionally, the supplemented diets enriched the meat with essential minerals, particularly phosphorus and zinc, and increased the levels of vitamin E and coenzyme Q10, enhancing both meat quality and its nutritional value. Meat analyses performed on *Longissimus thoracis et lumborum* (LTL) muscle and thigh meat (THM) revealed notable findings. While no significant differences in LTL physical quality were observed in the first production cycle, LTL brightness was significantly higher in the second cycle. Thigh meat in the second cycle also exhibited higher fat content. Importantly, the supplementation with ELS and PP extract demonstrated a consistent positive impact on the polyunsaturated fatty acid (PUFA) profile and the overall nutritional value of the meat. Although some variability was observed between the two breeding cycles, the results highlight the potential of these dietary strategies to produce healthier, nutrient-rich rabbit meat, offering significant benefits for human health.

Author Contributions: Conceptualization, N.E.F., M.C. (Michela Contò), S.F. and G.B.; methodology, M.C. (Michela Contò), M.C. (Marta Castrica), S.F., S.D.G., N.E.F., G.B. and G.R.; software, A.Q., S.A. and L.M.; validation, G.B., D.V., G.C., C.M.B. and S.F.; formal analysis, M.C. (Marta Castrica), S.A., L.M., S.F. and E.A.; investigation, N.E.F., M.C. (Michela Contò), M.C. (Marta Castrica) and S.F.; data curation, E.A., D.B., A.Q., M.C. (Michela Contò) and S.A.; writing—original draft preparation, N.E.F., M.C. (Michela Contò), S.F., M.C. (Marta Castrica), A.Q., G.R. and S.D.G.; writing—review and editing, N.E.F., G.C., S.F., G.B. and C.M.B.; visualization, N.E.F., M.C. (Michela Contò) and S.F.; supervision, N.E.F., M.C. (Michela Contò), S.F. and A.Q.; project administration, G.B., D.V., S.F. and C.M.B.; funding acquisition, G.B. and D.V. All authors have read and agreed to the published version of the manuscript.

Funding: This research was funded by the PRIMA project "Omega rabbit: food for health benefIT" (ORABBIT), G15F21000810001.

Data Availability Statement: The original contributions presented in the study are included in the article.

Acknowledgments: The authors would like to thank Giovanni Migni and Azienda Agricola Brachino Patrizia for their contribution.

Conflicts of Interest: The authors declare no conflicts of interest.

References

1. Cullere, M.; Dalle Zotte, A. Rabbit Meat Production and Consumption: State of Knowledge and Future Perspectives. *Meat Sci.* **2018**, *143*, 137–146. [CrossRef]
2. Hoffman, L.; Nkhabutlane, P.; Schutte, D.; Vosloo, C. Factors Affecting the Purchasing of Rabbit Meat: A Study of Ethnic Groups in the Western Cape. *J. Fam. Ecol. Consum. Sci.* **2010**, *32*, 26–35. [CrossRef]
3. Dalle Zotte, A.; Szendro, Z. The Role of Rabbit Meat as Functional Food. *Meat Sci.* **2011**, *88*, 319–331. [CrossRef]
4. Kumar, S.A.; Kim, H.J.; Jayasena, D.D.; Jo, C. On-Farm and Processing Factors Affecting Rabbit Carcass and Meat Quality Attributes. *Food Sci. Anim. Res.* **2023**, *43*, 197. [CrossRef] [PubMed]
5. Kumar, P.; Sharma, N.; Narnoliya, L.K.; Verma, A.K.; Umaraw, P.; Mehta, N.; Ismail-Fitry, M.R.; Kaka, U.; Yong-Meng, G.; Lee, S.J.; et al. Improving Quality and Consumer Acceptance of Rabbit Meat: Prospects and Challenges. *Meat Sci.* **2025**, *219*, 109660. [CrossRef]
6. Siddiqui, S.A.; Gerini, F.; Ikram, A.; Saeed, F.; Feng, X.; Chen, Y. Rabbit Meat—Production, Consumption and Consumers' Attitudes and Behavior. *Sustainability* **2023**, *15*, 2008. [CrossRef]
7. Szendro, K.; Szabó-Szentgróti, E.; Szigeti, O. Consumers' Attitude to Consumption of Rabbit Meat in Eight Countries Depending on the Production Method and Its Purchase Form. *Foods* **2020**, *9*, 654. [CrossRef]
8. Andoni, E.; Curone, G.; Agradi, S.; Barbato, O.; Menchetti, L.; Vigo, D.; Zelli, R.; Cotozzolo, E.; Ceccarini, M.R.; Faustini, M.; et al. Effect of Goji Berry (*Lycium barbarum*) Supplementation on Reproductive Performance of Rabbit Does. *Animals* **2021**, *11*, 1672. [CrossRef]
9. Vlaicu, P.A.; Untea, A.E.; Varzaru, I.; Saracila, M.; Oancea, A.G. Designing Nutrition for Health—Incorporating Dietary By-Products into Poultry Feeds to Create Functional Foods with Insights into Health Benefits, Risks, Bioactive Compounds, Food Component Functionality and Safety Regulations. *Foods* **2023**, *12*, 4001. [CrossRef] [PubMed]
10. Delis-Hechavarria, E.A.; Guevara-Gonzalez, R.G.; Ocampo-Velazquez, R.V.; Gomez-Soto, J.G.; Vargas-Hernandez, M.; Parola-Contreras, I.; Torres-Pacheco, I. Functional Food for Rabbits. Current Approaches and Trends to Increase Functionality. *Food Rev. Int.* **2023**, *39*, 2057–2074. [CrossRef]
11. Decker, E.A.; Park, Y. Healthier Meat Products as Functional Foods. *Meat Sci.* **2010**, *86*, 49–55. [CrossRef] [PubMed]
12. Metzler-Zebeli, B.U. The Role of Dietary and Microbial Fatty Acids in the Control of Inflammation in Neonatal Piglets. *Animals* **2021**, *11*, 2781. [CrossRef]
13. Manessis, G.; Kalogianni, A.I.; Lazou, T.; Moschovas, M.; Bossis, I.; Gelasakis, A.I. Plant-Derived Natural Antioxidants in Meat and Meat Products. *Antioxidants* **2020**, *9*, 1215. [CrossRef] [PubMed]
14. Agradi, S.; Sulce, M.; Menchetti, L.; Vigo, D.; Castrica, M.; Barbato, O.; Andoni, E.; Quattrone, A.; Munga, A.; Marongiu, M.L.; et al. Dietary Supplementation with N-3 Polyunsaturated Fatty Acids: Effects on Reproductive and Productive Performance and Meat Quality in Rabbit Breeding. *Anim. Nutr.* **2023**, *14*, 70–78. [CrossRef] [PubMed]
15. Cavani, C.; Petracci, M.; Trocino, A.; Xiccato, G. Advances in Research on Poultry and Rabbit Meat Quality. *Ital. J. Anim. Sci.* **2009**, *8*, 741–750. [CrossRef]
16. Xiccato, G. Feeding and Meat Quality in Rabbits: A Review. *World Rabbit. Sci.* **1999**, *7*, 75–86. [CrossRef]
17. Simopoulos, A.P. The Importance of the Omega-6/Omega-3 Fatty Acid Ratio in Cardiovascular Disease and Other Chronic Diseases. *Exp. Biol. Med.* **2008**, *233*, 674–688. [CrossRef]
18. Bianchi, M.; Petracci, M.; Cavani, C. Effects of Dietary Inclusion of Dehydrated Lucerne and Whole Linseed on Rabbit Meat Quality. *World Rabbit. Sci.* **2006**, *14*, 247–258. [CrossRef]
19. Siudak, Z.; Kowalska, D. Dietary Supplements Used in Rabbit Nutrition and Their Effect on the Fatty Acid Profile of Rabbit Meat—A Review. *J. Anim. Feed. Sci.* **2023**, *33*, 159–169. [CrossRef]
20. Bianchi, M.; Petracci, M.; Cavani, C. The Influence of Linseed on Rabbit Meat Quality. *World Rabbit. Sci.* **2009**, *17*, 97–107. [CrossRef]
21. Rodríguez, M.; Carro, M.D.; Valiente, V.; Formoso-Rafferty, N.; Rebollar, P.G. Supplementation with Fish Oil Improves Meat Fatty Acid Profile Although Impairs Growth Performance of Early Weaned Rabbits. *Animals* **2019**, *9*, 437. [CrossRef] [PubMed]
22. Danuta, J.; Marian, C.; Wiesław, P.; Anna, R. The Effect of Fish Oil, Lycopene and Organic Selenium as Feed Additives on Rabbit Meat Quality. *J. Appl. Anim. Res.* **2020**, *48*, 476–483. [CrossRef]
23. Rossi, R.; Vizzarri, F.; Chiapparini, S.; Ratti, S.; Casamassima, D.; Palazzo, M.; Corino, C. Effects of Dietary Levels of Brown Seaweeds and Plant Polyphenols on Growth and Meat Quality Parameters in Growing Rabbit. *Meat Sci.* **2020**, *161*, 107987. [CrossRef] [PubMed]
24. Al-Soufi, S.; García, J.; Muíños, A.; López-Alonso, M. Marine Macroalgae in Rabbit Nutrition—A Valuable Feed in Sustainable Farming. *Animals* **2022**, *12*, 2346. [CrossRef] [PubMed]

25. Peiretti, P.G.; Meineri, G. Effects of Diets with Increasing Levels of Golden Flaxseed on Carcass Characteristics, Meat Quality and Lipid Traits of Growing Rabbits. *Ital. J. Anim. Sci.* **2010**, *9*, 372–377. [CrossRef]
26. Peiretti, P.G.; Meineri, G. Effects of Diets with Increasing Levels of Spirulina Platensis on the Carcass Characteristics, Meat Quality and Fatty Acid Composition of Growing Rabbits. *Livest. Sci.* **2011**, *140*, 218–224. [CrossRef]
27. Agradi, S.; Fehri, N.E.; Quattrone, A.; Vigo, D.; Brecchia, G.; Menchetti, L.; Barbato, O.; Bosco, A.D.; Failla, S.; Contò, M.; et al. Effect of Extruded Linseed and Algae *Padina pavonica*'s Dietary Supplementation on Meat Fatty Acid Profile in Fattening Rabbits. In Proceedings of the 13th World Rabbit Congress, IRTA, ASESCU, Tarragone, Spain, 2–4 October 2024.
28. Makkar, H.P.S.; Tran, G.; Heuzé, V.; Giger-Reverdin, S.; Lessire, M.; Lebas, F.; Ankers, P. Seaweeds for Livestock Diets: A Review. *Anim. Feed. Sci. Technol.* **2016**, *212*, 1–17. [CrossRef]
29. Quattrone, A.; Belabbas, R.; Fehri, N.E.; Agradi, S.; Mazzola, S.M.; Barbato, O.; Dal Bosco, A.; Mattioli, S.; Failla, S.; Abdel-Kafy, E.S.M.; et al. The Effect of Dietary Plant-Derived Omega 3 Fatty Acids on the Reproductive Performance and Gastrointestinal Health of Female Rabbits. *Vet. Sci.* **2024**, *11*, 457. [CrossRef]
30. Galindo, A.; Reis, D.B.; Rodríguez, I.; Pérez, J.A.; Abdul-Jalbar, B.; Zárate, R.; Nunes, N.; Pinheiro De Carvalho, M.A.A.; Acosta, N.G.; Rodríguez, C. Lipid Characterization of 14 Macroalgal Species from Madeira Archipelago: Implications for Animal and Human Nutrition. *Bot. Mar.* **2022**, *65*, 51–67. [CrossRef]
31. Rocha, C.P.; Pacheco, D.; Cotas, J.; Marques, J.C.; Pereira, L.; Gonçalves, A.M.M. Seaweeds as Valuable Sources of Essential Fatty Acids for Human Nutrition. *Int. J. Environ. Res. Public Health* **2021**, *18*, 4968. [CrossRef] [PubMed]
32. Morais, T.; Inácio, A.; Coutinho, T.; Ministro, M.; Cotas, J.; Pereira, L.; Bahcevandziev, K. Seaweed Potential in the Animal Feed: A Review. *J. Mar. Sci. Eng.* **2020**, *8*, 559. [CrossRef]
33. Fatma, C.A.F.; Yilmaz, O.; Durucan, F.; Şen Özdem, N. Biochemical Components of Three Marine Macroalgae (*Padina pavonica, Ulva lactuca* and *Taonia atomaria*) from the Levantine Sea Coast of Antalya, Turkey. *J. Biol. Environ. Sci.* **2015**, *6*, 401–411.
34. Mekinić, I.G.; Šimat, V.; Botić, V.; Crnjac, A.; Smoljo, M.; Soldo, B.; Ljubenkov, I.; Čagalj, M.; Skroza, D. Bioactive Phenolic Metabolites from Adriatic Brown Algae *Dictyota dichotoma* and *Padina pavonica* (Dictyotaceae). *Foods* **2021**, *10*, 1187. [CrossRef] [PubMed]
35. Pla, M.; Zomeño, C. Effect of the Dietary N-3 and n-6 Fatty Acids on Rabbit Carcass and Meat Quality | Request PDF. In Proceedings of the 9th World Rabbit Congress, Verona, Italy, 10–13 June 2008.
36. Blasco, A.; Ouhayoun, J. Harmonization of Criteria and Terminology in Rabbit Meat Research. Revised Proposal. *World Rabbit. Sci.* **1996**, *4*, 93–99. [CrossRef]
37. Ripoll, G.; Albertí, P.; Panea, B.; Failla, S.; Hocquette, J.F.; Dunner, S.; Sañudo, C.; Olleta, J.L.; Christensen, M.; Ertbjerg, P.; et al. Colour Variability of Beef in Young Bulls from Fifteen European Breeds. *Int. J. Food Sci. Technol.* **2018**, *53*, 2777–2785. [CrossRef]
38. Honikel, K.O.; Hamm, R. Measurement of Water-Holding Capacity and Juiciness. In *Quality Attributes and Their Measurement in Meat, Poultry and Fish Products*; Springer: Berlin/Heidelberg, Germany, 1994; pp. 125–161. [CrossRef]
39. Contò, M.; Cifuni, G.F.; Iacurto, M.; Failla, S. Effect of Pasture and Intensive Feeding Systems on the Carcass and Meat Quality of Buffalo. *Anim. Biosci.* **2021**, *35*, 105. [CrossRef] [PubMed]
40. *AOAC Official Methods of Analysis*, 22nd ed.; Association of Official Analytical Chemists: Gaithersburg, MD, USA, 2023.
41. Gąsior, R.; Pieszka, M. VALIDATION OF A RAPID METHOD FOR SIMULTANEOUS DETERMINATION OF VITAMINS A AND E IN MILK USING HPLC. *Pol. J. Food Nutr. Sci.* **2007**, *57*, 151–155.
42. Qian, H.; Sheng, M. Simultaneous Determination of Fat-Soluble Vitamins A, D and E and pro-Vitamin D2 in Animal Feeds by One-Step Extraction and High-Performance Liquid Chromatography Analysis. *J. Chromatogr. A* **1998**, *825*, 127–133. [CrossRef] [PubMed]
43. Mattila, P.; Lehtonen, M.; Kumpulainen, J. Comparison of In-Line Connected Diode Array and Electrochemical Detectors in the High-Performance Liquid Chromatographic Analysis of Coenzymes Q9 and Q10 in Food Materials. *J. Agric. Food Chem.* **2000**, *48*, 1229–1233. [CrossRef] [PubMed]
44. Maraschiello, C.; Díaz, I.; García Regueiro, J.A. Determination of Cholesterol in Fat and Muscle of Pig by HPLC and Capillary Gas Chromatography with Solvent Venting Injection. *J. High. Resol. Chromatogr.* **1996**, *19*, 165–168. [CrossRef]
45. IUPAC. *Standard Methods for The Analysis of Oils, Fats and De-Rivatives. Method 2.411*, 7th ed.; International Union of Pure and Applied Chemistry, Blackwell Scientific Publications: Cation, London, 1987.
46. Failla, S.; Buttazzoni, L.; Zilio, D.M.; Contò, M.; Renzi, G.; Castellini, C.; Amato, M.G. An Index to Measure the Activity Attitude of Broilers in Extensive System. *Poult. Sci.* **2021**, *100*, 101279. [CrossRef]
47. Jastrzębska, A. Modifications of Spectrophotometric Methods for Total Phosphorus Determination in Meat Samples. *Chem. Papers* **2009**, *63*, 47–54. [CrossRef]
48. de Blas, C.; Wiseman, J. *Nutrition of the Rabbit*, 2nd ed.; CABI: Oxfordshire, UK, 2010.

49. Abreu, P.; Leal-Cardoso, J.H.; Ceccatto, V.M.; Hirabara, S.M. Regulation of Muscle Plasticity and Trophism by Fatty Acids: A Short Review. *Rev. Assoc. Med. Bras. (1992)* **2017**, *63*, 148–155. [CrossRef] [PubMed]
50. Dal Bosco, A.; Castellini, C.; Bianchi, L.; Mugnai, C. Effect of Dietary α-Linolenic Acid and Vitamin E on the Fatty Acid Composition, Storage Stability and Sensory Traits of Rabbit Meat. *Meat Sci.* **2004**, *66*, 407–413. [CrossRef]
51. Trocino, A.; Filiou, E.; Tazzoli, M.; Birolo, M.; Zuffellato, A.; Xiccato, G. Effects of Floor Type, Stocking Density, Slaughter Age and Gender on Productive and Qualitative Traits of Rabbits Reared in Collective Pens. *Animal* **2015**, *9*, 855–861. [CrossRef] [PubMed]
52. Mattioli, S.; Castellini, C.; Mancini, S.; Roscini, V.; Cartoni Mancinelli, A.; Cotozzolo, E.; Pauselli, M.; Dal Bosco, A. Effect of Trub and/or Linseed Dietary Supplementation on In Vivo Oxidative Status and Some Quality Traits of Rabbit Meat. *Meat Sci.* **2020**, *163*, 108061. [CrossRef] [PubMed]
53. Li, R.G.; Wang, X.P.; Wang, C.Y.; Ma, M.W.; Li, F.C. Growth Performance, Meat Quality and Fatty Acid Metabolism Response of Growing Meat Rabbits to Dietary Linoleic Acid. *Asian-Australas. J. Anim. Sci.* **2012**, *25*, 1169–1177. [CrossRef] [PubMed]
54. Deng, Y.; Nie, Q.; Zhou, Y.; Wang, W.; Wu, Z.; Ji, L.; Zhang, J.; Zhou, D. Research Progress in Processing Technology of Rabbit Meat. *Food Bioeng.* **2024**, *3*, 314–322. [CrossRef]
55. Gamboa, M.D.; Santiago, D.M.O.; Sulabo, A.S.L.; Torio, M.A.O. Microbiological and Physicochemical Changes in Rabbit Meat During Ambient Storage. *Philipp. Agric. Sci.* **2024**, *107*, 89–101. [CrossRef]
56. Benatmane, F.; Kouba, M.; Youyou, A.; Mourot, J. Effect of a Linseed Diet on Lipogenesis, Fatty Acid Composition and Stearoyl-CoA-Desaturase in Rabbits. *Animal* **2011**, *5*, 1993–2000. [CrossRef]
57. Rossi, R.; Vizzarri, F.; Ratti, S.; Palazzo, M.; Casamassima, D.; Corino, C. Effects of Long-Term Supplementation with Brown Seaweeds and Polyphenols in Rabbit on Meat Quality Parameters. *Animals* **2020**, *10*, 2443. [CrossRef] [PubMed]
58. Dalle Zotte, A. Perception of Rabbit Meat Quality and Major Factors Influencing the Rabbit Carcass and Meat Quality. *Livest. Prod. Sci.* **2002**, *75*, 11–32. [CrossRef]
59. Fadlilah, A.; Rosyidi, D.; Susilo, A. Chemical Quality of Fresh New Zealand White Rabbit Meat in Batu, Indonesia. *IOP Conf. Ser. Mater. Sci. Eng.* **2020**, *811*, 012024. [CrossRef]
60. Prates, J.A.M. Enhancing Meat Quality and Nutritional Value in Monogastric Livestock Using Sustainable Novel Feed Ingredients. *Foods* **2025**, *14*, 146. [CrossRef]
61. Ribeiro, D.M.; Martins, C.F.; Costa, M.; Coelho, D.; Pestana, J.; Alfaia, C.; Lordelo, M.; de Almeida, A.M.; Freire, J.P.B.; Prates, J.A.M. Quality Traits and Nutritional Value of Pork and Poultry Meat from Animals Fed with Seaweeds. *Foods* **2021**, *10*, 2961. [CrossRef]
62. Semeniuc, C.A.; Ranga, F.; Podar, A.S.; Ionescu, S.R.; Socaciu, M.I.; Fogarasi, M.; Fărcaș, A.C.; Vodnar, D.C.; Socaci, S.A. Determination of Coenzyme Q10 Content in Food By-Products and Waste by High-Performance Liquid Chromatography Coupled with Diode Array Detection. *Foods* **2023**, *12*, 2296. [CrossRef]
63. Vlaicu, P.A.; Untea, A.E.; Turcu, R.P.; Saracila, M.; Panaite, T.D.; Cornescu, G.M. Nutritional Composition and Bioactive Compounds of Basil, Thyme and Sage Plant Additives and Their Functionality on Broiler Thigh Meat Quality. *Foods* **2022**, *11*, 1105. [CrossRef]
64. Martins, C.F.; Pestana Assunção, J.; Ribeiro Santos, D.M.; Madeira, M.S.M.d.S.; Alfaia, C.M.R.P.M.; Lopes, P.A.A.B.; Coelho, D.F.M.; Cardoso Lemos, J.P.; de Almeida, A.M.; Mestre Prates, J.A.; et al. Effect of Dietary Inclusion of Spirulina on Production Performance, Nutrient Digestibility and Meat Quality Traits in Post-Weaning Piglets. *J. Anim. Physiol. Anim. Nutr.* **2021**, *105*, 247–259. [CrossRef] [PubMed]
65. Poławska, E.; Cooper, R.G.; Jóźwik, A.; Pomianowski, J. Meat from Alternative Species—Nutritive and Dietetic Value, and Its Benefit for Human Health—A Review. *CYTA—Food* **2013**, *11*, 37–42. [CrossRef]
66. Ivanova, P.; Petrov, P.B.; Gerzilov, V.; Keranova, N.; Petkova, Z.; Teneva, O.; Antova, G.; Hristakieva, P.; Penchev, I.G. Effect of Flaxseed Oil Supplemented Diets on Growth Performance and Meat Quality Traits in Broilers. *Bulg. J. Agric. Sci.* **2022**, *28*, 101–110.
67. Peiretti, P.G.; Mussa, P.P.; Prola, L.; Meineri, G. Use of Different Levels of False Flax (*Camelina sativa* L.) Seed in Diets for Fattening Rabbits. *Livest. Sci.* **2007**, *107*, 192–198. [CrossRef]
68. Kouba, M.; Benatmane, F.; Blochet, J.E.; Mourot, J. Effect of a Linseed Diet on Lipid Oxidation, Fatty Acid Composition of Muscle, Perirenal Fat, and Raw and Cooked Rabbit Meat. *Meat Sci.* **2008**, *80*, 829–834. [CrossRef]
69. Marino, R.; Della Malva, A.; Caroprese, M.; De Palo, P.; Santillo, A.; Sevi, A.; Albenzio, M. Effects of Whole Linseed Supplementation and Treatment Duration on Fatty Acid Profile and Endogenous Bioactive Compounds of Beef Muscle. *Animal* **2019**, *13*, 444–452. [CrossRef] [PubMed]
70. Atti, N.; Methlouthi, N.; Saidi, C.; Mahouachi, M. Effects of Extruded Linseed on Muscle Physicochemical Characteristics and Fatty Acid Composition of Lambs. *J. Appl. Anim. Res.* **2013**, *41*, 404–409. [CrossRef]

71. Peiretti, P.G.; Meineri, G. Effects on Growth Performance, Carcass Characteristics, and the Fat and Meat Fatty Acid Profile of Rabbits Fed Diets with Chia (*Salvia hispanica* L.) Seed Supplements. *Meat Sci.* **2008**, *80*, 1116–1121. [CrossRef] [PubMed]
72. Bernardini, M.; Dal Bosco, A.; Castellini, C. Effect of Dietary N-3/n-6 Ratio on Fatty Acid Composition of Liver, Meat and Perirenal Fat in Rabbits. *Anim. Sci.* **1999**, *68*, 647–654. [CrossRef]
73. Kowalska, D.; Bielański, P. Meat Quality of Rabbits Fed a Diet Supplemented with Fish Oil and Antioxidant. *Anim. Sci. Pap. Rep.* **2009**, *27*, 139–148.
74. Al-Soufi, S.; García, J.; Nicodemus, N.; Lorenzo, J.M.; Cegarra, E.; Muíños, A.; Losada, A.P.; Miranda, M.; López-Alonso, M. Marine Macroalgae in Rabbit Feed—Effects on Meat Quality. *Meat Sci.* **2024**, *216*, 109584. [CrossRef]
75. Hermida, M.; Gonzalez, M.; Miranda, M.; Rodríguez-Otero, J.L. Mineral Analysis in Rabbit Meat from Galicia (NW Spain). *Meat Sci.* **2006**, *73*, 635–639. [CrossRef]

foods

MDPI

Article

Antioxidant Activity of *Bougainvillea spectabilis* Bracts as an Alternative to Nitrites in Cooked Pork Ham

T. Alexandra Ferreira [1], Jose A. Rodriguez [2], Irais Sánchez-Ortega [2], Jose M. Lorenzo [3,4] and Eva M. Santos [2,*]

1 Campus Puebla, Universidad del Valle de Mexico, Camino Real a San Andrés Cholula No. 4002, Emiliano Zapata, San Andrés Cholula 72810, Mexico; thania.ferreira@uvmnet.edu
2 Área Académica de Química, Universidad Autónoma del Estado de Hidalgo, Carr. Pachuca-Tulancingo Km. 4.5, Mineral de la Reforma 42184, Mexico; josear@uaeh.edu.mx (J.A.R.); irais_sanchez5498@uaeh.edu.mx (I.S.-O.)
3 Centro Tecnológico de la Carne de Galicia, Rúa Galicia n° 4, Parque Tecnológico de Galicia, San Cibrao das Viñas, 32900 Ourense, Spain; jmlorenzo@ceteca.net
4 Área de Tecnoloxía dos Alimentos, Facultade de Ciencias, Universidade de Vigo, 32004 Ourense, Spain
* Correspondence: emsantos@uaeh.edu.mx

Abstract: In this study, the impact of incorporating *Bougainvillea spectabilis* powder into ham formulation as a potential color replacement for nitrites was evaluated. Three drying methods were proposed to preserve the antioxidant properties of bougainvillea: foam-mat drying, air drying, and oven drying. Antioxidant assays (DPPH, ABTS, and FRAP) assays revealed that the presence of bougainvillea powders enhanced the antioxidant properties and maintained the stability of the ham over 8 weeks of storage at 4 °C. In addition, total polyphenolic content and presence of thiobarbituric acid reactive substances (TBARS) were evaluated and showed higher and lower scores, respectively, in the samples with the incorporation of bougainvillea compared to the control samples, suggesting their potential to replace nitrite salts by providing natural antioxidant protection. Sensorial analysis also revealed no significant differences in sensory attributes in hams with 0.1% bougainvillea powder compared to nitrite samples. The incorporation of the bougainvillea powders in the ham formulation improved the sensorial attributes and consumer overall acceptance even after 8-week cold storage at 4 °C.

Keywords: *Bougainvillea spectabilis*; antioxidant activity; ham; edible flowers; sensory analysis

1. Introduction

Meat and meat products play an important role in human nutrition, acting as a valuable source of bioavailable nutrients, including proteins, iron, minerals, and vitamins [1]. Nitrites (NO_2^-) and nitrates (NO_3^-) are commonly used as meat formulation additives to improve meat quality. The nitrites not only contribute to the appealing color of meat through their interaction with muscle myoglobin but also improve flavor, exhibit antioxidant properties, and enhance antimicrobial characteristics, thereby extending the shelf life of meat products [2–4]. However, the International Agency for Research on Cancer (IARC) has stated that red and processed meat consumption is likely to improve the cancer risk associated with the presence of nitrites, among other issues [5]. These compounds induce the formation of N-nitrosamines and N-nitrosamides, recognized for their impact on carcinogenic and genotoxic processes [3,4]. However, the multiple functions of nitrite and nitrate salts in meat and meat products have prevented their substitution by other food additives, although their presence is generally regulated.

In recent decades, an effort has been made to develop "clean label" products by exploring additive alternatives to reduce the use of nitrites in meat processing. Some vegetables, herbs, spices, fruits, and flowers with antioxidant properties have been studied [4,6]. Pan et al. [7] incorporated bamboo leaf extract into pork ham to prevent nitrite transformation into N-nitrosamines [7]. Ozaki et al. [8] investigated the addition of radish powder and oregano essential oil in fermented cooked pork and beef sausages to enhance the color and inhibit the mesophilic bacteria. However, it did not prevent lipid oxidation effectively. Other examples include the use of celery juice concentrated in ham [9], tomato processing byproducts [10], red wine or red wine and garlic [11], beet root powder [12], pomegranate peel extract [13], and cranberry powder in sausages [14]. However, organoleptic properties need to be further studied.

Bougainvillea spectabilis, commonly known as bougainvillea, is renowned for its colorful bracts (Figure 1) due to the presence of natural coloring pigments such as betalains (betacyanin and betaxanthins), with interesting chemical properties. On the other hand, bougainvillea also contains compounds such as flavonoids, alkaloids, phenols, and tannins, which contribute to its potential as a natural additive [15,16]. These components show strong antioxidant activity and potential anti-inflammatory and anticancer properties and liver-protective effects [17]. Recently, Abdelrahman et al. [18] have reported the antioxidant and antimicrobial activity of phenolic acids, specifically anthocyanins, from bougainvillea and their in vitro inhibitory effect on the viability of certain cancer cells. This flower has been used in Mexican folk medicine as a tea for treating respiratory diseases [16,19–23], suggesting potential benefits for food preservation. Kaushik et al. have described the applications of *Bougainvillea spectabilis* in food products. It has been mentioned that bougainvillea has been incorporated into noodles, pasta, juice, macaroni, frozen desserts, beverages, milk products, tablets, and syrup as the coloring agent. It has been described that this plant could be considered for the food and pharmacy industries [24]. However, it is important to note that comprehensive studies on the safety and efficacy of incorporating bougainvillea extracts into meat products are limited. Further research is essential to understand the specific advantages and potential challenges associated with utilizing *Bougainvillea spectabilis* in the context of meat and meat products.

Figure 1. Bougainvillea spectabilis morphology.

In this context, an innovative proposal emerges to use *Bougainvillea spectabilis* as an additive in cooked ham, aiming to replace conventional nitrite salts. This study aims to evaluate the impact of different drying methods (air-drying, foam-mat drying, and oven drying) on the preparation of bougainvillea powder and their effect on the physicochemical and sensory properties of the ham, including antioxidant characteristics.

2. Materials and Methods

2.1. Reagents, Additives, and Solutions

All the reagents employed were of analytical grade and used without further purification. Reagents used in the preparation of the foam mat from bougainvillea, including maltodextrin, hydroxyethyl cellulose, and egg albumin (EA), were obtained from Food Technologies Trading (Mexico City, Mexico). Tween-80 was acquired from J.T. Baker (Phillipsburg, NJ, USA). Additives for ham formulation were coarse marine salt (Altamar, Mexico), polyphosphates (Bekafos Ambsa, Bekarem, Mexico), dextrose (marca), carrageenan (Gelybekam, Bekarem), sodium erythorbate (Bekarem, Iztapalapa, Mexico), sodium nitrite (Sigma-Aldrich, Saint Louis, MO, USA), and dextrose (Food Technologies Trading, Mexico City, Mexico).

For antioxidant activity assays, reagents such as 2,2-Diphenyl-1-picrylhydrazyl (DPPH), 6-hydroxy-2,5,7,8-tetramethylchroman-2-carboxylic acid (Trolox), potassium persulfate, 2,4,6-tris-2-pyridyl-s-triazine (TPTZ), hydrochloric acid, acetic acid, Folin–Ciocalteu reagent, gallic acid, thiobarbituric acid (TBA), 1,1,3,3-tetraethoxypropane (TEP), and methanol (MeOH) were sourced from Sigma Aldrich (St. Louis, MO, USA). ABTS was obtained from Roche Diagnostics (Indianapolis, IN, USA), while trichloroacetic acid (TCA) was acquired from Meyer (Estado de Mexico, Mexico). Iron (II) sulfate, sodium carbonate, and sodium acetate were provided by J.T. Baker (Phillipsburg, NJ, USA), and ferric (III) chloride was purchased from Merck (Darmstadt, Hesse, Germany). For nitrite determination, reagents were as follows: sodium tetraborate, potassium ferrocyanide, zinc acetate, sulfanilamide, sodium nitrite, acetic acid, and naphthyl ethylenediamine (NED). A NED solution was prepared by dissolving 0.2 g of NED in 150 mL of acetic acid solution (15% v/v). A sulfanilamide solution was prepared by dissolving 0.5 g of the reagent in 150 mL of acetic acid solution (15% v/v).

All the solutions were prepared with deionized water (Milli-Q Merck, Millipore Darmstadt, Hesse, Germany) with a resistivity of 18.2 MΩ cm or greater.

2.2. Preparation of the Bougainvillea Ingredients

Bougainvillea bracts and flowers were carefully collected from *Bougainvillea spectabilis* plants in Pachuca, Hidalgo, Mexico, from January to April 2023. For the preparation of bougainvillea powders, both the flowers and bracts were considered since it has been described that these parts contain the highest concentration of bioactive compounds of interest [15].

In the case of air drying (BA), approximately 100 g of bougainvillea bracts and flowers were weighed and allowed to air dry at room temperature (20–25 °C) in a dark and dry place with good ventilation. These conditions are necessary to prevent and minimize color loss. This drying process typically takes 5 to 7 days. For oven drying (BO), around 100 g of bougainvillea bracts and flowers were weighed and placed in a refractory container. The drying process was carried out in an oven at 65 °C for 4 h.

Finally, the foam-mat drying was applied as the previously described procedure [15]. For this purpose, 25 g of bougainvillea bracts and flowers were combined with 100 mL of distilled water and blended in a food processor. Then, this mixture was whipped for 15 min until frothing using a hand mixer in the presence of 15 g of albumin, 10.0 g of maltodextrin, 2.0 g of hydroxyethyl cellulose, and 2.0 g of Tween-80. The resulting foam was then dried in an oven at 60 °C for 4 h and called BF. After the drying processes, the dried materials BA, BO, and BF were milled in a UDY cyclone sample mill (UDY Corp., Fort Collins, CO, USA) to a particle size of 0.5 mm and preserved in hermetic polyethylene bags in darkness at room temperature until its use.

2.3. Cooked Ham Manufacturing

Five ham formulations were designed with the addition of bougainvillea powders plus two control formulations as follows: C, the control with no nitrites nor bougainvillea addition; C-NO_2, the control with the addition of 150 mg of nitrites kg^{-1} ham; F1, F2, and F3 formulations with 0.05%, 0.1%, and 0.25% of bougainvillea powder in the form of foam BF, respectively; F4 with 0.1% of BA; and F5 with 0.1% of BO. Hams of 2 kg (3 units per formulation) were elaborated at the pilot plant of the Food Chemistry Area in the Universidad Autónoma del Estado de Hidalgo according to regular procedures. Pork leg meat was purchased from a local provider coming from a single farm, manually cleaned from fat and connective tissue, cut into pieces around 5 × 5 cm, mixed, and frozen at −20 °C. The meat was thawed overnight and manually injected with brine prepared to achieve the designed formulation. The formulation of ham consisted of 71.43% pork meat, 2% salt, 0.5% polyphosphates, 0.7% dextrose, 0.75% carrageenan, 0.05% sodium erythorbate, and 24.57% water and the bougainvillea additive or nitrites according to the formulation. Brine was prepared using the following formulation (86% cold water, 7% salt, 1.75% polyphosphates, 2.45% dextrose, 2.62% carrageenan, and 0.175% sodium erythorbate) and injected at 40%. Bougainvillea powders were added to the brine before injection to reach the designed concentrations in the final product. After injection, the hams were tumbled and massaged for 10 min every hour at refrigeration temperature for 14 h. After that, the meat was placed in polyethylene bags inside cylindrical plastic molds. The hams were cooked until a core temperature of 67 °C in a Rational 20-2/1 oven unit (Landsberg am Lech, Germany), cooled in ice water for 2 h, and refrigerated at 4 °C. After 24 h, hams were unmolded and sampled for physicochemical and sensorial analysis, while the remaining product was vacuum packed and stored refrigerated at 4 °C for 8 weeks to follow color and antioxidant evolution at 4 and 8 weeks.

2.4. Physicochemical Properties

The cooking yield of the process was gravimetrically calculated as the percentage of ham weight after unmolding related to the weight before cooking. The pH of each sample of each formulation was measured using a digital pH meter (HI 99161, Hanna Instruments, Ronchi Di Villafranca Padovana, Italy) equipped with a glass probe for penetration. Water activity was determined in an AquaLab 3TE (Decagon, WA, USA). To determine drip loss, a 100 g sample was placed in a plastic bag vacuum-sealed and kept at 4 °C. After 8 weeks, the sample was dried with absorbent paper and weighed. The amount of drip was expressed as the percentage of drained water related to the initial weight. The water holding capacity (WHC) was measured based on the filter paper press method described by Steen et al. [25]. A weight of 0.3 g of ham was placed on a Whatman No. 2 filter paper between two plexiglass plates, and a weight of 1 Kg was placed over it for 5 min. The expelled fluid was absorbed by the filter paper, forming an outer circle around the inner circle formed by the meat. Both areas were measured with the help of the software ImageJ ver. 1.54f (NIH, 2023) [26], and the WHC was expressed as the ratio area of water expelled/area of meat (cm^2/cm^2) [27].

Moisture content in hams was determined using the moisture determination method outlined by the Association of Analytical Communities (AOAC, 2003) [28] in their standard 23.003:2003. This involved weighing samples of 2.000 ± 0.001 g each, which were subsequently dried in an oven at 105 °C until a constant weight was reached.

Crude fat was extracted by the Soxhlet procedure with petroleum ether at 80 °C and determined gravimetrically according to the AOAC 960.39 standard [28]. Protein content was also determined by the International Organization for Standardization standard ISO Kjeldahl method 937:1978 (ISO, 1978) [29]. Nitrogen was determined after digestion of a 1 g sample with sulfuric acid, distilled with NaOH, recovering the liberated ammonia in boric acid in a Gerhardt distillation unit (Königswinter, Germany), and titrated with HCl 0.1 N. Protein was calculated from total nitrogen concern using the conversion factor of 6.25. These above physicochemical determinations were obtained in triplicate.

Nitrite content in the ham formulations was determined using the Griess test [30,31]. Before analysis, the ham samples were homogenized and deproteinized following the method described by Belluci et al. [32]. Briefly, a 5.0 g sample was mixed with 2.5 mL of 5% sodium tetraborate and 25 mL of hot water (85 °C), then heated for 15 min. After transferring to a 100 mL flask, an additional 25 mL of hot water was added, and the mixture was cooled. Moreover, 2.5 mL of 15% potassium ferrocyanide and 2.5 mL of a 30% zinc acetate solution were incorporated into the mixture, and the volume was adjusted to 100 mL. Afterwards, sulfanilamide and NED were added to 5.0 mL of the filtrate, and the absorbance was measured at 540 [31,32]. Measurements were performed in triplicate. Spectrophotometric measurements were conducted with a UV/Vis spectrometer, the Perkin Elmer Lambda 40 (Waltham, MA, USA), employing the Perkin Elmer UVWinLab software. Results were expressed as mg NO_2 kg^{-1}.

CieLab color parameters L*, a*, and b* (where L* represents color lightness, a* denotes redness, and b* indicates yellowness) were measured at four randomly selected points on the sample surface after cutting on the middle of the sample with a portable colorimeter Hunter Lab miniScan EZ 4500L (HunterLab, Reston, VA, USA) under D65 illuminant and 10° observer angle. Tiles in black and white were used to calibrate the equipment. The mean of the four measurements was used per sample. The three hams per formulation were measured, and the means per ham were used for statistical analysis. The color was measured at 0, 4, and 8 weeks.

2.5. Sensory Evaluation

A 7-point hedonic test was employed for the sensory evaluation of ham formulations. Samples were sliced 2 h before the tasting (2 mm thick), wrapped in aluminum foil, and kept at room temperature. Thirty minutes prior to the sensorial session, slices were cut into rectangular pieces of 6 × 3 cm. Seventeen untrained but accustomed to participating in meat products sensory analysis panelists conducted the test in one session with the seven samples labeled by three-digit numbers and presented them randomly. The hedonic scores ranged from 1 to 7, as follows: very unpleasant (1), quite unpleasant (2), slightly unpleasant (3), acceptable (4), slightly pleasant (5), quite good (6), and excellent (7). The test included color, odor, taste, and overall acceptability. Hams were evaluated 48 h after the cooking process. Plastic dishes were used to present the samples to the panelists, and water and bread were provided to cleanse the palate from residual flavors between tastings [33].

2.6. Antioxidant Activity

Methanolic extracts from the samples were prepared to evaluate the antioxidant activity. Two hundred grams of ham sample were ground with a food processor, and 2.0 g were placed in a polypropylene tube with 5.0 mL of methanol. The mixture was vortexed for 10 min to obtain the methanolic extract [15]. The mixtures underwent ultrasound extraction for 10 min, followed by centrifugation at 2200 rpm and filtration using Whatman paper (Whatman 41). Each sample was obtained in triplicate.

The antioxidant activity of the ham formulations was evaluated considering DPPH and ABTS scavenging activity, as well as ferric-reducing antioxidant activity (FRAP). Additionally, the assessment included the determination of total polyphenolic content (TPC) and the measurement of lipid oxidation. This approach facilitated a comprehensive evaluation of the oxidative stability of the ham formulations throughout distinct storage intervals (weeks 0, 4, and 8). Spectrophotometric measurements were conducted with the UV/Vis spectrometer Perkin Elmer Lambda 40 (Waltham, MA, USA) employing the Perkin Elmer UVWinLab software(version 6.2).

The DPPH method was performed as described by Rivero-Perez et al. [34]. This procedure involved monitoring the decrease in absorbance at 515 nm when the DPPH radical is exposed to the presence of antioxidant species in the ham formulations. For DPPH radical scavenging activity, methanolic extracts of the samples were mixed with a DPPH solution, and the subsequent decrease in absorbance was quantified. ABTS radical scavenging activity was determined by measuring the reduction in the number of ABTS radical cations ($ABTS^{+\bullet}$) in the presence of the sample extracts; the change in absorbance was measured at 734 nm. The results of DPPH and ABTS methodologies were expressed as inhibition percentages. All extracts were prepared using the same sample amount, and the antioxidant activity evaluation was performed with a consistent extract volume of 20 μL and a radical solution volume of 980 μL in each case [33].

The FRAP method was assessed by quantifying the reduction of a ferric complex to its ferrous form according to Benzie et al. [35]. The concentration of antioxidant compounds is related to the increase in the absorbance at 593 nm. The results were expressed as mmol $FeSO_4 \bullet 100\ g^{-1}$.

In addition to the evaluation of antioxidant activity, the total polyphenolic content (TPC) was determined using the Folin–Ciocalteu method. In this process, a 2 g sample reacted with Folin–Ciocalteu reagent, and the resultant blue color was measured spectrophotometrically (750 nm), with gallic acid as the standard [34]. The results were expressed as $mg_{gallic\ acid}\ g^{-1}$.

Lipid oxidation was evaluated by the quantification of thiobarbituric acid reactive substances (TBARS). Following the methodology outlined by Vyncke with slight modifications [36]. Briefly, two grams of a ground sample are mixed with 10 mL of 5% TCA in a Falcon tube, then homogenized on ice for 2 min using an Ultraturrax T-18 (IKA, Wilmington, NC, USA). The samples are centrifuged at 3500 rpm for 10 min after freezing for 10 min to precipitate proteins. The supernatant is filtered, and 5 mL of extract is transferred to a Falcon tube, with dilutions made if necessary. After adding 5 mL of TBA, the sample is vortexed, incubated at 97 °C for 40 min, cooled, and placed in an ultrasonic bath for 15 min. The absorbance is measured at 532 nm. TBARS values were expressed as mg MDA kg^{-1}, and the progression of TBARS, indicative of lipid oxidation, was monitored during storage.

2.7. Statistical Analyses

The data from color and antioxidant activity were evaluated with a two-factor analysis of variance (ANOVA) (treatment and storage time), while data from the other physicochemical parameters were evaluated by a one-way ANOVA. Tukey's test was used to compare the mean values when the ANOVA was significant ($p < 0.05$). Regarding the sensory analysis, panelists were considered as a random effect (each panelist tasted samples from all formulations in a single session). Statistical analyses were performed using the Statgraphics Centurion XVI version 16.1.03 (32-bits) (StatPoint Technologies, Inc., Warrenton, VA, USA).

3. Results and Discussion

3.1. Physicochemical Results

The cooking yield ranged between 94.3 and 97.3% without significant differences ($p > 0.05$) between formulations. No significant differences ($p > 0.05$) were found in aw (0.976–0.983) and small differences were detected in pH (Table 1), while WHC and moisture were significantly affected ($p < 0.05$) by the inclusion of bougainvillea ingredient, especially in the form of foam. Water constitutes approximately 75% of the total weight in meat products, and its retention capacity, known as water holding capacity (WHC), is a crucial parameter for ensuring meat quality [37]. The sample with the higher proportion of bougainvillea accompanied by foam material (F3) presented the lowest significant moisture content (74.74 ± 1.86%) and the highest water holding capacity since these samples showed less water expelled after compression (0.59 ± 0.05). F1 and F2 samples, also with the colorant as foam but with less concentration of bougainvillea, also presented high water holding capacity despite the moisture being similar to the controls and samples with BA and BO colorant. The albumin added in the process of foaming also contributed to a significant increase ($p < 0.05$) in the protein percentage in formulations F2 and F3. Fat content ranged from 1.44 to 2.80%, with small differences between samples attributed to the heterogeneity of raw material.

Table 1. Physicochemical data obtained from ham formulations (N = 3).

Formulation	pH	WHC	Drip Loss (%)	Moisture Content (%)	Fat (%)	Protein (%)
C	5.60 ± 0.01^{a}	0.69 ± 0.01^{b}	2.83 ± 0.61	76.45 ± 1.12^{ab}	1.81 ± 0.34^{a}	14.73 ± 0.40^{a}
CNO_2	5.69 ± 0.01^{d}	0.69 ± 0.03^{b}	3.57 ± 0.91	76.33 ± 0.66^{ab}	2.80 ± 0.17^{c}	14.51 ± 0.76^{a}
F1	5.67 ± 0.02^{c}	0.62 ± 0.02^{a}	3.56 ± 0.29	77.39 ± 0.49^{b}	1.59 ± 0.08^{a}	14.74 ± 0.35^{a}
F2	5.67 ± 0.02^{c}	0.62 ± 0.01^{a}	2.59 ± 0.37	77.35 ± 0.13^{b}	1.89 ± 0.26^{ab}	15.92 ± 0.57^{b}
F3	5.59 ± 0.01^{a}	0.59 ± 0.05^{a}	2.91 ± 0.26	74.74 ± 1.86^{a}	2.36 ± 0.39^{bc}	17.07 ± 0.54^{c}
F4	5.66 ± 0.02^{c}	0.71 ± 0.02^{b}	3.16 ± 0.61	77.31 ± 1.38^{b}	1.66 ± 0.31^{a}	14.69 ± 0.70^{a}
F5	5.63 ± 0.02^{b}	0.72 ± 0.03^{b}	3.11 ± 0.38	77.75 ± 0.64^{b}	1.44 ± 0.39^{a}	14.28 ± 0.69^{a}

$^{a\text{–}d}$ means in the same column with different superscripts are significantly different ($p < 0.05$).

Regarding drip loss evaluation, protein oxidation has been implicated in reducing water-holding capacity in meat products [37]. After 8 weeks of cold storage, drip loss was measured, yielding values lower than 4% in all cases. F2 and F3 formulations presented the lowest values (2.59 and 2.91%, respectively), likely attributable to the addition of egg albumin during the foam-mat drying process used to produce the bougainvillea powder (BF).

After conducting the evaluation of nitrite content using the modified Griess method [32], it was determined that there was no presence of nitrites in the formulations containing the bougainvillea powders nor the control; just C-NO_2 presents a residual concentration of 41.60 ± 1.93 mg_{NO2} kg^{-1}. This finding confirms that the bougainvillea does not contribute to the nitrite levels in the ham formulations. This supports the potential use of bougainvillea as a safe and natural ingredient in food products aimed at reducing reliance on synthetic additives.

3.2. Instrumental Color and Sensory Results

The use of bougainvillea powders in the ham formulations provides a reddish color while offering antioxidant properties and replacing the presence of nitrites (see Figure 2).

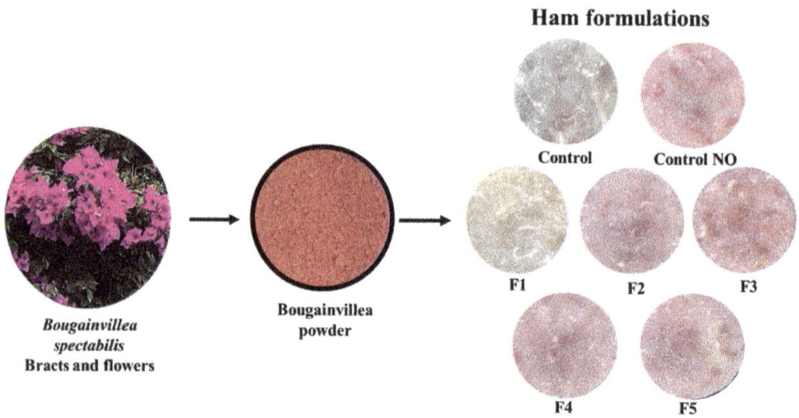

Figure 2. *Bougainvillea spectabilis* powder and ham formulations.

According to the color results, formulations F4 and F5 presented L values with no significant difference from the control sample with nitrites, although a* values were significantly lower while b* values were significantly higher than the recorded for C-NO_2. In fact, F4 and F5 presented the most similar color to the nitrite control sample (62.49 ± 0.53). The F3 formulation with the highest amount of bougainvillea colorant exhibited a similar a* value (8.30 ± 0.56) to the nitrite sample but with significantly higher values of L* and b*. On the contrary, F1 and F2 samples presented a paler color similar to control samples without nitrites (Table 2).

Table 2. Color parameter results (L, a*, and b*) during storage period (0, 4, and 8 weeks).

		Storage (Weeks)		
	Formulation	0	4	8
L	C	64.06 ± 0.51 ^{cd,X}	63.49 ± 0.26 ^{d,X}	66.62 ± 0.36 ^{d,Y}
	C-NO_2	62.49 ± 0.53 ^{a}	61.90 ± 0.56 ^{c}	62.55 ± 0.40 ^{b}
	F1	64.32 ± 0.23 ^{cd,Z}	62.70 ± 0.32 ^{cd,Y}	63.77 ± 0.25 ^{c,X}
	F2	64.40 ± 0.57 ^{d,Y}	62.38 ± 0.31 ^{c,X}	63.93 ± 0.08 ^{c,Y}
	F3	63.57 ± 0.34 ^{bc,Z}	59.22 ± 0.75 ^{a,X}	61.20 ± 0.40 ^{a,Y}
	F4	62.83 ± 0.40 ^{ab,Y}	61.07 ± 0.59 ^{b,X}	63.49 ± 0.47 ^{c,Y}
	F5	62.43 ± 0.47 ^{a,Z}	59.66 ± 0.23 ^{a,X}	61.00 ± 0.09 ^{a,Y}
a*	C	4.62 ± 0.42 ^{a,X}	5.90 ± 0.21 ^{a,Y}	4.97 ± 0.08 ^{a,X}
	C-NO_2	8.35 ± 0.39 ^{d}	8.60 ± 0.83 ^{c}	9.27 ± 0.93 ^{d}
	F1	5.24 ± 0.20 ^{a,X}	6.10 ± 0.32 ^{a,Y}	6.08 ± 0.04 ^{a,Y}
	F2	6.13 ± 0.15 ^{b,X}	7.12 ± 0.31 ^{b,Y}	8.64 ± 0.02 ^{cd,Z}
	F3	8.30 ± 0.56 ^{d,X}	10.63 ± 0.26 ^{d,Y}	10.32 ± 0.23 ^{e,Y}
	F4	6.54 ± 0.41 ^{b,X}	8.43 ± 0.28 ^{c,Y}	8.37 ± 0.24 ^{c,Y}
	F5	7.50 ± 0.29 ^{c,X}	8.61 ± 0.80 ^{c,Y}	11.44 ± 0.06 ^{f,Z}
b*	C	15.93 ± 0.56 ^{e}	15.56 ± 0.65 ^{d}	16.05 ± 0.40 ^{e}
	C-NO_2	11.19 ± 0.28 ^{a,X}	13.09 ± 0.38 ^{b,Y}	13.02 ± 0.81 ^{bc,Y}
	F1	14.97 ± 0.31 ^{cd,X}	16.01 ± 0.25 ^{d,Y}	14.89 ± 0.35 ^{d,X}
	F2	15.18 ± 0.59 ^{d,Y}	14.52 ± 0.31 ^{c,Y}	13.26 ± 0.42 ^{c,X}
	F3	14.41 ± 0.14 ^{c,Y}	14.17 ± 0.08 ^{c,Y}	12.32 ± 0.21 ^{b,X}
	F4	13.44 ± 0.38 ^{b,Y}	13.41 ± 0.53 ^{b,Y}	11.16 ± 049 ^{a,X}
	F5	12.92 ± 0.11 ^{b,Y}	12.36 ± 0.30 ^{a,Y}	10.74 ± 0.63 ^{a,X}

Results are expressed as mean value ± standard deviation (N = 3). $^{a–f}$ means in the same column with a different letter are significantly different ($p < 0.05$). $^{X–Z}$ means in the same row with a different letter are significantly different ($p < 0.05$).

During cold storage, there were consistent slight variations in color, especially in F1-F5 formulations. While a* and L values hardly changed in C-NO_2 and b* scores slightly increased, indicating that color remained stable during preservation, in the case of F1–F5 samples a significant increase in a* was noticed ($p > 0.05$). Also, a significant decrease in b* values was observed in F1–F5 samples, improving the reddish color of samples, making them more like C-NO_2, and proving that bougainvillea bracts and flowers can provide a reddish color to cooked ham, replacing nitrites. Similar efforts have been described to incorporate natural colorants with antioxidant activity, such as betacyanins and betaxanthins, in ham formulations. The incorporation of red radish, red beetroot, and hibiscus in cooked ham formulations was proposed by Dias et al. [38]. However, only red beetroot provided a color closer to the intended, but color stability was not evaluated. Natural pigments such as betalains tend to fade out with time because of oxidation reactions, but in this case, it seems that the antioxidant properties of the bougainvillea could be contributing to stabilizing the color [39].

These physical color properties were confirmed by the sensory panel results. According to Table 3, the formulations best evaluated in the visual aspect were samples with nitrites and the formulations with the bougainvillea colorant dried by air (F4) and oven (F5). This color perception was also transferred to the overall acceptance, which was higher in the same samples with no significant differences between them. The odor and the taste were not significantly affected ($p > 0.05$) by the inclusion of the bougainvillea. Odor and taste did not represent significant differences among the different formulations.

Table 3. Sensory panel results.

Formulation	Visual Aspect	Odor	Taste	Overall Acceptance
C	3.35 ± 1.37^{a}	4.53 ± 1.07^{a}	4.94 ± 1.14^{a}	4.88 ± 1.05^{ab}
C-NO_2	6.24 ± 1.15^{d}	4.53 ± 1.62^{a}	5.41 ± 1.42^{a}	5.47 ± 1.42^{b}
F1	4.29 ± 1.40^{b}	4.82 ± 1.19^{a}	5.12 ± 1.05^{a}	4.88 ± 0.83^{ab}
F2	4.65 ± 1.32^{bc}	4.29 ± 1.10^{a}	5.06 ± 1.25^{a}	4.88 ± 0.93^{ab}
F3	4.59 ± 1.33^{bc}	4.24 ± 1.56^{a}	4.59 ± 1.37^{a}	4.29 ± 1.65^{a}
F4	5.65 ± 1.41^{d}	4.65 ± 1.32^{a}	5.24 ± 1.09^{a}	5.24 ± 1.09^{b}
F5	5.41 ± 1.28^{cd}	4.76 ± 1.25^{a}	5.18 ± 1.63^{a}	5.18 ± 1.33^{b}

Results are expressed as mean value ± standard deviation. $^{a–d}$ means in the same column with a different letter are significantly different ($p < 0.05$).

3.3. Antioxidant Results

Although the antioxidant properties of bougainvillea have been previously studied [15,40–42], its incorporation as an additive in meat products has not been described. After the preparation of the formulations, analyses of antioxidant activity were conducted using DPPH, ABTS, and FRAP methods, along with the assessment of total polyphenolic content (TPC) and lipid oxidation (TBARS). This approach allowed the evaluation of the oxidative stability of the ham formulations over various storage periods (weeks 0, 4, and 8) at 4 °C.

Control samples showed the presence of antioxidant compounds due to the addition of sodium erythorbate. Antioxidative activity methods showed that ham samples with bougainvillea powders exhibited higher inhibition percentage values compared to the control sample (Figure 3). The incorporation of bougainvillea powder in ham formulations either matches or enhances the antioxidant capacity compared to the formulation containing nitrite salts. In the DPPH assay, the formulations containing the bougainvillea powders presented a significant difference ($p < 0.05$) compared with the control. The samples exhibited an inhibition percentage of 63.6–70.8%, while the control samples were 54.0%. Similar values were observed in ABTS (71.5–76.6%) during week 0.

Nevertheless, the antioxidant effect significantly diminished during cold storage ($p < 0.05$) in all formulations, independently of adding nitrites or the natural colorant. After 8 weeks of storage, DPPH values decreased to values between 24.12% and 32.98% (24.26% for the control sample and 26.48% for the nitrite sample). Samples F5 showed significant differences ($p < 0.05$) with the control and nitrite control showing an inhibition percentage of 32.64%. For the ABTS assay, a smaller decrease in inhibition percentages was observed, with values ranging from 48.98% to 62.89%. In this case, formulation F4 presented the highest ABTS value (62.89%), while the control and nitrite samples showed the lowest values (48.98% and 56.02%, respectively). All the formulations, including bougainvillea, presented significant differences with the control. DPPH and ABTS results indicate greater stability of polar compounds.

Related to FRAP results, formulations F1 to F5 exhibited more stable antioxidant activity over the 8-week storage period, with smaller decreases in activity than the control, suggesting that bougainvillea contributes to maintaining antioxidant stability. While storage typically leads to the degradation of natural reducing compounds, the control samples showed less antioxidant protection over time. These findings highlight the effectiveness of bougainvillea powders as a natural antioxidant alternative, offering extended oxidative protection compared to synthetic additives like sodium erythorbate (C) and nitrite salts (C-NO_2).

The antioxidant activity of plants is related to their bioactive content [15]. Orozco-Villafuerte et al. [43] observed a direct correlation between phenolic compounds and antioxidant activity of *Bougainvillea spectabilis*. In this case, the content of phenolic compounds (TPC) significantly increased ($p < 0.05$) in the presence of bougainvillea additives in ham. The higher content of TPC was found in the F4 formulation (0.831 ± 0.024 mg_{GAE} kg^{-1}). The lowest concentration was found in the control sample (0.211 ± 0.006 mg_{GAE} kg^{-1}).

In this case, the presence of polyphenolic compounds in the control and nitrite control was observed. This information agrees with the information described by Bešlo et al., who described a growing interest in the use of by-products in animal nutrition with high concentrations of polyphenols. They described this diet as contributing to greater stability of meat to fatty acid oxidation of meat products for human consumption [44,45].

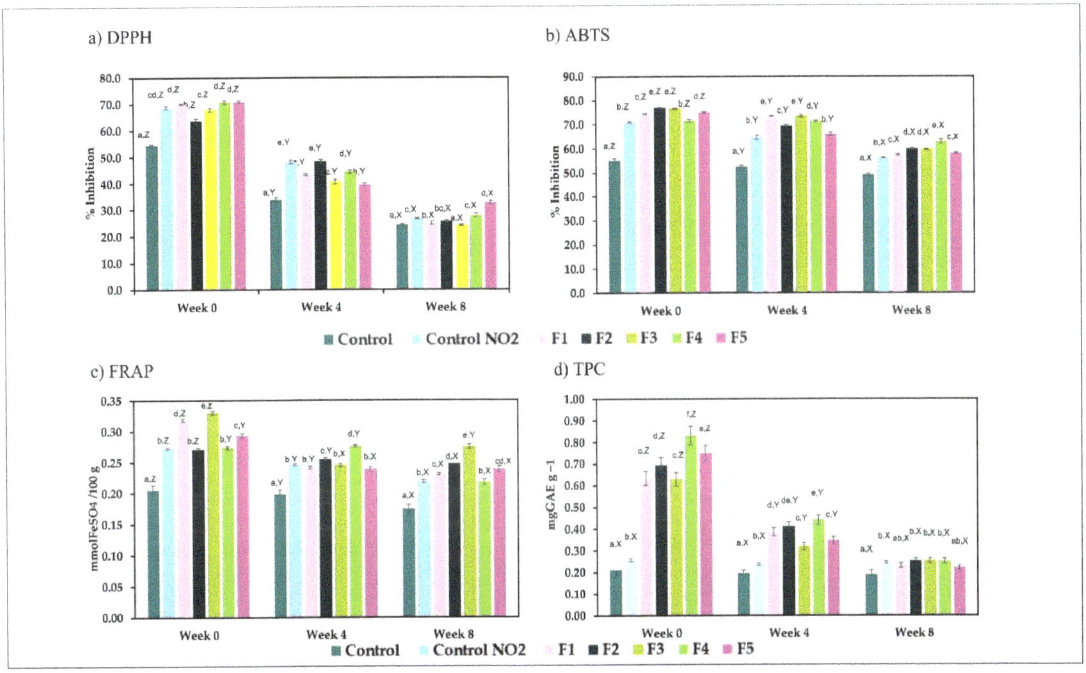

Figure 3. Effect of the addition of bougainvillea powder on the antioxidant activity measured in ham with (**a**) DPPH method, (**b**) ABTS method, (**c**) FRAP method, and (**d**) total polyphenolic content method. a–f mean values for each week with different letters differ significantly ($p < 0.05$), X–Z mean values for each formulation with different letters differ significantly ($p < 0.05$).

The presence of bioactive compounds with antioxidant activity influences lipid stability, which is related to meat quality since it prevents protein oxidation, discoloration, and rancidity [46]. This parameter was assessed with the thiobarbituric acid reactive substances method (TBARS). The addition of bougainvillea powders decreased lipid oxidation (F2 presented the significant lowest value of 0.0360 $\mu g_{MDA}\ g^{-1}$) compared with the control (0.1440 $\mu g_{MDA}\ g^{-1}$) and nitrite control samples (0.0691 $\mu g_{MDA}\ g^{-1}$). Generally, the TBARS values in ham increased with cool storage; however, because of the presence of bougainvillea, this effect was observed to a lesser extent (Figure 4). In this study, all the formulations have acceptable levels during the evaluation period (<0.500 $\mu g_{MDA}\ g^{-1}$).

The analysis of TBARS shows that the addition of bougainvillea in the formulations significantly reduces oxidative processes in meat. The TBARS test, which measures lipid peroxidation and thus oxidative rancidity, indicated that formulations with bougainvillea had lower values compared to the control, which exhibited the highest TBARS concentration; at week 0, all the formulations had values <0.150 $\mu g_{MDA}\ g^{-1}$. This suggests that bougainvillea is effective in mitigating oxidative damage, competing with the antioxidative effect provided by nitrites. Several plants, such as celery or Swiss chard powder and beetroot or barberry extract, have been considered alternative sources of nitrites, but some of them contain nitrate, which can be transformed into nitrites [3]. In this case, the absence of nitrite presence in the cooked ham with bougainvillea, the improvement of color, and antioxidant properties make *Bougainvillea spectabilis* a good candidate to be used in cooked meat products, but always as one more strategy within a hurdle technology that ensures a microbiologically safe product.

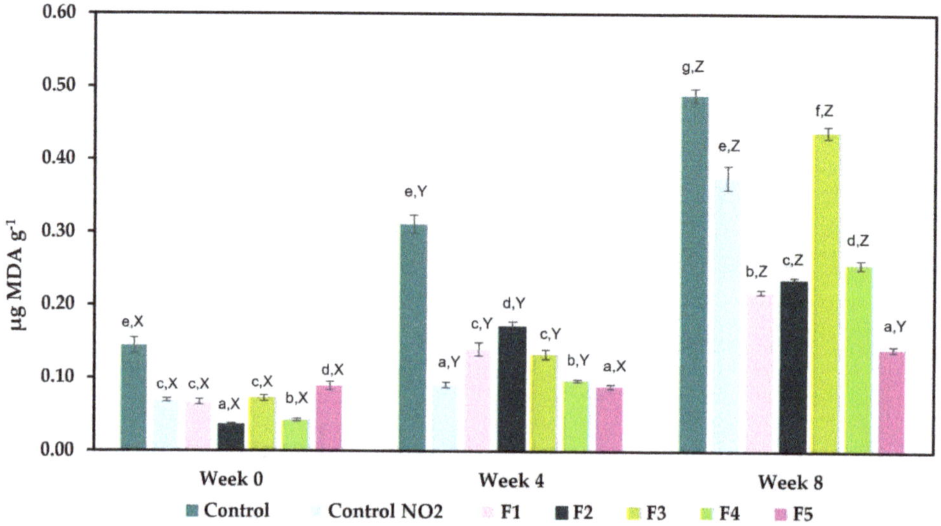

Figure 4. Effect of the addition of bougainvillea on the content of thiobarbituric acid reactive substances (TBARS assay) in mg MDA/kg. a–g mean values for each week with different letters differ significantly ($p < 0.05$). X–Z mean values for each formulation with different letters differ significantly ($p < 0.05$).

4. Conclusions

This study demonstrates that incorporating bougainvillea powder into ham formulations provides a viable natural color alternative to nitrites, maintaining cooking yield and physicochemical properties such as water holding capacity and moisture content. The bougainvillea powder, regardless of the drying process to obtain it, improved the antioxidant stability of the ham, as evidenced by increased total polyphenolic content and superior performance in DPPH, ABTS, and FRAP assays. Sensory evaluation confirmed that the bougainvillea-treated hams retained desirable color, odor, and taste, with formulations F4 and F5 achieving high visual and overall acceptance scores. Importantly, no nitrites were detected in any formulation, affirming the potential of *Bougainvillea spectabilis* as a natural color ingredient for healthier and more sustainable meat products within a set of measures that ensure the safety of the product.

Author Contributions: Conceptualization, E.M.S. and J.A.R.; methodology, T.A.F. and I.S.-O.; formal analysis, T.A.F., J.A.R. and I.S.-O.; writing—original draft preparation, T.A.F.; writing—review and editing, T.A.F. and J.M.L.; supervision, E.M.S. and J.M.L. All authors have read and agreed to the published version of the manuscript.

Funding: This research received no external funding.

Data Availability Statement: The original contributions presented in the study are included in the article, further inquiries can be directed to the corresponding author.

Acknowledgments: Eva M. Santos, Jose A. Rodriguez, and Jose M. Lorenzo, as members of the Healthy Meat network, funded by the CYTED (grant number: 119RT0568).

Conflicts of Interest: The authors declare no conflicts of interest.

References

1. Giromini, C.; Givens, D.I. Benefits and risks associated with meat consumption during key life processes and in relation to the risk of chronic diseases. *Foods* **2022**, *11*, 2063. [CrossRef] [PubMed]
2. Zhang, Y.; Zhang, Y.; Jia, J.; Peng, H.; Qian, Q.; Pan, Z.; Liu, D. Nitrite and nitrate in meat processing: Functions and alternatives. *Curr. Res. Food Sci.* **2023**, *6*, 100470. [CrossRef] [PubMed]
3. Shakil, M.H.; Trisha, A.T.; Rahman, M.; Talukdar, S.; Kobun, R.; Huda, N.; Zzaman, W. Nitrites in cured meats, health risk issues, alternatives to nitrites: A Review. *Foods* **2022**, *11*, 3355. [CrossRef] [PubMed]
4. Di Nunzio, M.; Loffi, C.; Montalbano, S.; Chiarello, E.; Dellafiora, L.; Picone, G.; Antonelli, G.; Tedeschi, T.; Buschini, A.; Capozzi, F.; et al. Cleaning the label of cured meat; Effect of the replacement of nitrates/nitrites on nutrients bioaccessibility, peptides formation, and cellular toxicity of in vitro digested salami. *Int. J. Mol. Sci.* **2022**, *23*, 12555. [CrossRef] [PubMed]
5. International Agency for Research on Cancer. Red meat and processed meat volume 114. In *IARC Monographs on the Evaluation of Carcinogenic Risks to Humans*; International Agency for Research on Cancer: Lyon, France, 2008.
6. Santos, E.M.; Rodriguez, J.A.; Lorenzo, J.M.; Mondragón, A.C.; Pateiro, M.; Gutiérrez, E.; Ferreira, T.A. Antioxidant effect of pumpkin flower (*Cucurbita maxima*) in chicken patties. *Foods* **2022**, *11*, 2258. [CrossRef]
7. Pan, J.; She, X.; Hu, L.; Shi, L.; Zhang, H. The application of bamboo leaf extracts in cured foods. *J. Bamboo Res.* **2019**, *38*, 57–61.
8. Ozaki, M.M.; Santos, M.; Ribeiro, W.O.; Ferreira, N.; Pollonio, M. Radish powder and oregano essential oil as nitrite substitutes in fermented cooked sausages. *Food Res. Int.* **2020**, *140*, 109855. [CrossRef]
9. Horsch, A.; Sebranek, J.G.; Dickson, J.S.; Niebuhr, S.E.; Larson, E.M.; Lavieri, N.A.; Ruther, B.L.; Wilson, L.A. The effect of pH and nitrite concentration on the antimicrobial impact of celery juice concentrate compared with conventional sodium nitrite on *Listeria monocytogenes*. *Meat Sci.* **2014**, *96*, 400–407. [CrossRef]
10. Cadariu, A.I.; Cocan, I.; Negrea, M.; Alexa, E.; Obistioiu, D.; Hotea, I.; Radulov, I.; Poiana, M.A. Exploring the potential of tomato processing byproduct as a natural antioxidant in reformulated nitrite-free sausages. *Sustainability* **2022**, *14*, 11802. [CrossRef]
11. Patarata, L.; Martins, S.; Silva, J.S.; Fraqueza, M.J. Red wine and garlic as a possible alternative to minimize the use of nitrite for controlling *Clostridium sporogenes* and *Salmonella* in a cured sausage: Safety and sensory implications. *Foods* **2020**, *9*, 206. [CrossRef] [PubMed]
12. Brzezinska-Rojek, J.; Sagatovych, S.; Malinowska, P.; Gadaj, K.; Prokopowicz, M.; Grembecka, M. Antioxidant capacity, nitrite and nitrate content in beetroot-based dietary supplements. *Foods* **2023**, *12*, 1017. [CrossRef] [PubMed]
13. Aliyari, P.; Kazaj, B.; Barzegar, M.; Gavlighi, H.A. Production of functional sausage using pomegranate peel and pistachio green hull extracts as natural preservatives. *J. Agric. Sci. Technol.* **2020**, *22*, 159–172.
14. Karwowska, M.; Dolatowski, Z.J. Effect of acid whey and freeze-dried cranberries on lipid oxidation and fatty acid composition of nitrite-/nitrate-free fermented sausage made from deer meat. Asian-Aust. *J. Anim. Sci.* **2017**, *30*, 85. [CrossRef] [PubMed]
15. Kenari, R.E.; Razavi, R. Encapsulation of bougainvillea (*Bougainvillea spectabilis*) flower extract in *Urtica dioica* L. seed gum: Characterization, antioxidant/antimicrobial properties, and in vitro digestion. *Food Sci. Nutr.* **2020**, *10*, 3436–3443. [CrossRef]
16. Abarca-Vargas, R.; Petricevich, V.L. *Bougainvillea* genus: A review on phytochemistry, pharmacology, and toxicology. *Evid.-Based Complement. Altern. Med.* **2018**, *2018*, 9070927. [CrossRef] [PubMed]
17. Domínguez, R.; Munekata, P.E.; Pateiro, M.; Maggiolino, A.; Bohrer, B.; Lorenzo, J.M. Red beetroot. A potential source of natural additives for the meat industry. *Appl. Sci.* **2020**, *10*, 8340. [CrossRef]
18. Abdelrahman, K.N.; Ghany, A.G.A.A.; Saber, R.A.; Osman, A.; Sitohy, B.; Sitohy, M. Anthocyanins from pomegranate peel (*Punica granatum*), chili pepper fruit (*Capsicum annuum*), and bougainvillea flowers (*Bougainvillea spectabilis*) with multiple biofunctions: Antibacterial, antioxidant, and anticancer. *Heliyon* **2018**, *10*, e32222. [CrossRef]
19. Petricevich, V.L.; Cedillo-Cortezano, M.; Abarca-Vargas, R. Chemical composition, antioxidant activity, cytoprotective and in silico study of ethanolic extracts of *Bougainvillea buttiana* (Var. Orange and Rose). *Molecules* **2022**, *27*, 6555. [CrossRef]
20. Soto, K.M.; Pérez Bueno, J.D.J.; Mendoza López, M.L.; Apátiga-Castro, M.; López-Romero, J.M.; Mendoza, S.; Manzano-Ramírez, A. Antioxidants in traditional Mexican medicine and their applications as antitumor treatments. *Pharmaceuticals* **2023**, *16*, 482. [CrossRef]
21. Ravikumar, P.M.; Pratheep, T. Exploring the therapeutic properties of *Bougainvillea glabra*: A study of its antimicrobial and antioxidant effects in methanol extract. *Ann. For. Res.* **2022**, *65*, 11285–11292.
22. García, I.G.O.; Barrera, A.L.G.; González, F.J.A.; Vela, N.A.C.; Montiel, D.G. *Bougainvillea glabra* Choisy (Nyctinaginacea): Review of phytochemistry and antimicrobial potential. *Front. Chem.* **2023**, *11*, 1276514. [CrossRef]
23. Javed, I.M.; Aleem, A.; Hamad, R.; Furqan, N.M.; Ammad, R. A review paper on foam-mat drying of fruits and vegetables to develop powders. *MOJ Food Process. Technol.* **2018**, *6*, 465–467. [CrossRef]
24. Kaushik, D.; Kumar, M.; Proestos, C.; Oz, F.; Gupta, P.; Kumar, A.; Kundu, P.; Kaur, J.; Kumar, V.; Anjali, A.; et al. A narrative review on the anti-inflammatory efficacy of *Bougainvillea spectabilis* Willd. and its various applications. *J. Agric. Food Res.* **2023**, *12*, 100570. [CrossRef]
25. Steen, L.; Neyrinck, E.; De Mey, E.; De Grande, A.; Telleir, D.; Raes, K.; Paelinck, H.; Fraeye, I. Impact of raw ham quality and tumbling time on the technological properties of polyphosphate-free cooked ham. *Meat Sci.* **2020**, *164*, 108093. [CrossRef] [PubMed]
26. Image J: Image Processing and Analysis in Java. Available online: https://imagej.net/ij/ (accessed on 15 July 2024).

27. Li, C.; Szczepaniak, S.; Steen, L.; Goemaere, O.; Impens, S.; Paelinck, H.; Zhou, G. Effect of tumbling time and cooking temperature on quality attributes of cooked ham. *Int. J. Food Sci. Tech.* **2011**, *46*, 2159–2163. [CrossRef]
28. Association of Official Agricultural Chemists. Official Methods of Analysis of the Association of Official's Analytical Chemists. In Proceedings of the 17th Edition, Association of Official Analytical Chemists (AOAC), Arlington, VA, USA, 17 March 2003.
29. *ISO 937:1978*; Determination of Nitrogen Content. International Standards Meat and Meat Products. International Organization for Standardization: Geneve, Switzerland, 1978.
30. Yucel, A.A.; Gulen, S.; Dincer, S.; Yucel, A.E.; Yetkin, G.I. Comparison of two different applications of the Griess method for nitric oxide measurement. *J. Exp. Integr. Med.* **2012**, *2*, 167–171. [CrossRef]
31. Sun, J.; Zhang, X.; Broderick, M.; Fein, H. Measurement of nitric oxide production in biological systems by using Griess reaction assay. *Sensors* **2003**, *3*, 276–284. [CrossRef]
32. Bellucci, E.R.B.; Souza, C.V.B.; Lorenzo, J.M.; Aleu, G.; Teixeira, A.; Domínguez, R.; da Silva-Barretto, A.C. Nitrate and Nitrite. In *Methods to Assess the Quality of Meat Products*; Springer: New York, NY, USA, 2022; pp. 85–95.
33. Hentati, F.; Barkallah, M.; Atitallah, A.B.; Dammak, M.; Louati, I.; Pierre, G.; Fendri, I.; Attia, H.; Michaud, P.; Abdelkafi, S. Quality characteristics and functional and antioxidant capacities of algae-fortified fish burgers prepared from common barbel (*Barbus barbus*). *BioMed Res. Int.* **2019**, *2019*, 11–14. [CrossRef] [PubMed]
34. Rivero-Perez, M.D.; Muñiz, P.; Gonzalez-Sanjosé, M.I. Antioxidant profile of red wines evaluated by total antioxidant capacity, scavenger activity, and biomarkers of oxidative stress methodologies. *J. Agric. Food Chem.* **2007**, *55*, 5476–5483. [CrossRef]
35. Benzie, I.F.F.; Strain, J.J. Ferric reducing/antioxidant power assay: Direct measure of total antioxidant activity of biological fluids and modified version for simultaneous measurement of total antioxidant power and ascorbic acid concentration. *Methods Enzymol.* **1999**, *299*, 15–27. [CrossRef]
36. Vyncke, W. Evaluation of the direct thiobarbituric acid extraction method for determining oxidative rancidity in mackerel (*Scomber scombrus* L.). *Fette Seifen Anstrichm.* **1975**, *77*, 239–240. [CrossRef]
37. Traore, S.; Aubry, L.; Gatellier, P.; Przybylski, W.; Jaworska, D.; Kajak-Siemaszko, K.; Santé-Lhoutellier, V. Higher drip loss is associated with protein oxidation. *Meat Sci.* **2012**, *90*, 917–924. [CrossRef] [PubMed]
38. Dias, S.; Castanheira, E.M.; Fortes, A.G.; Pereira, D.M.; Rodrigues, A.R.O.; Pereira, R.; Gonçalves, M.S.T. Application of natural pigments in ordinary cooked ham. *Molecules* **2020**, *25*, 2241. [CrossRef] [PubMed]
39. Singh, T.; Pandey, V.K.; Dash, K.K.; Zanwar, S.; Singh, R. Natural bio-colorant and pigments: Sources and applications in food processing. *J. Agric. Food Res.* **2023**, *12*, 100628. [CrossRef]
40. Abo-Elghiet, F.; Ahmed, A.H.; Aly, H.F.; Younis, E.A.; Rabeh, M.A.; Alshehri, S.A.; Alshahrani, K.S.A.; Mohamed, S.A. D-Pinitol content and antioxidant and antidiabetic activities of five *Bougainvillea spectabilis* Willd. cultivars. *Pharmaceuticals* **2023**, *16*, 1008. [CrossRef]
41. Kuhn, F.; de Azevedo, E.S.; Frazzon, J.; Noreña, C.P.Z. Evaluation of green extraction methods on bioactive compounds and antioxidant capacity from *Bougainvillea glabra* bracts. *Sustain. Chem. Pharm.* **2021**, *19*, 100362. [CrossRef]
42. Wu, Q.; Fu, X.; Chen, Z.; Wang, H.; Wang, J.; Zhu, Z.; Zhu, G. Composition, color stability and antioxidant properties of betalain-based extracts from bracts of *Bougainvillea*. *Molecules* **2022**, *27*, 5120. [CrossRef]
43. Orozco-Villafuerte, J.; Escobar-Rojas, A.; Buendía-González, L.; García-Morales, C.; Hernandez-Jaimes, C.; Alvarez-Ramirez, J. Evaluation of the protection and release rate of Bougainvillea (*Bougainvillea spectabilis*) extracts encapsulated in alginate beads. *J. Dispers. Sci. Technol.* **2018**, *40*, 1065–1074. [CrossRef]
44. Bešlo, D.; Golubi ´c, N.; Rastija, V.; Agi ´c, D.; Karnaš, M.; Šubarić, D.; Lučić, B. Antioxidant activity, metabolism, and bioavailability of polyphenols in the diet of animals. *Antioxidants* **2023**, *12*, 1141. [CrossRef]
45. Arshad, M.S.; Imran, A.; Nadeem, M.T.; Sohaib, M.; Saeed, F.; Anjum, F.M.; Kwo, J.; Hussain, S. Enhancing the quality and lipid stability of chicken nuggets using natural antioxidants. *Lipids Health Dis.* **2017**, *16*, 108. [CrossRef]
46. Campo, M.M.; Nute, G.R.; Hughes, S.I.; Enser, M.; Wood, J.D.; Richardson, R.I. Flavour perception of oxidation in beef. *Meat Sci.* **2006**, *72*, 303–311. [CrossRef] [PubMed]

foods

MDPI

Article

Incorporating Olive By-Products in Bísaro Pig Diets: Effect on Dry-Cured Product Quality

Ana Leite [1,2], Lia Vasconcelos [1,2], Sergio Lopez [3], Divanildo Outor-Monteiro [4], Victor Pinheiro [4], Sandra Rodrigues [1,2,*] and Alfredo Teixeira [1,2]

1 Centro de Investigação de Montanha (CIMO), Instituto Politécnico de Bragança, Campus de Santa Apolónia, 5300-253 Bragança, Portugal; anaisabel.leite@ipb.pt (A.L.); lia.vasconcelos@ipb.pt (L.V.); teixeira@ipb.pt (A.T.)
2 Laboratório para a Sustentabilidade e Tecnologia em Regiões de Montanha, Instituto Politécnico de Bragança, Campus de Santa Apolónia, 5300-253 Bragança, Portugal
3 IES Andrés de Valdelvira, 02006 Albacete, Spain; sergiologar97@gmail.com
4 Animal Science Department, Veterinary and Animal Research Centre (CECAV), University of Trás-os-Montes e Alto Douro (UTAD), 5000-801 Vila Real, Portugal; divanildo@utad.pt (D.O.-M.); vpinheir@utad.pt (V.P.)
* Correspondence: srodrigues@ipb.pt

Abstract: The objective of this study was to assess the impact of incorporating olive cake into the diet of indigenous Bísaro pigs on the quality of processed meat products. To this end, loins and "cachaços" were processed using a standardized manufacturing flowchart to produce dry-cured products. The two products were manufactured using the same formulation, ingredients, and curing process. Concerning the physicochemical composition, there were significant differences between the two products for the parameters of a_w ($p < 0.001$), moisture ($p < 0.001$), total fat ($p < 0.001$), protein ($p < 0.001$), and haem pigments ($p < 0.001$). The diet significantly impacted the NaCl content ($p < 0.05$). However, neither the product nor the diet affected the fractions of saturated fatty acids (SFA), monounsaturated fatty acids (MUFA), or polyunsaturated fatty acids (PUFA) ($p > 0.05$). However, a significant difference was observed for n-3 ($p < 0.05$). Adding olive cake increased these fatty acids, and the diet containing 25% centrifuged olive cake showed the highest levels for both products. Compared with the control, the diets containing olive cake had a higher content of n-3 fatty acids, resulting in a lower PUFA n-6/n-3 ratio ($p < 0.01$).

Keywords: indigenous Bísaro pig; processed meat products; olive cake; valorization

1. Introduction

The production of olive oil is a well-known phenomenon on a global scale, with an average output of approximately 3 million tons of olive oil per year. Of this total, 2 million tons (or over 67%) are produced by the European Union (EU). Among the EU countries, the largest share of olive production comes from Spain, Italy, Greece, and Portugal. More than half of the total produced in the European Union (66%) comes from Spain. Italy and Greece have very similar productions, with 15% and 13%, respectively. Lastly, Portugal is responsible for 5% of the total produced in the EU. Additionally, the EU is the biggest consumer and exporter of olive oil. The EU is responsible for consuming 50% of annual production, 1.5 million tons of olive oil, and exporting 570,000 tons annually [1].

As a result of the high production of olive oil, many by-products that are highly toxic to the environment are generated. Efficient management of these by-products is necessary to reduce their environmental impact and ensure economic profitability since most of these raw materials, if not disposed of correctly, result in excessively high treatment costs. According to Molina-Alcaide et al. [2], olive oil by-products can be categorized as olive leaves, olive cake, and other by-products. The olive leaves refer to a mixture of leaves and branches from pruning the olive trees and harvesting and cleaning the olives before the oil is extracted [2]. Olive cake consists of olive pulp, skin, stone, and water. The terminology for olive cake varies depending on the extraction methods, such as pressing, extracting, or centrifuging. In the case of centrifugal extraction, it is crucial to distinguish between three-phase and two-phase systems. The main difference is the quantity of oil and moisture, with the two-phase centrifuged olive cake being more efficient and environmentally friendly (with a higher moisture content and lower oil content) [3]. As far as pressed olive cake is concerned, extraction is achieved by a discontinuous press process, a more traditional extraction method. The nutritional valorization of these by-products has become a key focus for various research sectors. While they are widely used in ruminant feed, their application in other animal species remains limited.

Furthermore, these by-products are an energy resource due to their lignin content, providing a high calorific value with a low ash content [4]. Another potential application for this product is composting, which can be used as a fertilizer or a component of agricultural substrates [5]. Olive cake has had various uses but remains an under-exploited resource [2]. The market for commercializing raw materials for animal feed has been significantly constrained by price volatility.

Moreover, incorporating this by-product makes perfect sense in the northern region of Trás-os-Montes (Portugal), given the amount of olive cake produced and the number of farms with Bísara animals. In addition, the time of year when these by-products are obtained coincides with the finishing phase of these animals. Therefore, transporting this by-product is more accessible due to the proximity of the farms and oil mills, and storage is also more accessible. Bísaro pigs offer products characterized by high quality [6]. Due to the characteristics of their meat, native breeds are highly valued for producing high-quality dry-cured products [7].

Consequently, the main objectives of this work are to (1) evaluate the potential of olive cake (pressed and centrifuged) as a diet component for Bísaro pigs; (2) assess effects on the physicochemical composition of the two processed products obtained on an industrial scale; (3) analyze the impact of these products on the fatty acid profile; and (4) examine the influence of cutting the Longissimus thoracis lumborum (LTL) muscle in obtaining the two processed products.

2. Materials and Methods

2.1. Experimental Diets and Slaughter Procedure

The animals used were indigenous Bísaro pigs, kept in an extensive system on the farm belonging to the company Bísaro-Salsicharia Tradicional, Lda® (Gimonde-Bragança, Portugal). The first diet consisted of a traditional diet typical of these native breeds in an extensive system, thus considered the control diet. The second diet included the Base diet + 15% of centrifuged olive cake (BgCf15). The third diet consisted of a Base diet + 25% of centrifuged olive cake (BgCf25). The fourth diet consisted of a Base diet + 15% of pressed olive cake. The diets' chemical composition and fatty acid profile are presented in Table 1.

Table 1. Ingredient composition of the experimental diets (g/kg, as fed basis) and fatty acids composition (g/100 g).

	Diets			
	Control	**BgCf15**	**BgCf25**	**BgPr15**
	Chemical composition of the diet			
DM	86.35	83.76	90.77	88.40
OM	94.73	95.35	95.03	95.00
NDF	20.06	27.47	36.63	27.50
ADF	7.62	12.31	23.67	13.97
ADL	2.49	4.78	9.62	5.75
PB	12.31	11.18	13.78	10.94
GB	5.74	6.75	4.51	5.86

Table 1. *Cont.*

	Diets			
	Control	BgCf15	BgCf25	BgPr15
	Fatty acids (g/100 g)			
ΣSFA	15.27	12.88	16.28	11.91
ΣMUFA	23.80	59.32	41.71	56.69
ΣPUFA	22.25	9.18	25.11	7.27
n-6/n-3	17.18	7.77	14.08	7.57

DM—dry matter; OM—organic matter; NDF—neutral detergent fiber; ADF—acid detergent fiber; ADL: acid detergent lignin; PB—crude protein; GB—crude fat. C—control; BgCf15—Base diet + 15% olive cake two-phases; BgCf25—Base diet + 25% olive cake two-phases; and BgPr15—Base diet + 15% crude olive cake.

The Bísara animals were reared under the supervision of the University of Trás-os-Montes and Alto Douro in Vila Real, Portugal. A total of 48 animals were used for this study. The animals were separated into distinct groups to receive different diets (Figure 1). The control group was given a traditional diet for this indigenous breed, consisting of local vegetables, cereals, and feed appropriate for each growth stage. The other diets also had the same Base diet as the control diet, plus the type of olive cake with 15 and 25 percent (see Table 1). The animals were fed these diets during the finishing phase (90 days) and with an average feed of 3 kg daily. The University of Trás-os-Montes e Alto Douro team monitored the animals to ensure the diets were administered correctly throughout the finalization process.

Figure 1. Schematic of the diets applied to the Bísaro pigs.

Once the final finishing phase was complete, the animals were slaughtered at 12 months of age, with an average live weight of approximately 135 kg. The Bragança Municipal Slaughterhouse team was responsible for the slaughter. The methodology employed for the slaughter and carcass preparation has been previously described by Álvarez-Rodríguez and Teixeira [8]. All animals were provided with appropriate care and were slaughtered following EU Council Regulation (EC) No. 1099/2009 [9], which sets out animal welfare regulations at the time of slaughter. After the slaughter and cutting, the joints obtained were transported to Bísaro-Salsicharia Tradicional, Lda®, which was responsible for processing them.

2.2. *Dry-Cured Bísaro Loin and "Cachaço"*

The curing process was carried out at Bísaro-Salsicharia Tradicional, Lda®, using forty-eight loins and forty-eight "cachaços" from forty-eight slaughtered animals (Figure 2). The ingredients were added in decreasing order: 1.5% salt, 0.5% paprika, 0.5% garlic, and 0.1% oregano. Both the loins and "cachaços" were dry-cured for 60 days. After removing the muscles from the carcasses, the pieces were chilled in a chamber at 2–5 °C, and excess surface fat was removed from each piece. Selected pieces then underwent the curing process. In the salting and seasoning phase, the pieces were placed on a rotating drum for approximately 30 min after adding salt, paprika, garlic, and oregano. After mixing, the joints were transferred to a container and placed in a refrigeration chamber at a temperature range of 2–4 °C with approximately 90% relative humidity for 4 days to allow the ingredients to penetrate. Next, the pieces were stuffed into collagen casings. The final phase was drying and curing, during which significant biochemical changes occurred. The temperature and humidity were adjusted as the curing progressed: for the first 15 days, the cuts were kept at 4–8 °C with 80–90% relative humidity. Subsequently, the temperature was raised to 8–12 °C with 70–80% relative humidity for another 15 days. Finally, for the last 20 days, the product was maintained at 12–18 °C with 60–70% relative humidity. This curing process has been validated by the company Bísaro-Salsicharia Tradicional Lda®, considering all the quality and food safety standards. It should be noted that the company responsible for the entire manufacturing flowchart for this type of product has been certified for over seven years by extremely demanding quality benchmarks (IFS-Food).

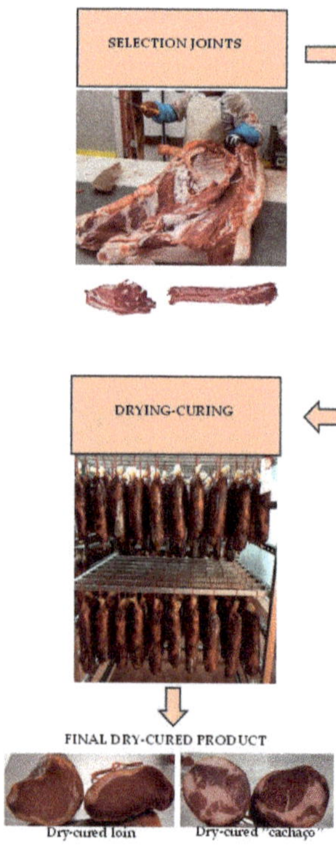

Figure 2. Process of obtaining Bísaro dry-cured loin and dry-cured "cachaço".

2.3. *Chemical Composition and Physicochemical Analysis of Dry-Cured Loin and Dry-Cured "Cachaço"*

All the physicochemical analyses on these two products were conducted using the following Portuguese standards. Water activity was assessed according to AOAC [10] using a HigroPalm Rotronic 8303 probe (Bassersdorf, Switzerland). Moisture content was determined according to the Portuguese standard NP 1614 [11]. For this, 5 mL of ethanol (96% v/v) was added to 3 g of sample, and the samples were dried in a drying oven (Raypa DO-150, Barcelona, Spain) at 103 ± 2 °C for 24 h. Ash content was determined according to the Portuguese standard [12]. To do this, 1 mL of magnesium acetate (15% w/v) was added to 3 g of sample in crucibles. The samples were then heated to 550 ± 25 °C for 5 h in a muffle furnace (Vulcan BOX Furnace Model 3-550, Yucaipa, CA, USA). Protein content was measured following the Portuguese standard [13] using the Kjeldahl Sampler System (K370, Flawil, Switzerland) and Digest System (K-437, Flawil, Switzerland). In 25 mL of sulfuric acid (97%), two catalyst tablets and 2 g of sample were placed in mineralization tubes. Following the completion of mineralization, the distillation procedure was performed. Subsequently, the distillate was titrated using a hydrochloric acid solution, and the necessary volume was registered. The determination of hydroxyproline and collagen content was carried out by Portuguese Standards NP 1987 [14]. The haem pigment content [15] was determined by measuring the reflectance on the exposed surface using spectroscopy with a Spectronic Unicam 20 Geneys instrument. The results are expressed as mg myoglobin/g fresh muscle. Additionally, the total chloride content was assessed following the methodology specified in the Portuguese Standard NP 1845 [16], expressed as a percentage by mass of sodium chloride.

2.4. Fatty Acid Analysis

Fatty acids in dry-cured loin and dry-cured "cachaço" samples were analyzed at the ESA-IPB Laboratory. Total lipids were extracted from 25 g of meat using the Folch procedure [17]. The fatty acid profile was determined from 50 mg of fat. Fatty acids were transesterified following the method described by Domínguez et al. [18]. After adding 4 mL of sodium methoxide solution and vortexing intermittently for 15 min at room temperature, 5 mL of H_2SO_4 solution (50% in methanol) was added. Then, 2 mL of distilled water was added, followed by additional vortexing. The organic phase, containing the methyl esters of fatty acids, was extracted with 2.35 mL of hexane. The separation and quantification of fatty acid methyl esters were carried out using a gas chromatograph (GC-Shimadzu 2010Plus; Shimadzu Corporation, Kyoto, Japan) equipped with a flame ionization detector and an automatic sample injector AOC-20i and using a Supelco SP^{TM}-2560 fused silica capillary column (100 m length, 0.25 mm i.d., 0.2 µm film thickness). Fatty acid contents were calculated using chromatogram peak areas and were expressed as g per 100 g of total fatty acid methyl esters. Additionally, the percentage of saturated fatty acids (ΣSFA), monounsaturated fatty acids (ΣMUFA), polyunsaturated fatty acids (ΣPUFA), the ratio PUFA n-6/n-3, and Σtrans were calculated according to Vieira et al. [19]. To assess lipid quality, the atherogenicity index (AI) and the thrombogenicity index (IT) were calculated following the Ulbricht and Southgate methods [20].

2.5. Statistical Analysis

The Shapiro–Wilk test was used to test data for normal distribution and homogeneity of variance. Next, the effect of diet and type of product, and the interaction between diet and product, on the physicochemical composition and fatty acid profile were examined using analysis of variance (ANOVA) with the general linear model (GLM) procedure, in which these parameters were defined as dependent variables and diet and type of product as fixed effects. The results were presented in terms of mean values and standard error of the mean (SEM). When there was a significant effect ($p < 0.05$), the means were compared using Student's t-test. To extract a few key combinations (called principal components) from the group of measured variables that capture most of the variability in those variables, we conducted a principal component analysis (PCA). Each principal component was determined by linearly combining the correlation matrix's eigenvectors. The eigenvalues indicate how much variance each component holds. Moreover, a multiple factor analysis (MFA) related to principal components analysis (PCA) was performed to produce a table of eigenvalues, summary plots, and a consensus map. All analyses were performed using the statistical package JMP® Pro 17.0.0 by 2023 SAS Institute Inc.© (Cary, NC, USA).

3. Results and Discussion

3.1. Physicochemical Composition

The results of the chemical composition of dry-cured loin and dry-cured "cachaço" are presented in Table 2. This table shows the effect of the product on each of the parameters studied, the impact between the diets, and the interaction between the type of product and the diet applied. Regarding the interaction between the product and the diets, the parameters a_w, ash, and NaCl content showed significant differences ($p < 0.05$). The other parameters (moisture, total fat, protein, collagen, and haem pigments) did not show significant results between the type of product and the diet applied. The product type significantly affected ($p < 0.001$) the parameters of a_w, moisture, total fat, protein, and haem pigments. On the other hand, collagen, NaCl content, and ash content were not significantly influenced ($p > 0.05$) by the product type. We can also see that the diet did not affect most of the physicochemical parameters studied, except for the NaCl content.

Table 2. Physicochemical composition of dry-cured Bísaro loin and dry-cured Bísaro "cachaço".

	Physicochemical Composition (g/100 g)															
	a_w		Moisture		Ash		Total Fat		Protein		Collagen		Haem Pigments		NaCl	
	L	C	L	C	L	C	L	C	L	C	L	C	L	C	L	C
Control	0.893	0.857	41.27	31.24	5.07	4.97	21.40	41.09	32.61	24.52	3.44	3.15	2.31	4.17	3.75	4.24
Cf15	0.873	0.839	42.73	33.43	4.98	5.96	19.03	36.59	32.24	24.99	3.35	2.55	2.15	4.33	3.55	5.01
Cf 25	0.866	0.862	40.02	31.84	7.21	5.36	24.62	40.03	29.73	23.40	3.66	2.74	2.72	4.04	6.53	4.01
Pr15	0.896	0.815	41.69	30.99	4.25	6.13	16.90	43.37	34.26	24.39	2.83	2.87	2.41	3.84	2.76	5.36
SEM	0.01		1.37		0.57		3.46		1.13		0.56		0.29		0.53	
Significance product	***		***		ns		***		***		ns		***		ns	
Significance diet	ns		ns		ns		ns		ns		ns		ns		*	
Significance product x diet	*		ns		**		ns		ns		ns		ns		***	

ns—not significant, * $p < 0.05$; ** $p < 0.01$; *** $p < 0.001$. SEM (Standard Error of the Mean). L—dry-cured loin; C—dry-cured "cachaço"). Haem pigments in mg myoglobin/g fresh muscle. Cf15—Base diet + 15% olive cake centrifuged; Cf25—Base diet + 25% olive cake centrifuged; Pr15—Base diet + 15% olive cake pressed; CT—Base diet.

Concerning water activity (a_w), the values obtained for the dry-cured loin ranged from 0.896 to 0.856, while those for the dry-cured "cachaço" ranged from 0.862 to 0.815. A statistically significant correlation was observed between the product and the diet ($p <$ 0.05) for the water activity parameter. The highest value for this parameter (0.896) was observed for the dry-cured loin with the 15% pressed olive cake diet (Pr15), while the lowest value for this parameter was obtained for the dry-cured "cachaço" with the same diet. The a_w value was significantly higher ($p < 0.001$) in all diets (including the control) for the dry-cured loin compared to the dry-cured "cachaço". These differences in water activity values align with the results obtained for the moisture parameter. The dry-cured loin, influenced by fat content (discussed later), has a lower water activity and moisture content than other forms of dry-cured loin.

The a_w value for the dry-cured loin control was the highest, exceeding the values reported for the same product in a study where the olive cake was not added to the diet of the Bísaro pig [21]. In contrast, the a_w values observed in this work for diets containing olive cake are consistent with those reported in another study [21] (for the same product type and adding olive cake in other percentages). In previous studies, different authors reported average values of 0.841 for dry-cured Celta loin [22], 0.830 for dry-cured Korean loin [23], 0.838 for dry-cured Polish neck [24], and 0.838 for dry-cured foal loin [25]. Similar water activity values have been reported for other dry-cured products, including dry-cured shoulder [26–29]. The water activity of a food product is a significant factor that affects the safety of dry-cured meat products. It serves as a fundamental indicator of the shelf life, ensuring nutritional stability and identifying the types of microorganisms present [30]. The combination of water and salt creates osmotic changes that result in dehydration, which removes water from the meat [31]. Water can facilitate the attachment of a radical species and the removal of hydrogen from the fatty acids, thereby initiating the oxidation process [22]. Therefore, both products exhibited water activity values that align with the criteria for dry-cured products with the desired microbiological stability. Furthermore, adding olive cake to the Bísaro pig feed did not exert any discernible influence, whether positive or negative. Another crucial parameter that serves as an indicator of product ripeness is moisture content.

The moisture content of the dry-cured loin was found to be significantly higher ($p < 0.001$) than that of the dry-cured "cachaço". The mean moisture content of the dry-cured loin ranged from 40.02 to 42.73%. Meanwhile, the dry-cured "cachaço" range was 30.99 to 33.43%. Including olive cake in the animals' diet did not result in any observable change in moisture content. The moisture content of dry-cured products is inversely proportional to the maturation time [32]. The values obtained in this study were higher for both products than those observed by other authors [21]. Nevertheless, comparable values to those observed in the dry-cured "cachaço" have been documented in a conventional dry-cured product from Spain's Mediterranean coast [33] and in the traditional Italian product "coppa" [34]. Similar values (42%) were obtained for Iberian dry-cured loins [34,35], as well as for other dry-cured loins [22,25,36].

The observed variations in ash content are primarily attributed to the sodium chloride content. The ash values exhibited a range of 4.25 to 7.21 for dry-cured loin and 4.97 to 6.13 for dry-cured "cachaço." The interaction between the product and diet was statistically significant ($p < 0.01$). Nevertheless, introducing the olive cake diet or the type of product used did not result in significant differences in ash content. Other studies have reported higher ash values for dry-cured foal loin [25], ", Bísaro dry-cured shoulder [29], and Celta dry-cured ham [37]. Similar values have been observed in other studies involving Iberian dry-cured loin [38] and Turkish dried meat [39].

The total fat content exhibited a statistically significant difference ($p < 0.001$) between the dry-cured loin and dry-cured "cachaço", with values ranging from 16.90 to 24.62% and from 36.59 to 43.37%, respectively. Concerning dietary factors, no impact was observed on the total fat content of the processed products. Although these two products originate from the same *Longissimus thoracis lumborum* (LTL) muscle, they are derived from disparate sections. The "cachaço" is obtained from the proximal region of the LTL muscle, situated in the cervical area of the column and extending to the fifth thoracic vertebra, as evidenced by its location beneath the scapula. The loin is obtained from the lumbar region of the LTL muscle. The high complexity of the meat matrix results in products derived from the same muscle exhibiting markedly disparate values for the total fat parameter.

A reduction in protein content is anticipated to accompany an elevated fat content. As demonstrated in Table 2, the protein content was significantly higher ($p < 0.001$) in the dry-cured loin, with mean values ranging between 29.73 and 34.26%. In contrast, the dry-cured "cachaço" exhibited lower protein values, ranging from 23.40% to 24.99%. The application of the olive by-product diets did not result in any discernible impact on the protein content of the products under investigation. Other studies have also observed this difference between dry-cured loin and "cachaço" [21]. A product designated as dry-cured coppa exhibited similar values to those observed in the dry-cured "cachaço" [34]. The same values have also been reported by other authors for Turkish dried meat [39], dry-cured Celta ham [37], and Bísaro dry-cured shoulder [29]. Higher values were observed in the case of the dry-cured foal [25].

No significant differences ($p > 0.05$) were observed between diets with olive cake or between the two product types concerning collagen content, which ranged from 2.83 to 3.66% and from 2.55 to 3.15% in the dry-cured loin and dry-cured "cachaço", respectively.

The total haem pigments, expressed as myoglobin concentration, exhibited a statistically significant difference between the two products (Bísaro dry-cured loin and dry-cured "cachaço"). The myoglobin content for these products ranged from 2.15 to 2.72 mg/g and 3.84 to 4.33 mg/g in the case of the dry-cured loin and dry-cured "cachaço", respectively. The incorporation of pressed and centrifuged olive cake did not affect this parameter, as evidenced by the findings of other researchers [21].

Significant differences were observed in the chloride content for the interaction between diet and type of product ($p < 0.001$) and for the addition of olive cake ($p < 0.05$). The sodium chloride content exhibited variability, ranging from 2.76 to 6.53% in the dry-cured loin and from 4.01 to 5.36% in the dry-cured "cachaço". The highest salt content was observed in the dry-cured loin with the centrifuged olive cake diet at 25%. This value represents the ash content observed in the Cf25 diet in the dry-cured loin. Although a difference in the salt content of the dry-cured loin was observed, it cannot be definitively concluded that the 25% olive cake inclusion was the sole factor responsible for this value. This same pattern was not observed in the dry-cured "cachaço" case. The average NaCl value for the dry-cured loin was lower than that of the dry-cured "cachaço". The dry-cured loin (Cf25) exhibited a markedly elevated value, resulting from a notable diet impact. It is imperative to reiterate that these products were not manufactured in a laboratory setting. Therefore, it is not possible to attribute the high levels of sodium chloride observed solely to the diet. Similar NaCl content values were observed in Salame Milano (4.3%), Coppa (5.9%), Parma Ham (6.1%) [40], and dry-cured foal "Cecina" [41].

3.2. Fatty Acids Composition

The fatty acid (FA) composition of the dry-cured loin and dry-cured "cachaço" is presented in Table 3 for analysis purposes. The table illustrates the impact of the product and the effect between diets on the fatty acid profile. For both products, the predominant fatty acids were oleic acid as monounsaturated fatty acid (MUFA), palmitic acid, stearic acid as saturated fatty acid (SFA), and linoleic acid as polyunsaturated fatty acid (PUFA). These four acids account for more than 93% of the total fatty acids in both products. These results align with pork's typical fatty acid composition and are consistent with the fatty acid profile described by Leite et al. [42]. A comparable pattern in FA distribution has been documented in various types of dry-cured products [21,34,43–47]. Except for C18:0 fatty acid, no significant differences were observed in the remaining predominant fatty acids between the two products studied. The addition of olive cake to the animal diets had no discernible impact. Significant differences were observed between the products for the following FA: C15:0, C17:0, C18:0, C20:0, C18:3n-6, C22:0, C23:0, and n-3. As for the diet, the inclusion of olive cake in the animal diet of Bísaro pigs had a significant influence on the following FA: C16:1n-7, 9t-C18:1, C20:1n-9, C18:3n-3, C20:2n-6, n-3, and ratio n-6/n-3. The FA fractions were not affected by the type of product studied. Similarly, adding centrifuged and pressed olive cake in 15 and 25% percentages did not influence the lipid quality of dry-cured loin and dry-cured "cachaço". Therefore, despite numerous factors that can affect the lipid composition of meat, such as diet and genetic lineage, the fatty acid profile of pork remains consistent with the typical pork profile.

Table 3. Effect of diet in fatty acids profile in Bísaro dry-cured Loin and dry-cured "cachaço".

Fatty Acids	L				C				SEM	*p*	D
	Control	Cf15	Cf25	Pr15	Control	Cf15	Cf25	Pr15			
C10:0	0.02	0.02	0.02	0.02	0.02	0.01	0.02	0.01	0.004	ns	ns
C12:0	0.05	0.05	0.05	0.05	0.04	0.05	0.04	0.05	0.003	ns	ns
C14:0	1.14	1.15	1.13	1.11	1.13	1.13	1.12	1.11	0.03	ns	ns
C14:1	0.03	0.03	0.03	0.02	0.02	0.03	0.03	0.02	0.005	ns	ns
C15:0	0.05	0.05	0.05	0.07	0.02	0.03	0.02	0.01	0.008	***	ns
C16:0	26.12	25.98	26.05	25.48	26.38	26.17	26.37	25.80	0.30	ns	ns

Table 3. *Cont.*

Fatty Acids	L				C				SEM	p	D
	Control	Cf15	Cf25	Pr15	Control	Cf15	Cf25	Pr15			
C16:1n-7	2.50	2.32	2.29	2.36	2.53	2.34	2.32	2.15	0.08	ns	**
C17:0	0.17	0.20	0.21	0.12	0.24	0.23	0.24	0.22	0.02	**	ns
C17:1n-7	0.23	0.22	0.22	0.19	0.24	0.23	0.23	0.20	0.02	ns	ns
C18:0	13.1	13.41	13.30	12.96	12.68	12.72	12.93	12.84	0.32	*	ns
9t-C18:1	0.21	0.23	0.19	0.21	0.20	0.22	0.18	0.22	0.01	ns	**
C18:1n-9	47.82	47.40	47.49	48.43	47.97	48.01	47.43	48.49	0.55	ns	ns
C18:2n-6	6.50	6.79	6.76	6.74	6.62	6.81	7.02	6.90	0.26	ns	ns
C20:0	0.21	0.19	0.21	0.22	0.17	0.17	0.18	0.18	0.01	***	ns
C18:3n-6	0.011	0.011	0.013	0.006	0.003	0.006	0.004	0.003	0.002	***	ns
C20:1n-9	0.77	0.80	0.91	0.82	0.76	0.80	0.86	0.79	0.03	ns	***
C18:3n-3	0.20	0.24	0.24	0.22	0.21	0.24	0.24	0.25	0.01	ns	***
C20:2n-6	0.24	0.24	0.2	0.25	0.23	0.24	0.28	0.24	0.01	ns	**
C22:0	0.04	0.05	0.03	0.05	0.03	0.04	0.03	0.03	0.03	*	ns
C20:3n-6	0.05	0.05	0.05	0.06	0.04	0.05	0.05	0.04	0.006	ns	ns
C22:1n-9	0.03	0.04	0.03	0.04	0.03	0.03	0.04	0.02	0.004	ns	ns
C23:0	0.35	0.40	0.33	0.43	0.27	0.32	0.22	0.29	0.05	***	ns
C24:1n-9	0.06	0.07	0.06	0.07	0.07	0.07	0.07	0.06	0.007	ns	ns
C22:6n-3	0.03	0.03	0.02	0.03	0.02	0.02	0.03	0.02	0.005	ns	ns
SFA	41.31	41.52	41.38	40.52	41.01	40.88	41.16	40.55	0.56	ns	ns
MUFA	51.65	51.11	51.23	52.15	51.83	51.73	51.15	51.95	0.56	ns	ns
PUFA	7.04	7.38	7.39	7.33	7.16	7.40	7.70	7.49	0.28	ns	ns
n-6	6.80	7.10	7.11	7.06	6.90	7.11	7.36	7.18	0.27	ns	ns
n-3	0.22	0.27	0.28	0.26	0.26	0.28	0.33	0.31	0.02	**	*
n-6/n-3	31.75	27.39	27.37	28.24	30.85	26.74	24.83	24.64	1.88	ns	**
IA	0.52	0.52	0.52	0.50	0.53	0.52	0.53	0.51	0.01	ns	ns
IT	1.35	1.36	1.35	1.30	1.33	1.32	1.34	1.3	0.03	ns	ns
h/H	2.00	2.01	2.01	2.09	2.00	2.02	2.00	2.07	0.04	ns	ns

ns—not significant, * $p < 0.05$; ** $p < 0.01$; *** $p < 0.001$. SEM (Standard Error of the Mean). Cf15—Base diet + 15% olive cake centrifuged; Cf25—Base diet + 25% olive cake centrifuged; Pr15—Base diet + 15% olive cake pressed; CT—Base diet. SFA, saturated fatty acids; MUFA, monounsaturated fatty acids; PUFA, polyunsaturated fatty acids; PUFA n-6/n-3 (∑ omega-6)/(∑ omega-3); IA, index of atherogenicity; IT, index of thrombogenicity; h/H = (C18:1n-9 + C18:2n-6 + C20:4n-6 + C18:3n-3 + C20; 5-n3 + C22:5n-3 + C22:6n-3)/C14:0 + C16:0; only fatty acids which represented more than 0.1% are presented in the table, although all detected fatty acids were used for calculating the totals and the indices.

Concerning MUFAs, oleic acid was the most prevalent FA, representing approximately 93% of the total MUFA content. This high proportion is significant for both products and represents an essential nutritional aspect, as this fatty acid fraction has been demonstrated to reduce cardiovascular risk factors [48]. Furthermore, MUFA has been shown to reduce plasma LDL cholesterol levels without compromising the anti-atherogenic properties of HDL cholesterol lipoproteins [49]. Oleic acid levels ranged from 51.11 to 52.15% in the dry-cured loin and from 51.15 to 51.95% in the dry-cured "cachaço", and these levels remained unchanged with the introduction of olive cake. Although the chemical composition of these two products showed significant differences in total fat content (Table 2), there were no differences between the two products regarding oleic acid. Similar MUFA were observed in Bísaro dry-cured loin and dry-cured "cachaço" [21]. Other authors have reported higher values of this acid in Iberian ham [50], Iberian dry-cured loin [34], dry-cured "coppa" [34], and Bísaro shoulder [29]. Lower MUFA values were reported in Iberian "lacón" [51], Korean dry-cured loin [23], Italian dry-cured loin [47], and Celta dry-cured ham [43].

Regarding SFA, the predominant FA was palmitic acid, showing levels of around 63% of the total SFA. Like oleic acid, palmitic acid levels were unaffected by the product type or olive cake's inclusion in the diet. We obtained values of between 40.52 and 41.52% for the dry-cured loin and 40.55–41.16% for the dry-cured "cachaço". The SFA fraction was also not influenced by adding olive cake, and there were no significant differences between the two products regarding palmitic acid content. Similar SFA was observed in Bísaro dry-cured loin and dry-cured "cachaço" [21], and Iberian dry-cured loin [52]. Lower values for this fraction were obtained in Bísaro shoulder [29], Celta dry-cured loin [44], Iberian lacón [51], Iberian and Serrano Ham, Bayonne and Corsican Ham, Parma, and San Daniele Ham, Jingua Ham [49], and dry-cured "coppa" [34]. Other authors have reported higher values of SFA in Korean dry-cured loin [23] and Croatian and Montenegrin dry-cured meat [46]. In agreement with other authors [53,54] about SFA, mortality rates correlated positively with the average percentage of dietary energy from saturated fatty acids.

For PUFAs, linoleic acid was the predominant FA, accounting for more than 95% of the total PUFA content. The levels of linoleic acid were unaffected by the product type or the olive cake's inclusion in the animal diet. Similar PUFA were observed in Bísaro dry-cured loin and dry-cured "cachaço" [21], Korean dry-cured loin [23], and Croatian and Montenegrin Prosciutto [46]. Other authors have reported higher values of PUFA in Bísaro shoulder [29], Celta dry-cured "lacón" [51], Iberian and Serrano dry-cured loin, Bayonne and Corsican Ham, Parma and San Daniele Ham, Jingua Ham [49], Pancetta and Croatian and Montenegrin dry-cured sirloin [46], and dry-cured "coppa" [34]. Lower values for this fraction were obtained in Iberian dry-cured loin [52]. It should be noted that PUFA is highly susceptible to oxidative degradation and is converted into other molecules [55]. For this reason, the curing process reduces the PUFA content in the final product. Adding a by-product such as olive cake did not counteract this reduction in unsaturated fatty acid.

Compared to previous studies' findings [21], including a higher percentage of olive cake (centrifuged and pressed) does not influence the MUFA, SFA, and PUFA fractions.

Regarding trans fatty acids, no significant differences were observed between the two product types ($p > 0.05$), with average values of 0.21 for dry-cured loin and 0.20 for dry-cured "cachaço". These values are below the recommended levels [56] and lower than those reported by other authors for processed Iberian dry-cured ham [57] and Bísaro shoulders [29]. Nevertheless, significantly lower values were observed in the group that consumed a diet including 25% centrifuged olive cake. Compared to the control diet, the diet containing 25% olive cake (Cf25) resulted in more favorable trans fatty acid values. In contrast, the diet with 15% centrifuged olive cake and 15% pressed olive cake exhibited detrimental outcomes concerning this category of trans fatty acid.

The PUFA/SFA and n-6/n-3 ratios, along with the IT and IA indices, are important indicators of the healthiness of fat in food. The recommended PUFA/SFA ratio for a healthy diet is 0.4 or less [58]. For the ratio n-6/n-3, the internationally recommended value for a healthy diet is 4 [59], with the optimal value being 1 [60–62]. Genetic modification and dietary changes have proven to be relatively effective strategies for achieving more desirable n-6/n-3 values in these products [43]. There was no significant difference in the n-6/n-3 ratio between the two products studied.

On the other hand, the diets applied with olive cake proved to be significantly different ($p < 0.05$). From the results shown in Table 3, adding olive cake to the animals' diets significantly reduces the n-6/n-3 ratio. This trend was not observed in other studies for Bísaro dry-cured loin and dry-cured "cachaço" with 10% olive cake added [21]. Therefore, we can say that introducing a percentage of 15 and 25% olive cake in the animal diet contributes to a decrease in this ratio. Higher values of this ratio were observed in Iberian and Parma dry-cured ham [49]. However, significantly lower values were observed in Serrano dry-cured ham, Bayonne dry-cured ham, and Corsican dry-cured ham [49]. Unsaturated fats, especially PUFAs, are well known as healthy fats with critical bodily functions, such as cell growth and development and disease prevention [63]. Although linoleic acid is an essential fatty acid, it should be noted that an excessively high intake of this PUFA may not be beneficial. Studies [64,65] have shown that excessive intake of linoleic acid has a pro-inflammatory effect. Fatty acids n-6 promote vasoconstriction and the formation of blood clots, while n-3 acids have the opposite effect [66]. Among the PUFA n-3, docosahexaenoic acid (C22:6n-3, DHA) is preventive and therapeutic in some chronic inflammatory diseases [67]. In small quantities, these FAs are crucial for the correct cerebral and visual development of the fetus and the maintenance of neural and visual tissues throughout life [67]. The human body cannot synthesize this n-3 FA (DHA) type, which is only obtained through food [68]. This study revealed significant variations in the level of PUFA n-3 based on both the type of product and the diet used. The n-3 content increased with olive cake in the diet, containing 25% centrifuged olive cake, resulting in the highest n-3 values for both products. For dry-cured "cachaço", n-3 values ranged from 0.26 to 0.33, whereas for dry-cured loin, the values were lower, ranging from 0.22 to 0.28.

A lower IA value is indicative of a reduced saturated-to-unsaturated fatty acid ratio, whereas a lower IT value is associated with a diminished risk of developing blood clots [69]. No significant differences were observed between the two product types regarding the IA and IR indices, nor was there any impact from the olive cake diets. Although no entity or organization provides reference values for these indices [70], it is generally understood that lower IA and IT values indicate better nutritional quality, potentially reducing the risk of coronary heart disease. Lower values for these indices have been reported in dry-cured "coppa" of Nero Siciliano pig [34] and Bísaro shoulder [29]. Similar values were obtained for Bísaro dry-cured loin and "cachaço" in animals fed with olive cake [21]. Other authors have reported that the intramuscular fat and backfat indices were affected by including 10% olive cake in the pig's diet, increasing the MUFA and PUFA content and improving the IA and IT quality indices [71]. As reported by Cava et al. [72], an elevated h/H ratio indicates enhanced nutritional adequacy of the fat content in the food. In this study, the ratio h/H was not influenced by the diets applied or the product type ($p > 0.05$). The dry-cured loin obtained values between 2.00 and 2.09 and the dry-cured "cachaço" between 2.00 and 2.07. Lower values of this ratio were obtained in Bísaro dry-cured loin and "cachaço" [21]. Higher h/H values were observed in Celtic ham with a chestnut-based diet [43]. In the case of fresh pork, different cuts of meat obtained different h/H ratio values [73]. Therefore, not only the type of feed but also the joint influences this ratio. Considering the value obtained for fresh loin (2.170) [73], the curing process does not affect this nutritional index.

3.3. *Principal Component Analysis (PCA)*

Principal component analysis (PCA) is a statistical technique used to simplify a dataset by transforming the original variables into a smaller set of new variables, called principal components, which capture the essential characteristics of the data. These principal components are linear combinations of the original variables that maximize the total variance. The primary graphical output of PCA is often a biplot, which maps the cases using the principal components and includes the original variables to help interpret the distances between case positions [74]. Figure 3 displays the results of the principal component analysis in the form of a biplot. The first principal component ($p < 0.001$) explained 40% of the total variance, and the second principal component ($p < 0.001$) explained 24.7% of the total variance. The two principal components explained the total variance of 64.7%. This principal component analysis shows a separation between the dry-cured loin and the dry-cured "cachaço". The physicochemical characteristics that best explain the dry-cured loin are salt content, total fat, and haem pigments. On the other hand, water activity, protein content, and moisture are the factors that best explain the dry-cured loin. That said, the principal component analysis shows that although the two products (dry-cured loin and dry-cured "cachaço") come from the same muscle (*Longissimus thoracis Lumborum*), their chemical composition is very different.

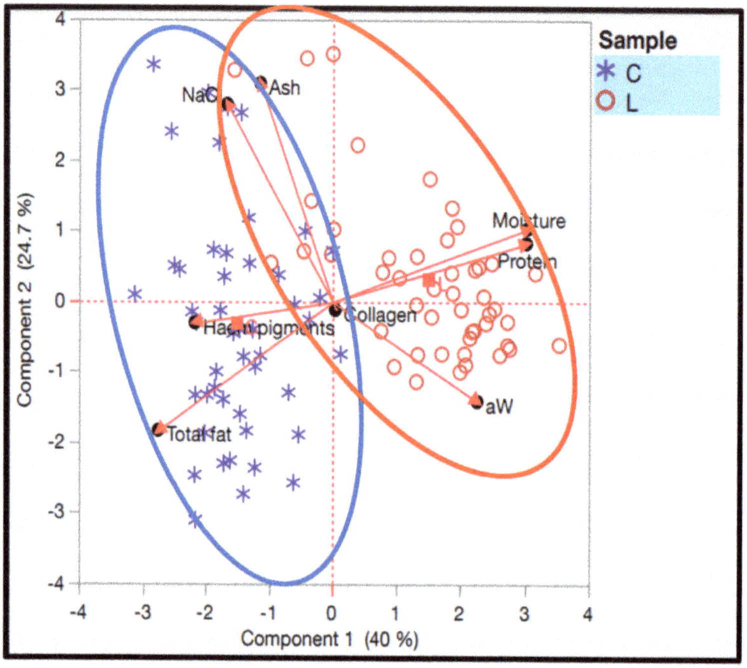

Figure 3. Biplot principal component analysis (C-dry-cured "cachaço"; L-dry-cured loin).

4. Conclusions

The results indicate that incorporating centrifuged and pressed olive cake in different percentages does not impact the physicochemical properties of dry-cured products. However, significant differences were observed between the two products, particularly in total fat, protein, a_w, moisture, and haem pigments. Significant interactions between the product and diet were also observed for a_w, ash, and NaCl parameters. This suggests that, despite both products originating from the same muscle (LTL), they are distinct, which can enhance their appeal to consumers. Adding olive cake significantly improved the n-3 content, with the olive cake with the highest percentage (Cf25) obtaining the highest n-3 value. These values influenced the PUFA n-6/n-3 ratio, showing the same trend, with significantly lower results in olive cake diets. There are no further differences in the different fatty acid fractions and nutritional quality indices. Still, it would be necessary for future work to understand how n-3 can influence the stability of the product and how these differences can affect the final product's organoleptic properties and consumer acceptability.

Author Contributions: Conceptualization, A.T.; methodology, A.L. and L.V.; formal analysis, A.L., L.V., S.L. and D.O.-M., investigation, A.L., L.V. and S.R.; data curation, A.L., L.V., S.L., D.O.-M., V.P., S.R. and A.T.; validation: A.L., L.V. and S.R.; writing—original draft preparation, A.L.; writing—review and editing S.R. and A.T.; supervision, A.T. All authors have read and agreed to the published version of the manuscript.

Funding: The authors are grateful to the Foundation for Science and Technology (FCT, Portugal) for financial support through national funds FCT/MCTES (PIDDAC) to CIMO (UIDB/00690/2020 and UIDP/00690/2020) and SusTEC (LA/P/0007/2020). The authors are also grateful to GAIN (Axencia Galega de Innovación) for supporting this article (grant number IN607A2019/01) and to Laboratory of Carcass and Meat Quality of Agriculture School of Polytechnic Institute of Bragança 'Cantinho do Alfredo'. Grants of A.L. and L.V. are due to NORTE-01-0247-FEDER-072234. The authors (S.R., J.M.L., R.D, and A.T.) are members of the Healthy Meat network, funded by CYTED (ref: 119RT0568). This study is part of a project between a research center (Carcass and Meat Quality and Technology

Laboratory of Agrarian School of Bragança) and a meat manufacturing industry (Bísaro Salsicharia Tradicional®) to develop and add value to animals reared in the extensive system and to create new processed meat products.

Institutional Review Board Statement: All animals were cared for and slaughtered in compliance with the welfare regulations and respecting EU Council Regulation (EC) No. 1099/2009. The study was conducted according to the guidelines of the protocol approved by the ORBEA (Animal Welfare Body) of the University Trás-os-Montes e Alto Douro (2253-e-DZ-2022).

Informed Consent Statement: Not applicable.

Data Availability Statement: The original contributions presented in the study are included in the article, further inquiries can be directed to the corresponding author.

Conflicts of Interest: The authors declare no conflict of interest.

References

1. European Commission. Producing 69% of the World's Production, the EU Is the Largest Producer of Olive Oil. 2020. Available online: https://ec.europa.eu/info/food-farming-fisheries/plants-and-plant-products/plant-products/olive-oil_en (accessed on 18 April 2024).
2. Molina-Alcaide, R.; Yáñez-Ruiz, D.R. Potential use of olive by-products in ruminant feeding: A review. *Anim. Feed. Sci. Technol.* **2008**, *147*, 247–264. [CrossRef]
3. Albuquerque, J.A.; Gonzalez, J.; García, D.; Cegarra, J. Agrochemical characterization of "alperujo", a solid by-product of the two-phase centrifugation method for olive oil extraction. *Bioresour. Technol.* **2004**, *91*, 195–200. [CrossRef]
4. Oktay, Z. Olive Cake as a Biomass Fuel for Energy Production. *Energy Sources Part A Recovery Util. Environ. Eff.* **2006**, *28*, 329–339. [CrossRef]
5. Coelho, L.; Portela, C.; Cravo, A.; Reis, M. Valorização do bagaço de azeitona por compostagem, para utilização agrícola. V Simpósio Nacional de Olivicultura, 14:133-140. Escola Superior Agrária de Santarém, 25 e 26 de Setembro. *Actas Port. Hortic.* **2009**, *14*, 133–140.
6. Teixeira, A.; Rodrigues, S. Consumer perceptions towards healthier meat products. *Curr. Opin. Food Sci.* **2021**, *38*, 147–154. [CrossRef]
7. Lorenzo, J.M.; Montes, R.; Purriños, L.; Cobas, N.; Franco, D. Fatty acid composition of Celta pig breed as influenced by sex and location of fat in the carcass. *J. Sci. Food Agric.* **2012**, *92*, 1311–1317. [CrossRef] [PubMed]
8. Álvarez-Rodríguez, J.; Teixeira, A. Slaughter weight rather than sex affects carcass cuts and tissue composition of Bísaro pigs. *Meat Sci.* **2019**, *154*, 54–60. [CrossRef]
9. Council Regulation (EC)-Official Journal of the European Communities No 1099/2009 of 24 September 2009 on the Protection of Animals at the Time of Killing. 2009, pp. 1–30. Available online: https://eur-lex.europa.eu/legal-content/EN/ALL/?uri=celex:32009R1099 (accessed on 23 April 2024).
10. AOAC International; Cunniff, P. *AOAC Official Methods of Analysis of AOAC International*, 16th ed.; The Association: Washington, DC, USA, 1995; ISBN 9780935584547.
11. *NP 1614/2002*; Determination of Moisture Content. Reference Method (ISO 1442:1197). In Portuguese Norm–Meat and Meat Products. Portuguese Institute of Quality, Ministry of Economy and Innovation: Caparica, Portugal, 2002.
12. *NP-ISO-1615/2002*; Determination of Total Ashes. Reference Method. In Portuguese Norm–Meat and Meat Products. Portuguese Institute of Quality, Ministry of Economy and Innovation: Caparica, Portugal, 2002.
13. *NP 1612/2002*; Determination of Total Nitrogen Content. Reference Method (ISO 937:1978). In Portuguese Norm–Meat and Meat Products. Portuguese Institute of Quality, Ministry of Economy and Innovation: Caparica, Portugal, 2002.
14. *NP 1987/2002*; Carnes e Produtos Cárneos; Determinação do teor em Hidroxiprolina. (Método de Referência). In Portuguese Norm–Meat and Meat Products. Portuguese Institute of Quality, Ministry of Economy and Innovation: Caparica, Portugal, 2002.
15. Hornsey, H.C. The colour of cooked cured pork. I-Estimation of the nitric oxide-haem pigments. *J. Sci. Food Agric.* **1956**, *7*, 534–540. [CrossRef]
16. *NP 1845/1982*; Meat and Meat Products—Determination of Chloride Content—Standard Method. Portuguese Institute of Quality, Ministry of Economy Innovation: Caparica, Portugal, 1982.
17. Folch, J.; Lees, M.; Stanley, G.H.S. A simple method for isolation and purification of total lipids from animal tissues. *J. Biol. Chem.* **1957**, *226*, 497–509. [CrossRef]
18. Dominguez, R.; Borrajo, P.; Lorenzo, J.M. The effect of cooking methods on nutritional value of foal meat. *J. Food Compos. Anal.* **2015**, *43*, 61–67. [CrossRef]
19. Vieira, C.; Sarmiento-García, A.; García, J.J.; Rubio, B.; Martínez, B. Quality and Shelf Life of Fresh Meat from Iberian Pigs as Affected by a New Form of Presentation of Oleic Acid and an Organic-Acid Mix in the Diet. *Foods* **2021**, *10*, 985. [CrossRef]
20. Ulbricht, T.L.V.; Southgate, D.A.T. Coronary heart disease: Seven dietary factors. *Lancet* **1991**, *338*, 985–992. [CrossRef]

21. Leite, A.; Vasconcelos, L.; Ferreira, I.; Domínguez, R.; Pereira, E.; Rodrigues, S.; Lorenzo, J.M.; Teixeira, A. Effect of the inclusion of olive cake in the diet on the physicochemical characteristics of dry-cured loin and dry-cured "cachaço" of Bísaro pig. *Appl. Sci.* **2023**, *13*, 1439. [CrossRef]
22. Pateiro, M.; Franco, D.; Carril, J.A.; Lorenzo, J.M. Changes on physico-chemical properties, lipid oxidation and volatile compounds during the manufacture of Celta dry-cured loin. *J. Food Sci. Technol.* **2015**, *52*, 4808–4818. [CrossRef]
23. Seong, P.N.; Park, K.M.; Kang, G.H.; Cho, S.H.; Park, B.Y.; Ba, H.V. The impact of ripening time on Technological quality traits, chemical change and Sensory characterisictics of dry-cured loin. *Asian Australas. J. Animal Sci.* **2015**, *28*, 677–687. [CrossRef] [PubMed]
24. Libera, J.; Kononiuk, A.; Keska, P.; Wójciak, K. Use of grape seed extract as a natural antioxidant additive in dry-cured pork neck technology. *Biotechnol. Food Sci.* **2018**, *82*, 141–150.
25. Lorenzo, J.M.; Carballo, J. Changes in physico-chemical properties and volatile compounds throughout the manufacturing process of dry-cured foal loin. *Meat Sci.* **2015**, *99*, 44–51. [CrossRef] [PubMed]
26. Asensio, M.; Silva, A.; Armenteros, M.; Caballero, D.; Martín, N.; Lorido, L.; Sánchez-Montero, L.; Fernandez, C.; Noguera, J.L.; Ramos, M. Quality evaluation of dry-cured shoulder from different genetic lines of Iberian pigs. *Arch. Zootec.* **2018**, *67*, 155–160. [CrossRef]
27. Caballero, D.; Asensio, M.; Fernández, C.; Reina, R.; García, M.J.; Noguera, J.L.; Silva, A. Effects of genotypes and crossbreeding on the quality parameters of dry-cured shoulders from different Iberian genetic pig lines. *J. Food Meas. Charact.* **2020**, *14*, 818–829. [CrossRef]
28. Cardoso-Toset, F.; Luque, I.; Morales-Partera, A.; Galán-Relaño, A.; Barrero-Domínguez, B.; Hernández, M.; Gómez-Laguna, J. Survival of Streptococcus suis, Streptococcus dysgalactiae and Trueperella pyogenes in dry-cured Iberian pork shoulders and loins. *Food Microbiol.* **2017**, *61*, 66–71. [CrossRef]
29. Leite, A.; Vasconcelos, L.; Ferreira, I.; Sarmiento-García, A.; Domínguez, R.; Santos, E.M.; Campagnol, P.C.B.; Rodrigues, S.; Lorenzo, J.M.; Teixeira, A. Chemical, physicochemical and sensorial characterization of nitrite-free dry Bísaro shoulders. *Foods* **2022**, *11*, 3079. [CrossRef]
30. Barbora-Cánovas, G.V.; Fontana, A.J.; Schmidt, S.-J.; Labuza, T.P. *Water Activity in Foods: Fundamentals and Aplications*, 2nd ed.; John Wiley & Sons: Hoboken, NJ, USA, 2020; p. 640.
31. Dimakopoulou-Papazoglou, D.; Katsanidis, E. Effect of maltodextrin, sodium chloride, and liquid smoke on the mass transfer kinetics and storage stability of osmotically dehydrated beef meat. *Food Bioprocess Tech.* **2017**, *10*, 2034–2045. [CrossRef]
32. Andres, A.I.; Ventanas, S.; Ventanas, J.; Cava, R.; Ruiz, J. Physicochemcial changes throughout the ripening of dry-cured hams with different salt content and processing conditions. *Eur. Food Res. Technol.* **2005**, *221*, 30–35. [CrossRef]
33. Sánchez-Parra, M.; Ordónez-Díaz, J.L.; Pérez-Aparicio, J.; Moreno-Rojas, J. Physicochemical and microbiological changes associated with processing in dry-cured Tuna. *Appl. Sci.* **2023**, *13*, 5900. [CrossRef]
34. Di Rosa, A.; Zumbo, A.; Furfaro, M.E.; Carcione, G.; D'Angelo, G.; Chiofalfo, V. Effect of two different rearing systems on quality of dry cured "Coppa" of Nero Siliciano pig. In Proceedings of the 8th International Symposium on the Mediterranean Pig, Ljubljana, Slovenia, 10–12 October 2013.
35. Ortiz, A.; Tejerina, D.; Contador, R.; Andrés, A.I.; Petrón, M.J.; Cáceres-Nevado, J.M.; García-Torres, S. Quality traits of dry-cured loins from Iberian pigs reared in *Montanera* system as affected by pre- freezing cure. *Foods* **2021**, *10*, 48. [CrossRef]
36. Vargas-Ramella, M.; Domínguez, R.; Pateiro, M.; Franco, D.; Barba, J.F.; Lorenzo, J.M. Chemical and physico-chemical changes during the dry-cured processing of deer loin. *Int. J. Food Sci. Technol.* **2020**, *55*, 1025–1031. [CrossRef]
37. Bermúdez, R.; Franco, I.; Franco, D.; Carballo, J.; Lorenzo, J.M. Physochemical changes during manufacture and final sensory characteristics of dry-cured Celta ham. Effect of muscle type. *Food Control* **2014**, *43*, 263–269. [CrossRef]
38. Soto, E.; Hoz, L.; Ordóñez, J.A.; Hierro, E.; Herranz, B.; López-Bote, C.; Cambero, M.I. Impact of feeding and rearing systems of Iberian pigs ion volatile profile and sensory characteristics of dry-cured loin. *Meat Sci.* **2008**, *79*, 666–676. [CrossRef]
39. Gençcelep, H.; Ihtiyar, B.; Yüzer, M.O. Determination of quality properties of kastamonu pastirma: A dry-cured meat product. *Harran Tarim. ve Gida Bilimleri Derg.* **2022**, *26*, 491–500. [CrossRef]
40. Zanardi, E.; Novelli, E.; Ghiretti, G.P.; Chizzolini, R. Oxidative stability of lipids and cholesterol in salame Milano, coppa and Parma ham: Dietary supplementation with vitamin E and oleic acid. *Meat Sci.* **2000**, *55*, 169–175. [CrossRef] [PubMed]
41. Cittadini, A.; Domínguez, R.; Gómez, B.; Pateiro, M.; Pérez-Santaescolástica, C.; López-Fernández, O.; Sarriés, M.V.; Lorenzo, J.M. Effect of NaCl replacement by other chloride salts on physicochemical parameters, proteolysis and lipolysis of dry-cured foal "Cecina". *J. Food Sci. Technol.* **2019**, *57*, 1628–1635. [CrossRef]
42. Leite, A.; Domínguez, R.; Vasconcelos, L.; Ferreira, I.; Pereira, E.; Pinheiro, V.; Outor-Monteiro, D.; Rodrigues, S.; Lorenzo, J.M.; Santos, E.M.; et al. Can the Introduction of Different Olive Cakes Affect the Carcass, Meat and Fat Quality of Bísaro Pork? *Foods* **2022**, *11*, 1650. [CrossRef]
43. Bermúdez, R.; Franco, I.; Franco, D.; Carballo, J.; Lorenzo, J.M. Influence of inclusion of chestnut in the finishing diet on fatty acid profile of dry-cured ham from Celta pig breed. *Meat Sci.* **2012**, *92*, 394–399. [CrossRef]
44. Gómez, M.; Cachaldora, A.; Fonseca, S.; Domínguez, R.; Carballo, J.; Franco, I. Biochemical, oxidative, and lipolytic changes during vacuum-packed storage of dry-cured loin: Effect of chestnuts intake by Celta pigs. *J. Food Qual.* **2018**, *2018*, 7690501. [CrossRef]

45. Contador, R.; Ortiz, A.; Ramírez, M.R.; García-Torres, S.; López-Parra, M.M.; Tejerina, D. Physico-chemical and sensory qualities of Iberian sliced dry-cured loins from various commercial categories and the effect of the type of packaging and refrigeration time. *LWT Food Sci. Technol.* **2021**, *141*, 110876. [CrossRef]
46. Pleadin, J.; Lesic, T.; Vujacic, V.; Milicevic, D.; Buneta, A.; Susnic, S.; Lukanic, I.; Kresic, G. Comparison of Chemical composition and fatty acid profile traditional meat products from Croatian and Montenegro. *J. Food Qual.* **2021**, *2021*, 5586436. [CrossRef]
47. Colonna, M.A.; Tarricone, S.; Giannico, F.; Selvaggi, M.; Carriero, F.; Crupi, P.; Corbo, F.; Clodoveo, M.L. Dietary effects of extra virgin olive oil extracted by ultrasound technology or refined olive oil on the quality traits of pork and "Capocollo di Martina Franca" dry-cured met. *Animals* **2021**, *11*, 954. [CrossRef]
48. Baum, S.J.; Kris-Etherton, P.M.; Willett, W.C.; Lichtenstein, A.H.; Rudel, L.L.; Maki, K.C.; Whelan, J.; Ramsden, C.E.; Block, R.C. Fatty acids in cardiovascular health and disease: A comprehensive update. *J. Clin. Lipidol.* **2012**, *6*, 216–234. [CrossRef]
49. Jiménez-Colmenero, F.; Ventanas, J.; Toldrá, F. Nutritional composition of dry-cured ham and its role in a healthy diet. *Meat Sci.* **2010**, *84*, 585–593. [CrossRef] [PubMed]
50. Cavas, R.; Ruiz, J.; Ventanas, J.; Antequera, T. Effect of α-tocopheryl acetate supplementation and the extensive feeding of pigs on the volatile aldehydes during the maturation of Iberian ham. *Food Sci. Technol. Int.* **1999**, *5*, 235–241. [CrossRef]
51. De Jesús, M.C.; Domínguez, R.; Cantalapiedra, J.; Iglesias, A.; Lorenzo, J.M.; Lorenzo, J.M. Effect of the amount of chestnuts in the diet of Celta pigs on the fatty acid profile of dry-cured lacon. *Grasas Y Aceites* **2016**, *67*, 119.
52. Ortiz, A.; González, E.; García-Torres, S.; Gaspar, P.; Tejerina, D. Do animal slaughter age and pre-cure freezing have a significant impact on the quality of Iberian dry-cured pork loin? *Meat Sci.* **2021**, *179*, 108531. [CrossRef]
53. Keys, A.; Mienotti, A.; Karvonen, M.J.; Aravanis, C.; Blackburn, H.; Buzina, R.; Djordjevic, B.S.; Dontas, A.S.; Fidanza, F.; Keys, M.H. The Diet and 15-Year Death Rate in the Seven Countries Study. *Am. J. Epidemiol.* **1986**, *124*, 903–915. [CrossRef]
54. Keys, A. *Seven Countries: A Multivariate Analysis of Death and Coronary Heart Disease*; Harvard University Press: Cambridge, MA, USA, 1980; pp. xi+381. ISBN 0-674-80237-3.
55. Domínguez, R.; Pateiro, M.; Gagaoua, M.; Barba, F.J.; Zhang, W.; Lorenzo, J.M. A comprehensive review on lipid oxidation in meat and meat products. *Antioxidants* **2019**, *8*, 429. [CrossRef] [PubMed]
56. WHO and FAO. Diet, Nutrition, and the Prevention of Chronic Diseases (Report of a joint WHO and FAO Expert Consulation). *WHO Tech. Rep. Ser.* **2003**, *916*, 1–160.
57. Bayés-García, L.; Tres, A.; Vichi, S.; Calvet, T.; Cuevas-Diarte, M.A.; Codony, R.; Boatella, J.; Caixach, J.; Ueno, S.; Guardiola, F. Authentication of Iberian dry-cured ham: New approaches by polymorphic fingerprint and ultrahigh resolution mass spectrometry. *Food Control* **2016**, *60*, 370–377. [CrossRef]
58. UK Department of Health. Report on health and social subjects No. 46. In *Nutritional Aspects of Cardiovascular Disease*; HMSO: London, UK, 1994; p. 51.
59. World Health Organization. Interim summary of conclusions and dietary recommendations on total fat and fatty acids. In *Joint FAO/WHO Expert Consultation on Fats and Fatty Acids in Human Nutrition, 10 to 14 November*; World Health Organization: Geneva, Switzerland, 2008.
60. British Nutrition Foundation. *Unsaturated Fatty Acids: Nutritional and Physiological Significance: The Report from the British Nutrition Foundation's Task Force*; Chapman and Hall, Ltd.: London, UK, 1992.
61. Simopoulos, A. An Increase in the Omega-6/Omega-3 Fatty Acid Ratio Increases the Risk for Obesity. *Nutrients* **2016**, *8*, 128. [CrossRef]
62. Simopoulos, A.P. Evolutionary aspects of the dietary Omega-6:Omega-3 fatty acid ratio: Medical implications. *World Rev. Nutr. Diet.* **2009**, *100*, 1–21.
63. Simopoulos, A.P. ω-3 fatty acids in the prevention management of cardiovascular disease. *Can. J. Physiol. Pharmacol.* **1997**, *75*, 234–239. [PubMed]
64. Maekawa, S.; Takada, S.; Nambu, H.; Furihata, T.; Kakutani, N.; Setoyama, D.; Ueyanagi, Y.; Kang, D.; Sabe, H.; Kinugawa, S. Linoleic acid improves assembly of the CII subunit and CIII2/CIV complex of the mitochondrial oxidative phosphorylation system in heart failure. *Cell Commun. Signal.* **2019**, *17*, 128. [CrossRef] [PubMed]
65. Zong, G.; Liu, G.; Willett, W.C.; Wanders, A.J.; Alssema, M.; Zock, P.L.; Hu, F.B.; Sun, Q. Associations between Linoleic Acid Intake and Incident Type 2 Diabetes among U.S. Men and Women. *Diabetes Care* **2019**, *42*, 1406–1413. [CrossRef] [PubMed]
66. Lands, W.E.M. Fish, Omega-3, and Human Health, 2nd ed.AOCS Press: Urbana, IL, USA, 2005.
67. Singh, M. Essential fatty acids, DHA and human brain. *Indian J. Pediatr.* **2005**, *72*, 239–242. [CrossRef]
68. Wertz, P.W. Essential fatty acids and dietary stress. *Toxicol. Ind. Health* **2009**, *25*, 279–283. [CrossRef] [PubMed]
69. Razmaite, V.; Švirmickas, G.J.; Šiukšcius, A.U. Effect of Weight, Sex and Hunting Period on Fatty Acid Composition of Intramuscular and Subcutaneous Fat from Wild Boar. *Ital. J. Anim. Sci.* **2012**, *11*, 174–179. [CrossRef]
70. Chen, J.; Liu, H. Nutritional Indices for Assessing Fatty acids: A Mini-Review. *Int. J. Mol. Sci.* **2020**, *21*, 5695. [CrossRef]
71. Liotta, L.; Chiofalo, V.; Presti, L.V.; Chiofalo, B. In vivo performances, carcass traits, and meat quality of pigs fed olive cake processing waste. *Animals* **2019**, *9*, 1155. [CrossRef]
72. Cava, R.; Estévez, M.; Ruiz, J.; Morcuende, D. Physicochemical characteristics of three muscles from free-range reared Iberian pigs slaughtered at 90 kg live weight. *Meat Sci.* **2003**, *63*, 533–541. [CrossRef]

73. Covaciu, F.D.; Feher, I.; Cristea, G.; Dehelean, A. Nutritional quality and safety assessment of pork meat cut from Romania: Fatty acids and Elemental profile. *Foods* **2024**, *13*, 804. [CrossRef]
74. Greenacre, M.; Groenen, P.J.F.; Hastie, T.; d'Enza, A.I.; Markos, A.; Tuzhilina, E. Principal component analysis. *Nat. Rev. Methods Primers* **2022**, *2*, 100. [CrossRef]

foods

MDPI

Article

Dietary Effect of Curcumin on Amino Acid, Fatty Acid, and Volatile Compound Profiles of Chicken Meat

Ying Shu [1,2], Fengyang Wu [1], Wei Yang [3], Wenhui Qi [1], Runyang Li [1] and Zhisheng Zhang [1,*]

[1] College of Food Science and Technology, Hebei Agricultural University, Lekai South Avenue, Baoding 071000, China; hebaushuying@hebau.edu.cn (Y.S.); fengyangwu2020@163.com (F.W.); wenhui406@yeah.net (W.Q.); runyangli@163.com (R.L.)
[2] Hebei Layer Industry Technology Research Institute, Economic Development Zone, Handan 545000, China
[3] Institute of Animal Husbandry and Veterinary Medicine of Hebei Province, Dongguan Avenue, Baoding 071030, China; realyangvv@163.com
* Correspondence: zhangzhisheng@hebau.edu.cn

Abstract: This study investigated the dietary effect of curcumin (CUR) on amino acid, 5′-nucleotides, fatty acid, and volatile compound profiles of chicken meat. A total of 400 healthy 1-day-old broiler male chicks were divided into 4 groups ($n = 10$) and fed either a basal diet or a diet with the addition of CUR with concentrations of 100 mg/kg, 150 mg/kg, and 200 mg/kg for 43 days. The results show that the addition of CUR in chicken diets is conducive to promoting the deposition of amino acids and increasing the content of 5′-nucleotides in chicken meat, reducing the contents of saturated fatty acid (SFA) and C20:4 n6 but increasing the ratio between polyunsaturated fatty acid (PUFA) and SFA. In addition, the volatile compound profile shows that the main volatile compounds in chicken meat are aldehydes (including hexanal, heptanal, octanal, and nonanal), with significant increases in their contents observed among chickens in the CUR-intake group. Moreover, it has been found that (E, E)-2,4-nonadienal, trans-2-decenal, benzaldehyde, and trans-2-octenal in chicken meat can significantly increase its overall aroma, and the addition of CUR with 150 mg/kg had the best effect on improving nutritional quality and flavor of chicken meat. This study provides a basis for the comprehensive utilization of CUR as a feed additive with the potential to substitute antibiotics.

Keywords: curcumin; meat; amino acid; fatty acid; volatile compounds

1. Introduction

Poultry meat is viewed as one of the largest sources of animal protein in people's diets around the world [1]. With the advantages of high-quality nutritional components, low cholesterol content, diverse amino acid compositions, and relatively high digestibility, it has been greatly favored by consumers. Antibiotics have been widely used in the poultry industry for disease prevention and treatment as well as production promotion [2,3]. Nevertheless, the overuse of antibiotics has led to various adverse consequences, such as drug residues, bacterial resistance, and environmental pollution. In 2006, 2017, and 2020, Europe, the United States, and China banned the addition of antibiotics in poultry and livestock feed in succession [4,5]. Now, natural plant extracts rich in phenolic compounds have been widely studied because of their good effects in delaying the oxidation of fats and proteins, enhancing antioxidation, improving growth performance, and improving quality [6,7].

Curcumin (CUR) is a hydrophobic phenolic compound extracted from the rhizomes of *Curcuma longa* L. [8]. This compound exhibits a wide range of pharmacological effects, including anti-inflammation [9,10], anti-oxidation [11,12], anti-tumor [13,14], anti-bacteria [15,16], anti-virus [17], digestion promotion [18], and immune regulation [19,20]. In recent years, some studies have been performed on the effects of CUR in the poultry farming industry. These studies showed that the use of CUR as a feed additive to chicken diets can affect the growth performance of poultry and various quality indices of poultry meat, improving the feed conversion efficiency, effectively reducing the deposition of abdominal and liver fats, and consequently improving meat quality [21–24]. The addition of CUR in the diets of Cherry Valley ducks improved their muscle quality and enhance their antioxidant activities [25]. In addition, the addition of CUR and acetylsalicylic acids in diets of heat-stressed broiler chickens significantly reduced the content of MDA in their breast muscle tissue, improving their antioxidant capacities [26]. The results of the previous experiment of our team showed that adding CUR to the diet of broilers could reduce the content of MDA and improve the processing quality and antioxidant activity of chicken meat [27], so we hypothesized that the use of CUR as a feed additive could affect nutritional components and flavor by influencing the antioxidant capacity of chicken meat. Therefore, a comprehensive evaluation was conducted in this study on the dietary effects of CUR on amino acid, 5′-nucleotides, fatty acid, and volatile compound profiles of chicken meat.

2. Materials and Methods

2.1. Experimental Design, Animal, and Diet

This research was approved by the Ethics Committee of Hebei Agricultural University (Number: 2022003). All animal experiments that comply with the REACH guidelines were conducted in accordance with the British Animals (Scientific Procedures) Act 1986 and its relevant guidelines, as well as the EU Directive 2010/63/EU on the protection of animals used for scientific purposes. The CUR (with a purity of ≥95%) used in this research was purchased from Chenguang Biotech Group Co., Ltd., Handan, China. The basal ration of a corn–soybean meal-type ration was prepared according to the Feeding Standard of Chickens (NY/T 33-2004) in China [28]. Its composition and nutritional levels are listed in Table 1. A total of 400 healthy 1-day-old Arbor Acres broiler male chickens with similar body weights (44.18 ± 1.05 g) were selected and randomly divided into four groups, with 10 replicates in each group and 10 chicks in each replicate. Chicks were reared on soft bedding cages and housed in a well-ventilated room. Artificial light was provided for 12 h until the end of experiment. Heaters were installed in the experimental chicken room to regulate the environmental temperature according to body requirement of the chicks. An experimental period of 43 days was applied in this research. Chickens in the control group (C0) were fed with a basal diet, while chickens in the CUR-addition group were individually fed with basal diets with the addition of curcumin with concentrations of 100 mg/kg (C1), 150 mg/kg (C2), and 200 mg/kg (C3). During the experiment, test subjects could freely eat, feed, and drink water.

Table 1. Ingredient composition and nutrient level of basal diets (dry-matter basis, %).

Daily Ration Composition %	1–21 Days of Age	22–43 Days of Age	Nutritional Level % [2]	1–21 Days of Age	22–43 Days of Age
Corn	48.20	51.45	Metabolic energy/(MJ/kg)	12.76	13.39
Soymeal	41.80	36.50	Crude protein	24.14	22.91
Vegetable oil	5.00	7.05	Crude fiber	3.48	3.15
Premix compound [1]	5.00	5.00	Calcium	0.83	0.81
Total	100.00	100.00	Total phosphorous	0.60	0.58
			Lysine	1.26	1.11
			Methionine	0.53	0.51

[1] Premix compound provides the following substances for per kilogram of daily ration: VA, 10,000 IU; VD_3, 4000 IU; VE, 20 IU; VK_3, 2 mg; VB_1, 2 mg; VB_2, 6 mg; VB_6, 3 mg; VB_{12}, 0.02 mg; nicotinamide, 40 mg; calcium pantothenate, 10 mg; folic acid, 1 mg; biotin, 0.12 mg; Cu, 16 mg; Fe, 80 mg; Zn, 110 mg; Mn, 120 mg; I, 1.5 mg; Se, 0.3 mg. [2] Metabolic energy, crude protein, calcium, and total phosphorous values presented in the nutritional level item are all measured values, while all the other values are calculated values.

2.2. Sample Collection

At the end of the experiment, one broiler chicken with a body weight close to the average level was selected in each replicate (a total of eight chickens in each group). After a 12-h period of no feeding, all broiler chickens that were selected underwent euthanasia through cervical dislocation. Left breast meat samples were collected from among these chickens. With the removal of extra fats, these meat samples were divided into small pieces and frozen in liquid nitrogen and vacuum-packed directly, then stored at −80 °C for further measurement.

2.3. Amino Acid Measurement

The content of amino acids in the sample meat was measured using high-performance liquid chromatography (HPLC, Waters e2695, Waters Corporation, Milford, MA, USA). A sample with a weight of 1 g was precisely weighed and then adjusted to volume after a 24-h hydrolysis period under 110 °C in 6 mol/L hydrochloric acid, followed by filtration and evaporation. The content of amino acids was measured through the derivation of 2,4-dinitrochlorobenzene. The corresponding mobile phases A and B were a 0.25% anhydrous sodium acetate solution and chromatography-grade acetonitrile, respectively.

The chromatographic conditions applied in the experiment are as follows: column temperature = 43 °C; detection wavelength = 360 nm; and mobile-phase flow rate = 1 mL/min. For the gradient elution program, during the period of 0–10 min, the volume ratio between mobile phases A and B was set at 82:18. During the period of 10–15 min, the volume ratio between mobile phases A and B was set at 71:29. During the period of 15–25 min, the volume ratio between mobile phases A and B was set at 66:34. During the period of 25–30 min, the volume ratio between mobile phases A and B was set at 45:55. During the period of 30–37 min, the volume ratio between mobile phases A and B was set at 40:60. During the period of 37–45 min, the volume ratio between mobile phases A and B was set at 82:18.

2.4. 5′-Nucleotide Measurement

A meat sample with a weight of 2.5 g was weighed and added into a centrifuge tube. The sample was homogenized using 20 mL of 5% $HClO_4$ and was then centrifuged at 12,000 rpm for 20 min. After collection of the supernatant, the precipitate was homogenized using a 5% $HClO_4$ solution and then centrifuged. After a combination of the secondary supernatant, the pH value of the solution was adjusted to 6.5, with its volume adjusted to 50 mL. Subsequently, a 0.22 μm filtration was conducted on the solution for further analysis. The filtrate was analyzed using HPLC (Waters e2695, Waters Corporation, USA) combined with a UV detector (254 nm) [29,30].

The analysis was conducted with the following conditions: chromatographic column = XBridge-C18-T (5 μm, 4.6 mm × 250 mm); column temperature = 25 °C; mobile phase A = phosphate saline buffer (1.40 mmol/L $C_{16}H_{37}NO_4S$; 0.01 mol/L K_2HPO_4; pH 3.2); mobile phase B = methanol; flow rate = 1 mL/min; injection volume = 10 μL; isocratic elution; A:B = 98:2, 25 min.

2.5. Fatty Acid Measurement

An NaOH-methanol solution (2%) with a volume of 8 mL was added to the extracted sample, which then underwent a water bath at 80 °C for 20 min. Subsequently, a 15% boron trifluoride-methanol solution with a volume of 7 mL was added to the sample, which then underwent a water bath at 80 °C for 2 min. After the sample cooled down, an n-heptane solution with a volume of 10 mL was added to the sample and mixed well. Anhydrous sodium sulfate with a weight of 2 g was added to the sample for water absorption. Subsequently, gas chromatography was performed. Sample separation and quantification were conducted using a gas chromatograph-mass spectrometer (GCMS-QP2020 NX, Shimadzu Corporation, Kyoto, Japan) with the following conditions. Chromatographic column = SH-RXI-5SIL MS (30 m × 0.25 μm, 0.25 μm, Shimadzu Corporation, Japan);

carrier gas = helium; flow rate = 1.0 mL/min; injection volume = 1.0 μL; injection port temperature = 270 °C; and injection method = splitless. The initial temperature of the column heater was set at 100 °C and maintained for 13 min. The temperature was then increased to 180 °C at a rate of 10 °C/min and was held there for 20 min. Subsequently, it was increased to 200 °C at a rate of 1 °C/min and was held there for 2 min. Lastly, the temperature was increased to 230 °C at a rate of 4 °C/min, and was held there for 12.5 min. Under 70 eV, quantitative and qualitative analyses of fatty acids were conducted using an external standard method with ion source and interface temperatures set at 250 °C and 280 °C, respectively [31].

2.6. Analysis of Volatile Flavor Components Using SPME-GC-MS

According to the literature [32], a sample with a weight of 2.00 g was precisely weighed and added into a 20 mL headspace vial. After that, 1.0 μL of 2-methyl-3-heptanone solution (0.408 μg/μL) was added into the vial, which was then sealed. The vial underwent a 55 °C water-bath equilibrium for 20 min. Subsequently, volatile components were adsorbed using a 75 μm CAR/PDMS fiber (Supelco, Inc., Bellefonte, PA, USA) at 55 °C for 40 min. After the completion of the adsorption, a 5-min desorption at 250 °C was immediately performed on the fiber. GC-MS analysis was conducted using a GCMS-QP2020 NX gas chromatograph-mass spectrometer (Shimadzu Corporation, Japan) with the following conditions: chromatographic column = SH-RXI-5SIL MS (30 m × 0.25 μm, 0.25 μm, Shimadzu Corporation, Japan); carrier gas = helium (≥99.999%); flow rate = 1.0 mL/min (constant); and injection port temperature = 250 °C (splitless). The initial temperature was set at 40 °C and maintained for 5 min. The temperature was then increased to 100 °C at a rate of 15 °C/min. After that, it was further increased to 220 °C at a rate of 5 °C/min and was held there for 2 min. Lastly, the temperature was increased to 260 °C and was held there for 5 min. Under 70 eV, electron impact mass spectrometry was performed with an ion-source temperature of 230 °C and an m/z scanning range of 40–450 μm. The obtained mass spectra were compared with the US National Institute of Standards and Technology (NIST) Mass Spectral Library (version 17.0) database to identify the volatile components. The gas chromatographic peak area of the volatile component was compared with the internal standard area to determine the concentration of the volatile component. Odor activity values (OAVs) were calculated using the following formula to determine the odor-active compounds:

$$OAV_i = \frac{C_i}{OT_i} \quad (1)$$

where C_i represents the concentration of the volatile compound, and OT_i represents the odor threshold in water.

2.7. Statistical Analysis

The experimental results are expressed as the mean ± standard error (SE). The difference significance test was conducted using an ANOVA in SPSS23.0. Duncan's method was used in multiple comparisons, with a p-value lower than 0.05 ($p < 0.05$) indicating a significant difference.

3. Results and Discussion

3.1. Analysis of Amino Acid Content

Amino acids are an important precursor substance of meat flavor compounds and a major indicator for assessing the quality of meat flavor [33]. They play a significant role in the formation of flavor compounds in chicken meat [34]. In this study, it was found that the addition of CUR with different concentrations in the diets of chickens presented different influences on the contents of amino acids in chicken meat (see Table 2). Compared to the chicken meat in the C0 group, the meat in the C2 group exhibited the highest concentrations of essential amino acids (EAA) and non-essential amino acids (NEAA) ($p < 0.05$). Except for glycine (Gly), chicken meat in the C2 group presented the highest contents of amino acids ($p < 0.05$). In addition, chicken meat in the C0 group exhibited the highest content of Gly ($p < 0.05$), while chicken meat in the C2 group presented the highest content of serine (Ser) ($p < 0.05$).

Table 2. Amino acid content of chicken breast meat samples ($n = 8$).

Amino Acid		Content (g/100 g)					
		C0 (0 mg/kg)	C1 (100 mg/kg)	C2 (150 mg/kg)	C3 (200 mg/kg)	SEM	*p*-Value
EAA	Thr	41.89 [b]	26.93 [d]	47.67 [a]	39.15 [c]	3.54	0.178
	Lys	29.98 [b]	14.41 [c]	34.77 [a]	28.56 [b]	2.89	0.012
	Val	35.02 [b]	22.51 [c]	41.12 [a]	33.51 [b]	3.15	0.187
	Met	23.47 [b]	13.16 [c]	26.63 [a]	21.69 [b]	2.05	0.05
	Ile	24.23 [b]	6.04 [d]	32.12 [a]	17.91 [c]	3.62	0.019
	Leu	3.10 [b]	0.97 [d]	3.98 [a]	2.62 [c]	0.42	0.018
	Phe	19.07 [b]	8.50 [c]	21.65 [a]	17.87 [b]	1.87	0.006
	Subtotal	176.76 [b]	92.53 [d]	207.93 [a]	161.33 [c]	16.79	0.034
NEAA	Asp	34.11 [b]	23.91 [c]	38.47 [a]	34.01 [b]	2.62	0.230
	Glu	45.01 [b]	32.24 [c]	49.90 [a]	45.08 [b]	3.19	0.218
	Ser	55.26 [b]	29.60 [d]	60.88 [a]	33.91 [c]	5.24	0.044
	Gly	22.11 [a]	15.06 [c]	17.07 [b]	21.56 [a]	2.27	0.681
	Ala	18.95 [b]	11.46 [c]	21.20 [a]	18.11 [b]	1.61	0.114
	Pro	22.45 [b]	15.37 [d]	26.50 [a]	20.67 [c]	2.01	0.283
	Cys	9.45 [b]	5.00 [c]	11.47 [a]	9.64 [b]	0.93	0.025
	His	14.89 [b]	10.92 [c]	16.40 [a]	15.24 [b]	1.19	0.417
	Arg	21.54 [b]	13.13 [d]	36.21 [a]	19.36 [c]	3.48	0.119
	Tyr	9.59 [b]	3.07 [c]	10.87 [a]	9.01 [b]	1.10	0.002
	Subtotal	253.37 [b]	159.78 [d]	288.99 [a]	226.59 [c]	20.68	0.112
TAA		430.13 [b]	252.32 [d]	496.93 [a]	387.92 [c]	37.27	0.068
SAA		200.09 [b]	117.84 [d]	219.57 [a]	171.61 [c]	16.80	0.104
BAA		141.32 [b]	75.24 [d]	178.11 [a]	128.21 [c]	14.45	0.035
FAA		79.13 [b]	56.15 [d]	88.38 [a]	79.09 [c]	5.81	0.223
EAA/TAA		41.09%	36.67%	41.84%	41.59%	0.00	0.034
EAA/NEAA		69.76%	57.91%	71.95%	71.20%	0.02	0.013

$n = 8$ replicates per treatment; different superscripts (a–d) means are significantly different in the same row. TAA = total amino acids; SAA = sweet amino acid (sum of Thr, Lys, Ser, Gly, Ala, and Pro); BAA = bitter amino acid (sum of Val, Met, Ile, Leu, Phe, Tyr, His, and Arg); FAA = flavor amino acid (sum of Asp and Glu).

EAA is an essential substance for the human body. However, it cannot be produced by the human body and must be obtained through food. Food with a higher EAA content exhibits a higher nutritional value [35]. At the same time, the contents of certain types of amino acids are closely associated with the normal metabolism of the human body. For instance, threonine (Thr) can promote muscle growth and facilitate fat metabolism. Leucine (Leu) can induce enhanced activities of antioxidases, improving the antioxidant capacity of the human body. Serine, as a metabolic intermediate, participates in the synthesis of methionine (Met), glycine (Gly), and cysteine (Cys), primarily playing a role in lipid metabolism, the immune system, and the central nervous system. Although Ser is not an essential amino acid, it plays a crucial role in a series of metabolic processes and the maintenance of bodily functions. The reason why chicken meat in the C2 group presents an increased Ser content could lie in that chickens fed with rations with the addition of CUR exhibit increased intake and turnover amounts of proteins. In addition, Gly can be directly converted into Ser. Therefore, the reason why there is a decreased Gly content could lie in the fact that Gly is converted into Ser through the catalysis of serine hydroxymethyltransferase.

The content of amino acids in muscle has a significant influence on the nutritional value of chicken meat. Moreover, some amino acids are key flavor compounds in chicken meat. Amino acids can react with reducing sugars, triggering the Maillard reaction, which constitutes one of the primary ways to form meat flavor [34]. In addition, amino acids can react with α-dicarbonyl compounds, triggering the Strecker degradation reaction, the product of which, Strecker aldehyde, is a primary source of meat flavor. Glutamate (Glu) and aspartic acid (Asp) are important flavor amino acids. During the hydrolysis process of the protein–peptide chain, these two amino acids can bind with sodium ions, leading to the formation of an umami taste [36,37]. Compared with the chicken meat in the C0 group, the chicken meat in the C2 group presented the highest contents of BAA, SAA, and FAA ($p < 0.05$). Sweet-tasting amino acids in chicken meat primarily include threonine (Thr), lysine (Lys), and serine (Ser). On the other hand, bitter-tasting amino acids in chicken meat are mostly important flavor enhancers involved in the Maillard reaction during the formation process of flavor. This study shows that bitter-tasting amino acids in chicken meat primarily include valine (Val), isoleucine (Ile), and methionine (Met). The results indicate that chickens fed with rations with the addition of CUR present a better overall flavor in their meat. In addition, it has been found in this research that chickens fed with rations with the addition of CUR exhibit EAA/TAA and EAA/NEAA ratios in their meat, which are close to those ideal ratios recommended by FAO/WHO. Changes in EAA, FAA, and TAA contents in chicken meat after the addition of CUR in chicken diets were investigated in this study. It indicated that chicken meat in the C2 group presented the most beneficial addition amount of CUR to the deposition of AA, and this amount can facilitate an improvement in the nutritional value and the formation of flavor in chicken meat [38].

3.2. Nucleotide Analysis

Flavor 5′-nucleotides include GMP, AMP, and IMP. These 5′-nucleotides and flavor amino acids can present a coordination effect of umami enhancement [39,40]. The effect data of different addition amounts of CUR on the nucleotide content in chicken meat are listed in Table 3. From this table, it can be seen that chickens fed with rations with the addition of CUR present significantly higher contents of 5′-nucleotides in their meat than those of chicken meat in the C0 group ($p < 0.05$). Specifically, chicken meat in the C2 group exhibited the highest contents of GMP, which were significantly higher than those of chicken meat in all other groups ($p < 0.05$). In addition, chicken meat in the C1 group presented the highest contents of AMP and IMP, which were significantly higher than those of chicken meat in all other groups ($p < 0.05$). The results show that the addition of CUR in chicken diets can significantly increase the contents of 5′-nucleotides in chicken meat, exhibiting an improvement effect on the contents of flavor 5′-nucleotides in chicken meat. This indicates that the addition of CUR in chicken diets can improve the contents of 5′-nucleotides in chicken meat, improving its flavor. This may be related to the increased level of IMP or GMP, which results from the improved antioxidant capacity of chicken meat [41].

Table 3. Nucleotide contents of chicken breast meat samples ($n = 8$).

Nucleotides	Content (g/100 g)				SEM	p-Value
	C0 (0 mg/kg)	C1 (100 mg/kg)	C2 (150 mg/kg)	C3 (200 mg/kg)		
GMP	0.38 ^{d}	0.58 ^{b}	0.86 ^{a}	0.41 ^{c}	0.07	<0.001
AMP	0.31 ^{d}	0.77 ^{a}	0.58 ^{c}	0.67 ^{b}	0.06	<0.001
IMP	0.05 ^{d}	0.08 ^{a}	0.07 ^{b}	0.06 ^{c}	0.00	<0.001

$n = 8$ replicates per treatment; different superscripts (a–d) means are significantly different on the same row.

3.3. Fatty Acid Analysis

From the perspective of human health, chicken meat can be regarded as a high-quality resource of essential fatty acids and proteins for the human body [42]. The tenderness, flavor, and nutritional value of chicken meat are all affected by the composition of fatty acids, which is greatly affected by the dietary control of chickens. The effect data of the addition of CUR in chicken diets on the contents of fatty acids in chicken meat are listed in Table 4. From the table, it can be seen that primary fatty acids detected in chicken meat included C16:0, C18:1 n-9 c, C18:2n-6, and C20:4n6.

Table 4. Fatty acid composition and content of chicken breast meat samples ($n = 8$).

Fatty Acid	Content (g/100 g)				SEM	*p*-Value
	C0 (0 mg/kg)	C1 (100 mg/kg)	C2 (150 mg/kg)	C3 (200 mg/kg)		
C10:0	0.01	0.01	0.01	0.01	0.00	<0.001
C12:0	0.02 ^{c}	0.03 ^{b}	0.02 ^{c}	0.03 ^{a}	0.00	<0.001
C13:0	0.01	0.01	0.01	0.01	0.00	<0.001
C14:0	0.32 ^{c}	0.53 ^{a}	0.29 ^{d}	0.39 ^{b}	0.03	<0.001
C15:0	0.10 ^{b}	0.16 ^{a}	0.10 ^{b}	0.15 ^{a}	0.10	<0.001
C16:0	6.55 ^{b}	9.46 ^{a}	6.09 ^{c}	4.41 ^{d}	0.89	0.261
C17:0	0.28 ^{b}	0.35 ^{a}	0.26 ^{c}	0.34 ^{a}	0.01	<0.001
C18:0	5.79 ^{c}	7.57 ^{a}	6.19 ^{bc}	6.76 ^{b}	0.26	0.005
C20:0	0.18 ^{b}	0.20 ^{a}	0.13 ^{c}	0.20 ^{a}	0.02	0.197
C21:0	0.03 ^{b}	0.03 ^{b}	0.07 ^{a}	0.05 ^{ab}	0.01	0.050
C22:0	0.09 ^{b}	0.08 ^{b}	0.12 ^{a}	0.11 ^{ab}	0.01	0.001
C23:0	0.10 ^{a}	0.06 ^{b}	0.09 ^{a}	0.09 ^{a}	0.01	0.171
C24:0	0.13 ^{a}	0.05 ^{a}	0.11 ^{a}	0.10 ^{a}	0.02	0.245
SFA	13.62 ^{b}	18.52 ^{a}	13.47 ^{b}	12.64 ^{c}	1.04	0.142
C14:1	0.10 ^{c}	0.23 ^{a}	0.09 ^{c}	0.10 ^{b}	0.02	<0.001
C16:1	3.28 ^{b}	5.84 ^{a}	2.98 ^{c}	3.28 ^{b}	0.44	<0.001
C18:1n9c	25.33 ^{a}	15.13 ^{b}	12.68 ^{c}	24.12 ^{a}	2.74	0.286
C20:1	0.08 ^{b}	0.14 ^{a}	0.08 ^{b}	0.10 ^{b}	0.01	0.013
C24:1	0.23 ^{bc}	0.18 ^{c}	0.50 ^{a}	0.29 ^{b}	0.05	0.001
MUFA	29.02 ^{a}	21.51 ^{b}	16.33 ^{c}	27.90 ^{a}	2.63	0.329
C18:2n6c	6.17 ^{a}	0.44 ^{d}	5.56 ^{b}	3.86 ^{c}	1.04	0.193
C18:3n6 (GLA)	0.38 ^{c}	0.74 ^{a}	0.34 ^{c}	0.63 ^{b}	0.06	<0.001
C18:3n3 (ALA)	3.86 ^{b}	3.62 ^{b}	3.07 ^{c}	4.73 ^{a}	0.23	0.001
C20:4n6 (AA)	5.98 ^{c}	5.99 ^{c}	12.16 ^{a}	9.07 ^{b}	1.02	0.018
C20:5n3 (EPA)	0.26 ^{c}	0.44 ^{b}	0.69 ^{a}	0.43 ^{b}	0.06	<0.001
C22:6n3 (DHA)	1.55 ^{b}	1.43 ^{b}	4.70 ^{a}	2.03 ^{b}	0.53	0.017
PUFA	18.21 ^{c}	12.67 ^{d}	26.52 ^{a}	20.75 ^{b}	2.06	0.048
PUFA/SFA	1.33 ^{ab}	0.68 ^{b}	1.97 ^{a}	1.73 ^{a}	0.21	0.069
n-3 PUFA	5.68 ^{b}	5.50 ^{b}	8.46 ^{a}	7.19 ^{ab}	0.49	0.031
n-6 PUFA	12.53 ^{a}	7.17 ^{d}	18.05 ^{a}	13.56 ^{c}	1.64	0.074

$n = 8$ replicates per treatment; different superscripts (a–d) means are significantly different on the same row. SFA—saturated fatty acid; PUFA—polyunsaturated fatty acid; PUFA/SFA—the ratio between contents of PUFAs and SFAs.

The addition of CUR in chicken diets exerts a certain impact on the total SFA contents in chicken meat. SFAs in chicken meat are mostly palmitic acids (C16:0) and stearic acids (C18:0), indicating that these two fatty acids play important roles in the formation of chicken meat flavor. Compared with the C0 group, different experimental groups presented significant differences in their contents of palmitic acids (C16:0) ($p < 0.05$). Chicken meat in the C3 group presented the lowest contents of palmitic acids, which were significantly lower than those of chicken meat in the C0 group ($p < 0.05$). After the addition of CUR in chicken diets, chicken meat in the C2 group presented reduced contents of SFA fatty acids, with more significantly reduced contents of SFA fatty acids among chicken meat in the C3 group. This indicates that an appropriate CUR addition amount can decrease the SFA content in chicken meat. It is worth noting that among different dosage groups, chicken meat in the C1 group presented significantly increased stearic acid (C18:0) and total SFA contents with decreased concentrations of oleic acids (C18:1n-9c) and monounsaturated fatty acid (MUFA). Analysis shows that the possible reason is that the addition of CUR in chicken diets leads to a reduced conversion efficiency of stearic acid (C18:0) to oleic acid (C18:1n-9c). However, its mechanism is not fully understood [43].

MUFAs in chicken meat primarily include oleic acids (C18:1n9c) and palmitoleic acids (C16:1), while PUFAs in chicken meat primarily include linoleic acids (C18:2n6c) and arachidonic acids (C20:4n6). Among all MUFAs, chicken meat in the C0 group presented the highest contents of oleic acids, which were significantly higher than those of chicken meat in the C1 and C2 groups ($p < 0.05$). Chicken meat in the C1 group exhibited the highest contents of C14:1, C16:1, and C20:1, which were significantly higher than those of chicken meat in all other groups ($p < 0.05$). The contents of C20:4n6 in chicken meat in the C1 group were not significantly different from those of chicken meat in the C0 group but significantly lower than those of chicken meat in the C2 and C3 groups ($p < 0.05$). Meanwhile, chicken meat in the C2 group presented the highest contents of C20:4n6 (AA), which were significantly higher than those of chicken meat in all other treatment groups ($p < 0.05$). Chicken meat in the C2 group exhibited the lowest contents of C14:1, C16:1, and C18:1n9c. Among them, the contents of C14:1 in chicken meat in the C2 group were significantly lower than those of chicken meat in the C1 and C3 groups ($p < 0.05$), but not significantly different from those of chicken meat in the C0 group. The contents of C16:1 and C18:1n9c in chicken meat in the C2 group were significantly lower than those of chicken meat in the C0, C1, and C3 groups. PUFA is an essential substance for the human body, and a certain intake of PUFA has many benefits for human health [42]. A relatively high intake of n-3PUFA through diet is conducive to controlling cardiovascular diseases among humans. Therefore, an increase in the content of n-3PUFA can more specifically reflect the improvement in the food health indicators [43,44]. n-6 PUFA has the effects of maintaining the blood–lipid balance and preventing cardiovascular diseases. The study results showed that chicken meat in the C2 group exhibited the highest contents of C20:5n3 (EPA), C20:4n6 (AA), C22:6n-3 (DHA), and total PUFA ($p < 0.05$), indicating the potential of CUR for enriching PUFA in chicken meat [39]. Chicken meat with a higher PUFA/SFA ratio presents better quality. The results indicate that an appropriate addition amount of CUR in rations of broiler chickens can improve the nutritional value of chicken meat by impacting the fatty-acid composition of chicken breast meat. The reason could lie in the fact that antioxidant components in CUR can prevent the oxidation of PUFA, thus reducing the oxidation of fatty acids in chicken breast meat and consequently improving the quality of chicken meat. Similar effects have also been reported for other additives with antioxidant activity [43,45].

3.4. *Analysis of Volatile Compounds*

It can be seen that a total of 78 volatile compounds were detected among all chicken breast meat samples (Table 5). These compounds can be divided into 7 chemical classes, including 21 aldehydes, 19 hydrocarbons, 12 ketones, 7 alcohols, 4 esters, 10 aromatic compounds, 3 acids, and 2 heterocyclic compounds. Meanwhile, changes in varieties and contents of flavor compounds in chicken meat can be observed after the addition of CUR in chicken diets.

Meat flavor is a key factor for consumers when judging the acceptability of meat. During the cooking process, a series of complex reactions, including the Maillard reaction, lipid oxidation degradation, and thiamine degradation, will occur in flavor precursor substances in chicken meat. Volatile odors and flavor compounds produced during these reactions collectively constitute the overall flavor of chicken meat. In addition, lipid oxidation and the Maillard reaction are two important reactions that produce volatile compounds in chicken meat [46]. Aldehydes are key flavor compounds produced through the Strecker reaction of amino acids and the oxidation degradation of fatty acids [47,48]. These compounds are one of the main sources of meat flavor, and food containing aldehydes presents the scents of lipid, grass [49], and citrus. Compared to chicken meat in the C0 group, after the addition of CUR in chicken diets, chicken meat in different experimental groups all presented increased contents of aldehydes. Six aldehyde compounds, including hexaldehyde, heptaldehyde, octyl aldehyde, and n-nonyl aldehyde, exhibited a dosage effect ($p < 0.05$). Alcohols in the chicken meat primarily come from the oxidation degradation of unsaturated fatty acids. Alcohols can be divided into straight-chain alcohols and branched-chain alcohols. Straight-chain alcohols primarily come from lipid oxidation [46]. Branched-chain alcohols are mostly produced through the reduction of branched-chain aldehydes and the Strecker degradation of amino acids. These alcohols normally contain low-carbon chains. With the increase in the number of carbon chains, the flavor of alcohols changes from an anesthetic odor to a scent of fruit and lipids. From Table 5, it can be seen that a total of seven alcohol compounds were detected among the chicken meats in the C2 and C3 groups, and only six alcohol compounds were detected among the chicken meats in the C0 and C1 groups, with no 1-octen-3-ol detected.

Table 5. Effects of dietary CUR on the volatile compounds of chicken breast meat samples ($n = 8$).

Volatile Compounds		Content (ng/g)					
		C0 (0 mg/kg)	C1 (100 mg/kg)	C2 (150 mg/kg)	C3 (200 mg/kg)	SEM	p-Value
Aldehydes	Hexanal	255.35 ^{c}	716.89 ^{b}	1514.93 ^{a}	1637.96 ^{a}	189.86	<0.001
	2-Hexenal,(E)-	0.25 ^{c}	0.64 ^{c}	3.07 ^{a}	2.10 ^{b}	0.37	<0.001
	Heptanal	9.57 ^{c}	36.08 ^{b}	45.74 ^{b}	59.84 ^{a}	6.27	<0.001
	2-Heptenal,(E)-	1.70 ^{d}	4.84 ^{c}	18.72 ^{a}	14.23 ^{b}	2.26	<0.001
	2,4-Heptadienal,(E,E)-	0.17 ^{d}	0.68 ^{c}	3.10 ^{a}	2.13 ^{b}	0.38	<0.001
	Octanal	6.89 ^{c}	30.44 ^{b}	38.76 ^{b}	55.59 ^{a}	5.94	<0.001
	2-methylundecanal	1.04 ^{b}	2.17 ^{a}	1.80 ^{a}	1.81 ^{a}	0.17	0.022
	2-Octenal,(E)-	0.90 ^{d}	3.61 ^{c}	14.22 ^{a}	11.00 ^{b}	1.77	<0.001
	Nonanal	8.84 ^{d}	43.94 ^{c}	62.78 ^{b}	97.87 ^{a}	10.70	<0.001
	1,3,4-trimethyl-3-Cyclohexene-1-carboxaldehyde,	1.32 ^{b}	2.2 ^{a}	0.09 ^{c}	0.74 ^{b}	0.27	0.002
	2-Nonenal,(E)-	0.15 ^{c}	0.69 ^{b}	2.28 ^{a}	2.25 ^{a}	0.31	<0.001
	Decanal	0.39 ^{d}	1.53 ^{c}	2.69 ^{b}	3.98 ^{a}	0.44	<0.001
	2,4-Nonadienal,(E,E)-	0.40 ^{d}	1.88 ^{c}	8.38 ^{a}	7.01 ^{b}	1.10	<0.001
	2-Decenal, (E)-	0.18 ^{c}	0.95 ^{b}	3.87 ^{a}	4.03 ^{a}	0.56	<0.001
	2,4-Decadienal,(E,Z)-	0.15 ^{d}	0.72 ^{c}	2.98 ^{a}	2.11 ^{b}	0.37	<0.001
	Undecanal	0.07 ^{d}	0.67 ^{a}	0.35 ^{b}	0.21 ^{c}	0.11	0.146
	2,4-Decadienal,(E,E)-	0.32 ^{d}	1.75 ^{c}	7.00 ^{a}	5.07 ^{b}	0.88	<0.001
	2-Undecenal,(E)-	0.09 ^{c}	0.38 ^{b}	1.61 ^{a}	1.58 ^{a}	0.23	<0.001
	Dodecanal	0.10 ^{d}	0.39 ^{c}	1.32 ^{a}	0.90 ^{b}	0.16	<0.001
	Tridecanal	-	0.45	0.84	0.64	0.12	0.018
	Pentadecanal	0.18 ^{c}	0.45 ^{b}	0.9 ^{a}	0.69 ^{ab}	0.09	0.002
	Subtotal	288.05 ^{c}	851.36 ^{b}	1735.39 ^{a}	1911.69 ^{a}	219.85	<0.001
Hydrocarbons	1,2-epoxyheptane	1.34 ^{c}	5.03 ^{b}	6.27 ^{a}	7.12 ^{a}	0.76	<0.001
	Bornyl bromide	1.29 ^{a}	2.00 ^{a}	0.09 ^{b}	0.12 ^{b}	0.28	0.002
	4,5-Epoxynonane,(E)-	-	0.09 ^{b}	0.27 ^{a}	0.12 ^{b}	0.04	0.076
	Tetradecane	0.29 ^{b}	0.75 ^{a}	0.80 ^{a}	0.94 ^{a}	0.09	0.013
	Heneicosane	0.73 ^{b}	1.62 ^{ab}	2.21 ^{a}	2.19 ^{a}	0.24	0.056
	Pentadecane	0.27 ^{b}	0.57 ^{b}	1.34 ^{a}	1.37 ^{a}	0.18	<0.001
	n-Hexadecane	0.40 ^{d}	0.66 ^{c}	0.99 ^{a}	0.87 ^{b}	0.11	0.017
	1,3-Octadiene	0.42 ^{c}	1.57 ^{b}	3.40 ^{a}	3.24 ^{a}	0.41	0.250
	β-myrcene	25.02 ^{c}	24.65 ^{c}	29.01 ^{b}	33.15 ^{a}	1.85	0.408
	2,6-Dimethyl-1,3,5,7-octatetrene,(3 E,5 E)-	1.96 ^{b}	3.22 ^{a}	0.13 ^{c}	0.18 ^{c}	0.44	<0.001
	2-Carene	1.39 ^{a}	1.55 ^{a}	0.03 ^{b}	0.07 ^{b}	0.26	0.008
	1,3-Hexadiene,3-ethyl-2-methyl-	275.71 ^{b}	689.55 ^{a}	17.46 ^{c}	13.23 ^{d}	101.38	0.008
	D-Limonene	0.61 ^{d}	0.93 ^{c}	3.99 ^{a}	3.60 ^{b}	0.50	<0.001
	beta.-Ocimene	2.83 ^{a}	2.84 ^{a}	0.08 ^{b}	0.17 ^{b}	0.49	0.010
	.gamma.-Terpinene	0.93 ^{c}	1.25 ^{b}	1.601 ^{a}	1.25 ^{b}	0.11	0.224
	Cyclohexene,1-methyl-4-(1-methylethylidene)-	1.06 ^{b}	1.65 ^{a}	0.13 ^{c}	0.16 ^{c}	0.23	0.010
	Limonene oxide,(E)-	2.42 ^{b}	5.69 ^{a}	0.02 ^{d}	1.04 ^{c}	0.74	<0.001
	3-Octadecene	0.8 ^{c}	1.62 ^{b}	2.21 ^{a}	2.19 ^{a}	0.21	0.006
	(E)-9-Octadecene	0.13 ^{d}	0.35 ^{c}	0.75 ^{a}	0.50 ^{b}	0.09	0.044
	Butanoic acid,anhydride	0.23 ^{b}	1.16 ^{a}	1.06 ^{a}	0.91 ^{a}	0.13	<0.001
	Subtotal	317.82 ^{b}	746.75 ^{a}	71.79 ^{c}	72.35 ^{c}	101.11	0.007

Table 5. *Cont.*

Volatile Compounds		Content (ng/g)					
		C0 (0 mg/kg)	C1 (100 mg/kg)	C2 (150 mg/kg)	C3 (200 mg/kg)	SEM	*p*-Value
Ketones	Hexanone,4-hydroxy-3-propyl-	0.58 ^{d}	1.49 ^{c}	2.03 ^{b}	1.80 ^{a}	0.21	0.023
	3-Heptanone	0.39 ^{c}	0.75 ^{b}	1.46 ^{a}	1.45 ^{a}	0.16	<0.001
	2-Heptanone	9.21 ^{c}	33.37 ^{b}	57.31 ^{a}	63.55 ^{a}	7.28	<0.001
	Heptanone,6-methyl-	0.90 ^{d}	6.19 ^{c}	10.52 ^{a}	8.49 ^{b}	1.22	<0.001
	3-Ethylcyclopentanone	0.50 ^{c}	1.87 ^{b}	3.95 ^{a}	3.31 ^{a}	0.45	<0.001
	2,5-Octanedione	37.27 ^{d}	173.13 ^{c}	254.94 ^{b}	288.56 ^{a}	32.98	<0.001
	3,5-Octadien-2-one	0.18 ^{c}	1.03 ^{b}	2.63 ^{a}	2.84 ^{a}	0.37	<0.001
	2-Nonanone	0.96 ^{c}	1.09 ^{bc}	1.15 ^{b}	1.36 ^{a}	0.06	0.087
	Cyclohexanone,2-methyl-5-(1-methylethenyl)-,trans-	3.32 ^{b}	5.80 ^{a}	0.91 ^{d}	1.05 ^{c}	0.68	<0.001
	2-Undecanone	0.24 ^{c}	1.03 ^{b}	1.61 ^{a}	1.58 ^{a}	0.21	0.012
	trans-3-Nonen-2-one	0.18 ^{d}	1.16 ^{c}	6.92 ^{a}	5.85 ^{b}	0.96	<0.001
	11-Heneicosanone	-	-	0.01 ^{b}	0.16 ^{a}	0.02	0.006
	Subtotal	53.72 ^{c}	226.92 ^{b}	343.42	379.96 ^{a}	43.22	<0.001
Alcohols	3,5-Octadien-2-ol	0.11 ^{d}	1.92 ^{c}	3.79 ^{a}	3.43 ^{b}	0.49	<0.001
	Benzyllinalool	1.24 ^{b}	1.25 ^{b}	2.3 ^{a}	2.48 ^{a}	0.21	0.010
	1,3-Dioxolane-2-butanol,2-methyl-	0.10 ^{d}	0.70 ^{c}	3.89 ^{b}	5.04 ^{a}	0.68	<0.001
	2,5-Dimethylcyclohexanol	0.31 ^{d}	2.18 ^{c}	11.85 ^{a}	10.04 ^{b}	1.62	<0.001
	Linalool	0.04 ^{c}	0.06 ^{bc}	0.11 ^{a}	0.07 ^{b}	0.01	<0.001
	2-Octen-1-ol,(E)-	0.05 ^{d}	0.42 ^{c}	1.75 ^{a}	0.64 ^{b}	0.20	<0.001
	1-Octen-3-ol	-	-	0.64 ^{a}	0.03 ^{b}	0.01	<0.001
	Subtotal	1.85 ^{d}	6.53 ^{c}	23.73 ^{a}	21.71 ^{b}	3.10	<0.001
Esters	Caproic acid vinyl ester	0.13 ^{d}	1.02 ^{c}	4.3 ^{b}	5.97 ^{a}	0.78	<0.001
	Butyrolactone	0.2 ^{d}	0.82 ^{a}	0.46 ^{c}	0.66 ^{b}	0.09	0.011
	Linalyl acetate	0.54 ^{b}	0.88 ^{a}	0.12 ^{c}	0.08 ^{c}	0.12	0.003
	Carbamodithioic acid,diethyl-,methyl ester	6.49 ^{b}	9.14 ^{a}	1.16 ^{c}	0.49 ^{d}	1.23	<0.001
	Subtotal	7.35 ^{b}	11.86 ^{a}	6.02 ^{b}	7.19 ^{b}	0.86	0.016
Aromatic	Toluene	8.82 ^{d}	11.04 ^{c}	33.46 ^{b}	68.51 ^{a}	7.74	<0.001
	Ethylbenzene	2.10 ^{c}	2.73 ^{c}	14.84 ^{b}	32.68 ^{a}	3.99	<0.001
	p-Xylene	5.76 ^{a}	6.41 ^{a}	1.95 ^{b}	6.93 ^{a}	0.66	0.011
	Styrene	1.05 ^{c}	1.13 ^{c}	46.29 ^{b}	113.95 ^{a}	14.82	<0.001
	Benzene,1,3-dimethyl-	2.87 ^{c}	3.82 ^{bc}	4.32 ^{b}	6.93 ^{a}	0.50	0.002
	Benzene,(1-methylethyl)-	0.05 ^{c}	0.1 ^{c}	1.93 ^{b}	4.49 ^{a}	0.58	<0.001
	Benzaldehyde	11.90 ^{c}	15.02 ^{bc}	21.92 ^{a}	18.39 ^{ab}	1.35	0.011
	o-Cymene	37.90 ^{b}	54.47 ^{a}	0.35 ^{c}	0.30 ^{c}	8.09	0.001
	Benzeneacetaldehyde	1.20 ^{b}	1.83 ^{ab}	2.30 ^{a}	2.48 ^{a}	0.20	0.044
	Benzaldehyde, 4-ethyl-	0.20 ^{c}	1.52 ^{b}	3.51 ^{a}	3.77 ^{a}	0.49	<0.001
	Subtotal	71.84 ^{d}	98.08 ^{c}	130.84 ^{b}	258.4 ^{a}	23.16	0.000
Acids	Pentanoic acid	0.29 ^{c}	1.18 ^{b}	2.08 ^{a}	2.34 ^{a}	0.29	0.007
	Hexanoic acid	2.54 ^{d}	15.02 ^{c}	38.86 ^{b}	51.99 ^{a}	6.47	<0.001
	Subtotal	2.83 ^{d}	16.20 ^{c}	40.94 ^{b}	54.33 ^{a}	6.75	0.000
Heterocyclic	1,2-Benzisothiazole	0.28 ^{d}	0.40 ^{c}	7.04 ^{b}	11.21 ^{a}	1.50	<0.001
	Hydantoin	0.13 ^{c}	1.50 ^{a}	0.21 ^{b}	0.12 ^{c}	0.21	<0.001
	Subtotal	0.40 ^{d}	1.90 ^{c}	7.24 ^{b}	11.33 ^{a}	1.42	0.000

n = 8 replicates per treatment; different superscripts (a–d) means are significantly different on the same row.

Hydrocarbons can be divided into straight-chain hydrocarbons and olefins. Among them, straight-chain hydrocarbons have a relatively high threshold [50], thus exhibiting a relatively low contribution to meat flavor. After the addition of CUR in chicken diets, changes in the varieties and contents of hydrocarbons in chicken meat in different CUR-dosage groups were observed, with no trans-4,5-epoxydecane detected among the chicken meats in the C0 group. Myrcene presents a faint, meaty aroma. Chicken meat in the C3 group presented the highest contents of myrcene, which were significantly higher than those of chicken meat in the C0, C1, and C2 groups ($p < 0.05$). Ketone compounds mostly come from the oxidation degradation of fats. Most ketone compounds present aromas of milk and fruit. Ketones present a much higher threshold than that of aldehydes. Therefore, ketone compounds make a lower contribution to meat flavor than aldehyde compounds. Compared with chicken meat in the C0 group, after the addition of CUR in chicken diets, chicken meat presented a significantly increased content of ketones. Among all ketone compounds, chicken meat in different treatment groups presented the highest contents of 2,5-octanedione. Among all ester compounds that were detected, chicken meat in the C3 group presented the highest contents of ethyl caproate, which were significantly higher than those of chicken meat in all other groups ($p < 0.05$). Meanwhile, among all chicken meat samples, only two acid compounds, valeric acid, and caproic acid, were detected. Compared with chicken meat in the C0 group, chicken meat after the addition of CUR in chicken diets presented a significantly increased content of total acids ($p < 0.05$). Among all chicken meat samples, two heterocyclic compounds, benzoisothiazole and hydantoin, were detected. With an increased CUR addition amount in chicken diets, the contents of benzoisothiazole in chicken meat significantly increased ($p < 0.05$). Aromatic compounds that were detected in this study present a dosage effect on their total contents. Among these compounds, toluene and styrene hold a large proportion. Significant differences in the contents of aromatic compounds were observed among chicken meat samples in different groups, with the content of ortho-isopropyl toluene presenting a negative correlation with the CUR addition amount.

3.5. Identification of Key Aroma Components Based on OAV

Odor activity value (OAV) is an index used in the selection and identification of key flavor compounds in meat products. It is generally believed that an OAV greater than 1 indicates that the compound in question has a contribution to meat flavor. Meanwhile, a higher OAV indicates a higher contribution level of the compound to meat flavor [51]. From Table 6, it can be seen that a total of 13 flavor compounds with an OAV greater than 1 were detected among chicken meat samples in different treatment groups. Among them, there were 11 aldehyde compounds and 2 aromatic compounds. This indicates that all these 13 compounds make a significant contribution to the flavor of chicken meat. With a very low threshold, aldehydes exhibit a relatively high OAV and a significant contribution to meat flavor. Chicken meat in the CUR groups presents higher OAVs of hexaldehyde, heptaldehyde, octyl aldehyde, nonaldehyde, capraldehyde, and trans-2-decenal than those of chicken meat in the C0 group. Compared with chicken meat in all other groups, only chicken meat in the C2 group presents OAVs of myrcene greater than one, and only chicken meat in the C3 group exhibits OAVs of ethylbenzene greater than one. Benzaldehyde has a relatively low threshold and a high content in chicken meat. Therefore, its OAV is relatively high, making it a compound that contributes to the sweet flavor of chicken meat. Ethylbenzene is an aromatic compound that presents a unique aroma. Ethylbenzene itself presents a low contribution to the flavor of chicken meat. However, its concentrations in chicken meat continuously increases with the CUR addition amount in chicken diets, leading to the continuous increase in its contribution to meat flavor. Based on the above analysis, it can be concluded that the addition of CUR in chicken diets presents the most significant impact on the content of volatile aldehyde compounds in chicken meat. Compared with chicken meat in the C0 group, chicken meat in the CUR-dosage groups all presents increased OAVs of those 13 flavor compounds. This indicates that, with the addition of CUR in chicken diets, the flavor of chicken meat could be impacted through its affected content of volatile compounds.

Table 6. The OAV of volatile flavor compounds in breast muscles of chicken (n = 8).

Compound	Odor Threshold (μg/kg)	Odor Description	OAV			
			C0 (0 mg/kg)	C1 (100 mg/kg)	C2 (150 mg/kg)	C3 (200 mg/kg)
Hexanal	5	Green, grassy, fat	51.07	143.38	302.99	327.59
Hexanal	3	Fresh, burnt fat	3.19	12.03	15.25	19.95
Octanal	1	Fatty, green	6.89	30.44	38.76	55.59
Trans-2-octenal	0.08	fruit, Fatty	11.28	45.11	177.67	137.50
Nonanal	1	Fatty, green	8.84	43.94	62.78	97.87
Trans-2-nonenal	0.08	Fatty	1.88	8.56	28.49	28.15
Decanal	0.1	Green, onion, yeast	3.90	15.33	26.91	39.75
2,4-nonadienal,(E,E)-	0.06	Flower, Fatty	6.61	31.29	139.55	116.84
Trans-2-decenal	0.3	Sweet orange	0.59	3.19	12.89	13.42
-2,4-Decadienal,(E,E)	0.7	Fatty, toasted, scallion	0.45	2.50	10.00	7.24
Lauric aldehyde	0.9	Flower	0.11	0.44	1.46	0.99
Benzaldehyde	3	Nutty, bitter almond, burnt sugar	3.96	5.01	7.31	6.13
Ethyl benzene	29	Fragrant	0.07	0.09	0.51	1.13

4. Conclusions

The results show that the addition CUR in chicken feed is conducive to promoting the deposition of amino acids in chicken meat and improving its content of 5′-nucleotides, thus reducing the contents of SFA and C20:4 n6 and increasing the PUFA/SFA ratio in chicken meat. Moreover, after the addition of CUR in chicken diets, the volatile flavor characteristics of chicken breast meat can be improved. Particularly, its content of aldehydes can be significantly increased. Among them, daily feeding with 150 mg/kg CUR has the best effect on improving the nutritional value and flavor of chicken.

Author Contributions: Conceptualization, Y.S. and Z.Z.; methodology, W.Y. and W.Q.; software, F.W.; validation, Y.S. and F.W.; formal analysis, R.L. and Z.Z.; investigation, W.Y.; resources, F.W.; data curation, W.Q.; writing—original draft preparation, Y.S. and W.Q.; writing—review and editing, Y.S. and Z.Z.; supervision, Y.S.; project administration, Y.S. and Z.Z.; funding acquisition, Y.S. and Z.Z. All authors have read and agreed to the published version of the manuscript.

Funding: This research was funded by Hebei Agricultural Research (HBCT2024270206), Natural Science Foundation of Hebei Province (C2023204010), Research on the Preservation Technology of Chicken (310513001) and Hebei Layer Industry Technology Research Institute (225676150H).

Institutional Review Board Statement: The study was conducted in accordance with the Declaration of Helsinki, and approved by the Ethics Committee of the Ethics Committee of Hebei Agricultural University (Number: 2022003).

Informed Consent Statement: Not applicable.

Data Availability Statement: The original contributions presented in the study are included in the article, further inquiries can be directed to the corresponding author.

Conflicts of Interest: The authors declare no conflicts of interest.

References

1. Wen, J. The current situation and development trend of broiler production in the world. *Poult. Husb. Dis. Control* **2023**, *2*, 31–39.
2. Begum, J.; Buyamayum, B.; Madhu, C.L.; Dev, K. Probiotics: Role in immunomodulation and consequent effects. *Lett. Anim. Biol.* **2021**, *1*, 1–6. [CrossRef]
3. Allen, H.K.; Trachsel, J.; Looft, T.; Casey, T.A. Finding alternatives to antibiotics. *Ann. N. Y. Acad. Sci.* **2014**, *1323*, 91–100. [CrossRef]
4. Valentini, J.; Da Silva, A.S.; Fortuoso, B.F.; Reis, J.H.; Gebert, R.R.; Griss, L.G.; Boiago, M.M.; Lopes, L.Q.S.; Santos, R.C.V.; Wagner, R.; et al. Chemical composition, lipid peroxidation, and fatty acid profile in meat of broilers fed with glycerol monolaurate additive. *Food Chem.* **2020**, *330*, 127187. [CrossRef]

5. Aminov, R.I. A Brief History of the Antibiotic Era: Lessons Learned and Challenges for the Future. *Front. Microbiol.* **2010**, *1*, 134. [CrossRef]
6. Ponnampalam, E.N.; Kiani, A.; Santhiravel, S.; Holman, B.W.B.; Lauridsen, C.; Dunshea, F.R. The Importance of Dietary Antioxidants on Oxidative Stress, Meat and Milk Production, and Their Preservative Aspects in Farm Animals: Antioxidant Action, Animal Health, and Product Quality—Invited Review. *Animals* **2022**, *12*, 3279. [CrossRef]
7. Álvarez-Rodríguez, J.; Urrutia, O.; Lobón, S.; Ripoll, G.; Bertolín, J.R.; Joy, M. Insights into the role of major bioactive dietary nutrients in lamb meat quality: A review. *J. Anim. Sci. Biotechnol.* **2022**, *13*, 20. [CrossRef]
8. Sahu, P.K. Design, structure activity relationship, cytotoxicity and evaluation of antioxidant activity of curcumin derivatives/analogues. *Eur. J. Med. Chem* **2016**, *121*, 510–516. [CrossRef]
9. Griesser, M.; Pistis, V.; Suzuki, T.; Tejera, N.; Pratt, D.A.; Schneider, C. Autoxidative and Cyclooxygenase-2 Catalyzed Transformation of the Dietary Chemopreventive Agent Curcumin. *J. Biol. Chem.* **2011**, *286*, 1114–1124. [CrossRef]
10. Nawab, A.; Li, G.; An, L.; Wu, J.; Chao, L.; Xiao, M.; Zhao, Y.; Birmani, M.W.; Ghani, M.W. Effect of curcumin supplementation on TLR4 mediated non-specific immune responses in liver of laying hens under high-temperature conditions. *J. Therm. Biol.* **2019**, *84*, 384–397. [CrossRef]
11. Khan, M.A.; Ali, S.; Yang, H.; Kamboh, A.A.; Ahmad, Z.; Tume, R.K.; Zhou, G. Improvement of color, texture and food safety of ready-to-eat high pressure-heat treated duck breast. *Food Chem.* **2019**, *277*, 646–654. [CrossRef] [PubMed]
12. Isabel Quiros-Fallas, M.; Vargas-Huertas, F.; Quesada-Mora, S.; Azofeifa-Cordero, G.; Wilhelm-Romero, K.; Vasquez-Castro, F.; Alvarado-Corella, D.; Sanchez-Kopper, A.; Navarro-Hoyos, M. Polyphenolic HRMS Characterization, Contents and Antioxidant Activity of Curcuma longa Rhizomes from Costa Rica. *Antioxidants* **2022**, *11*, 620. [CrossRef] [PubMed]
13. Lu, Y.; Wu, S.; Xiang, B.; Li, L.; Lin, Y. Curcumin Attenuates Oxaliplatin-Induced Liver Injury and Oxidative Stress by Activating the Nrf2 Pathway. *Drug Des. Dev. Ther.* **2020**, *14*, 73. [CrossRef] [PubMed]
14. Araiza-Calahorra, A.; Akhtar, M.; Sarkar, A. Recent advances in emulsion-based delivery approaches for curcumin: From encapsulation to bioaccessibility. *Trends Food Sci. Technol.* **2018**, *71*, 155–169. [CrossRef]
15. Gayani, B.; Dilhari, A.; Wijesinghe, G.K.; Kumarage, S.; Abayaweera, G.; Samarakoon, S.R.; Perera, I.C.; Kottegoda, N.; Weerasekera, M.M. Effect of natural curcuminoids-intercalated layered double hydroxide nanohybrid against *Staphylococcus aureus*, *Pseudomonas aeruginosa*, and *Enterococcus faecalis*: Abactericidal, antibiofilm, and mechanistic study. *MicrobiologyOpen* **2019**, *8*, 723. [CrossRef] [PubMed]
16. Shakeri, A.; Panahi, Y.; Johnston, T.P.; Sahebkar, A. Biological properties of metal complexes of curcumin. *Biofactors* **2019**, *45*, 304–317. [CrossRef] [PubMed]
17. Rattis, B.A.C.; Ramos, S.G.; Celes, M.R.N. Curcumin as a Potential Treatment for COVID-19. *Front. Pharmacol.* **2021**, *12*, 675287. [CrossRef] [PubMed]
18. Rahmani, M.; Golian, A.; Kermanshahi, H.; Bassami, M.R. Effects of curcumin or nanocurcumin on blood biochemical parameters, intestinal morphology and microbial population of broiler chickens reared under normal and cold stress conditions. *J. Appl. Anim. Res.* **2018**, *46*, 200–209. [CrossRef]
19. Rajput, N.; Naeem, M.; Ali, S.; Zhang, J.F.; Zhang, L.; Wang, T. The effect of dietary supplementation with the natural carotenoids curcumin and lutein on broiler pigmentation and immunity. *Poult. Sci.* **2013**, *92*, 1177–1185. [CrossRef] [PubMed]
20. Nawab, A.; Tang, S.; Li, G.; An, L.; Wu, J.; Liu, W.; Xiao, M. Dietary curcumin supplementation effects on blood immunological profile and liver enzymatic activity of laying hens after exposure to high temperature conditions. *J. Therm. Biol.* **2020**, *90*, 102573. [CrossRef]
21. Jin, S.; Yang, H.; Liu, F.; Pang, Q.; Shan, A.; Feng, X. Effect of Dietary Curcumin Supplementation on Duck Growth Performance, Antioxidant Capacity and Breast Meat Quality. *Foods* **2021**, *10*, 2981. [CrossRef] [PubMed]
22. Marcon, H.; Souza, C.F.; Baldissera, M.D.; Alba, D.F.; Favaretto, J.A.; Santos, D.S.; Borges, L.; Kessler, J.D.; Vedovatto, M.; Bianchi, A.E.; et al. Effect of curcumin dietary supplementation on growth performance, physiology, carcass characteristics and meat quality in lambs. *Ann. Anim. Sci.* **2021**, *21*, 623–638. [CrossRef]
23. Rajput, N.; Muhammad, N.; Yan, R.; Zhong, X.; Wang, T. Effect of Dietary Supplementation of Curcumin on Growth Performance, Intestinal Morphology and Nutrients Utilization of Broiler Chicks. *J. Poult. Sci.* **2013**, *50*, 44–52. [CrossRef]
24. Al-Muhammadawi, N.A.; Jassim Hammoudi, S. Effect of Adding Different Levels of Therapeutic Curcuma on Productive Traits in Broiler Chickens. *Arch. Razi Inst.* **2022**, *77*, 2059–2064. [PubMed]
25. Jin, S.; Pang, Q.; Liu, R.; Yang, H.; Liu, F.; Wang, M.; Wang, Y.; Feng, X.; Shan, A. Dietary curcumin decreased lipid oxidation and enhanced the myofibrillar protein structure of the duck (*Anas Platyrhynchos*) breast muscle when subjected to storage. *Lwt-Food Sci. Technol.* **2020**, *133*, 109986. [CrossRef]
26. Salah, A.S.; Mahmoud, M.A.; Ahmed-Farid, O.A.; El-Tarabany, M.S. Effects of dietary curcumin and acetylsalicylic acid supplements on performance, muscle amino acid and fatty acid profiles, antioxidant biomarkers and blood chemistry of heat-stressed broiler chickens. *J. Therm. Biol.* **2019**, *84*, 259–265. [CrossRef] [PubMed]
27. Zhou, Z.; Huang, M.; Shu, Y.; Qi, W.; Wu, F.; Zhang, Z.; Chen, B. Effect of Dietary Supplementation with Curcumin on the Chicken Quality. *Food Res. Dev.* **2024**, *45*, 42–47.
28. *NY/T 33-2004*; Nutrition Requirements of Chinese Chicken. The Ministry of Agriculture of the People's Republic of China: Beijing, China, 2020.

29. Liu, H.; Li, J.; Zhang, D.; Hamid, N.; Liu, D.; Hua, W.; Du, C.; Ma, Q.; Gong, H. The effect of thermal times of circulating non-fried roast technique on the formation of (non)volatile compounds in roasted mutton by multi-chromatography techniques and heat transfer analysis. *Food Res. Int.* **2023**, *174*, 113567. [CrossRef] [PubMed]
30. Kaewsatuan, P.; Poompramun, C.; Kubota, S.; Yongsawatdigul, J.; Molee, W.; Uimari, P.; Molee, A. Thigh muscle metabolic response is linked to feed efficiency and meat characteristics in slow-growing chicken. *Poult. Sci.* **2023**, *102*, 741. [CrossRef]
31. Gonzales-Barron, U.; Popova, T.; Bermúdez Piedra, R.; Tolsdorf, A.; Geß, A.; Pires, J.; Domínguez, R.; Chiesa, F.; Brugiapaglia, A.; Viola, I.; et al. Fatty acid composition of lamb meat from Italian and German local breeds. *Small Rumin. Res.* **2021**, *200*, 106384. [CrossRef]
32. Sun, Y.; Zhang, Y.; Song, H. Variation of aroma components during frozen storage of cooked beef balls by SPME and SAFE coupled with GC-O-MS. *J. Food Process. Preserv.* **2021**, *45*, 15036. [CrossRef]
33. Miao, J.; Lin, H.; Zhang, S.; Huo, J.; Deng, S. Effect on amino acid and mineral content of the loach (*Misgurnus anguillicaudatus*) by adding Fe (II) chelating hairtail protein hydrolysates (Fe (II)-HPH) to the feed. *Food Sci. Nutr.* **2020**, *8*, 1575–1582. [CrossRef] [PubMed]
34. Zhang, J.; Kaneko, G.; Sun, J.; Wang, G.; Xie, J.; Tian, J.; Li, Z.; Gong, W.; Zhang, K.; Xia, Y.; et al. Key Factors Affecting the Flesh Flavor Quality and the Nutritional Value of Grass Carp in Four Culture Modes. *Foods* **2021**, *10*, 2075. [CrossRef] [PubMed]
35. Su, Y.; Sun, X.; Zhao, S.; Hu, M.; Li, D.; Qi, S.; Jiao, X.; Sun, Y.; Wang, C.; Zhu, X.; et al. Dietary alfalfa powder supplementation improves growth and development, body health, and meat quality of Tibetan sheep. *Food Chem.* **2022**, *396*, 133709. [CrossRef] [PubMed]
36. He, P.; Lei, Y.; Zhang, K.; Zhang, R.; Bai, Y.; Li, Z.; Jia, L.; Shi, J.; Cheng, Q.; Ma, Y.; et al. Dietary oregano essential oil supplementation alters meat quality, oxidative stability, and fatty acid profiles of beef cattle. *Meat Sci.* **2023**, *205*, 109317. [CrossRef] [PubMed]
37. Ramalingam, V.; Song, Z.; Hwang, I. The potential role of secondary metabolites in modulating the flavor and taste of the meat. *Food Res. Int.* **2019**, *122*, 174–182. [CrossRef] [PubMed]
38. bin, H.; Liping, S.; Shuquan, M.; Peng, X. Effects of Four Chinese Herbal Preparations in the Feed on Body Composition and Muscle Quality of Micropterus Salmoide. *J. Yangtze Univ. (Nat. Sci. Ed.)* **2020**, *17*, 78–84.
39. Liu, C.; Meng, F.; Tang, X.; Shi, Y.; Wang, A.; Gu, Z.; Pan, Z. Comparison of nonvolatile taste active compounds of wild and cultured mud crab *Scylla paramamosain*. *Fish. Sci.* **2018**, *84*, 897–907. [CrossRef]
40. Madruga, M.S.; Elmore, J.S.; Oruna-Concha, M.J.; Balagiannis, D.; Mottram, D.S. Determination of some water-soluble aroma precursors in goat meat and their enrolment on flavour profile of goat meat. *Food Chem.* **2010**, *123*, 513–520. [CrossRef]
41. Zhang, Z.; Wu, Y.; Liu, Q.; Zhao, G.; Wei, L.; Zhang, C.; Huang, F. Comparative flavor precursors and volatile compounds of Wenchang chickens fed with copra meal based on GC–O–MS. *Food Res. Int.* **2023**, *174*, 113646. [CrossRef]
42. Xu, N.; Shi, J.; He, W.; Ye, J.; Han, M.; Xu, X. Research progress on the effect of post-mortem aging on chicken flavor. *J. Food Saf. Qual.* **2021**, *12*, 727–732.
43. Zhao, W.; Tian, Y.; Wang, Y.; Du, J.; Chen, L.; Gu, T.; Song, M.; Lu, L.; Sun, C. Dietary effect of *Dendrobium officinale* leaves on chicken meat quality, fatty acid composition, and volatile compounds profile. *Food Chem. X* **2024**, *22*, 101330. [CrossRef] [PubMed]
44. Zarate, R.; el Jaber-Vazdekis, N.; Tejera, N.; Perez, J.A.; Rodriguez, C. Significance of long chain polyunsaturated fatty acids in human health. *Clin. Transl. Med.* **2017**, *6*, 25. [CrossRef] [PubMed]
45. Kafantaris, I.; Kotsampasi, B.; Christodoulou, V.; Makri, S.; Stagos, D.; Gerasopoulos, K.; Petrotos, K.; Goulas, P.; Kouretas, D. Effects of Dietary Grape Pomace Supplementation on Performance, Carcass Traits and Meat Quality of Lambs. *In Vivo* **2018**, *32*, 807–812. [CrossRef] [PubMed]
46. Yao, H.; Jin, Y.; Zhang, X.; Yang, N.; Xu, X. Influence of pulsed electric field on thawing of frozen pork: Physical properties, fat oxidation and protein structure. *Food Biosci.* **2023**, *56*, 103175. [CrossRef]
47. Elmore, J.S.; Cooper, S.L.; Enser, M.; Donald, S.M.A.; Sinclair, L.A.; Wilkinson, R.G.; Wood, J.D. Dietary manipulation of fatty acid composition in lamb meat and its effect on the volatile aroma compounds of grilled lamb. *Meat Sci.* **2005**, *69*, 233–242. [CrossRef]
48. Yang, Y.; Zhang, X.; Wang, Y.; Pan, D.; Sun, Y.; Cao, J. Study on the volatile compounds generated from lipid oxidation of Chinese bacon (unsmoked) during processing. *Eur. J. Lipid Sci. Technol.* **2017**, *119*, 1600512. [CrossRef]
49. Marco, A.; Navarro, J.L.; Flores, M. The influence of nitrite and nitrate on microbial, chemical and sensory parameters of slow dry fermented sausage. *Meat Sci.* **2006**, *73*, 660–673. [CrossRef] [PubMed]
50. Ba, H.V.; Seo, H.-W.; Seong, P.-N.; Cho, S.-H.; Kang, S.-M.; Kim, Y.-S.; Moon, S.-S.; Choi, Y.-M.; Kim, J.-H. Live weights at slaughter significantly affect the meat quality and flavor components of pork meat. *Anim. Sci. J.* **2019**, *90*, 667–679. [CrossRef]
51. Zhu, J.; Chen, F.; Wang, L.; Niu, Y.; Chen, H.; Wang, H.; Xiao, Z. Characterization of the Key Aroma Volatile Compounds in Cranberry (*Vaccinium macrocarpon* Ait.) Using Gas Chromatography-Olfactometry (GC-O) and Odor Activity Value (OAV). *J. Agric. Food Chem* **2016**, *64*, 4990–4999. [CrossRef]

MDPI AG
Grosspeteranlage 5
4052 Basel
Switzerland
Tel.: +41 61 683 77 34

Foods Editorial Office
E-mail: foods@mdpi.com
www.mdpi.com/journal/foods

www.ingramcontent.com/pod-product-compliance
Lightning Source LLC
Chambersburg PA
CBHW061610210926
51850CB00009B/2273
9783725869329